HERMANN VON HELMHOLTZ'S MECHANISM: THE LOSS OF CERTAINTY

Archimedes

NEW STUDIES IN THE HISTORY AND PHILOSOPHY OF SCIENCE AND TECHNOLOGY

VOLUME 17

Archimedes has three fundamental goals; to further the integration of the histories of science and technology with one another: to investigate the technical, social and practical histories of specific developments in science and technology; and finally, where possible and desirable, to bring the histories of science and technology into closer contact with the philosophy of science. To these ends, each volume will have its own theme and title and will be planned by one or more members of the Advisory Board in consultation with the editor. Although the volumes have specific themes, the series itself will not be limited to one or even to a few particular areas. Its subjects include any of the sciences, ranging from biology through physics, all aspects of technology, broadly construed, as well as historically-engaged philosophy of science or technology. Taken as a whole, *Archimedes* will be of interest to historians, philosophers, and scientists, as well as to those in business and industry who seek to understand how science and industry have come to be so strongly linked.

Hermann von Helmholtz's Mechanism: The Loss of Certainty

A Study on the Transition from Classical to Modern Philosophy of Nature

GREGOR SCHIEMANN

Translated by Cynthia Klohr

Springer

Prof. Dr. Gregor Schiemann
Philosophisches Seminar
Bergische Universität
Gaußstrasse 20
42119 Wuppertal
Germany

Original title is:
Wahrheitsgewissheitsverlust: *Hermann von Helmholtz' Mechanismus im Anbruch der Moderne. Eine Studie zum Übergang von klassischer zu moderner Naturphilosophie*

ISBN 978-90-481-7413-3 e-ISBN 978-1-4020-5630-7

Softcover reprint of the hardcover 1st edition 2009

Printed on acid-free paper

9 8 7 6 5 4 3 2 1

springer.com

Preface to the English Edition

Focusing on Hermann von Helmholtz, this study addresses one of the nineteenth century's most important German natural scientists. Among his most well-known contributions to science are the invention of the ophthalmoscope and groundbreaking work towards formulating the law of the conservation of energy. The volume of his work, reaching from medicine to physiology to physics and epistemology, his impact on the development of the sciences far beyond German borders, and the contribution he made to the organization and popularization of research, all established Helmholtz's prominence both in the academic world and in public cultural life.

Helmholtz was also one of the last representatives of a conception of nature that strove to reduce all phenomena to matter in motion. In reaction to the increasingly insurmountable difficulties that program had in fulfilling its own standards for scientific explanation, he developed elements of a modern understanding of science that have remained of fundamental importance to this day.

This book is a translation of an abridged version of my German monograph *Wahrheitsgewissheitsverlust. Hermann von Helmholtz' Mechanismus im Anbruch der Moderne. Eine Studie zum Übergang von klassischer zu moderner Naturphilosophie.* Some passages and notes have been omitted to produce a condensed text. Bibliographical sources have been updated, English editions of Helmholtz's and other works added to the list. References to Helmholtz's works are page numbers in German editions. Lengthy, indented quotations have been taken from available translations whenever possible. The second page number refers to the corresponding English title listed along with the German title in the bibliography. Some of the English renderings have been tacitly improved.

I am grateful to Jed Z. Buchwald for including *Hermann von Helmholtz's Mechanism: The Loss of Certainty* in the Archimedes Series, to Springer's publishing manager Charles Erkelens for supervising the completion process, and to Cynthia Klohr for the translation. I also thank Felix Bräuer, Philip Flock, Uwe Schürmann, and Mirca Szigat for carefully organizing the footnotes and compiling the bibliography and the index.

Wuppertal
September 2008

Gregor Schiemann

Preface to the German Edition

Throughout the past two centuries, natural science has definitely contributed to revolutionizing social structures. Scientific findings exert sustained influence on people's minds. But in apparent antithesis to this enormous growth in significance, all the while the first signs were emerging, indicating that for various reasons, scientific knowledge was in the process of losing validity and heading ultimately towards progressive hypothesizing. Around the 1850s, as results from experimental research first came to be applied to large-scale industrial manufacturing and also accessible to a wider public, there seemed little reason to doubt that mankind could, basically, "comprehend the world entirely" (Hermann von Helmholtz). In subsequent decades, however, this changed fundamentally. Natural science's claim to knowledge underwent a crisis that peaked in early twentieth century physics. Today, striving for comprehensive and exclusively valid knowledge of nature has lost the esteem it once enjoyed. Today, such efforts represent merely one group of approaches within a complex spectrum of ways to establish theories in natural science.

The object of this book is to explore and understand features of the prehistory and formative phase of that transition, using Hermann von Helmholtz's doctrine of mechanism as an example. It focuses on claims to validity – some of which still seem familiar and others, which in many instances have meanwhile become obsolete. Historically, doubt about the scientific comprehensibility of the world, something that first prevailed in the twentieth century, can be traced well back into the past. Compared to an insight iterated since antiquity, namely that human knowledge is both limited in scope and essentially fallible, the pathos for science's claim to truth as proclaimed by nineteenth-century scientists seems difficult to follow. In fact, looking back, one might be inclined to presume that these scientists entertained motives other than an unselfish love of truth. But even if they perhaps primarily sought fame, social recognition, secure careers, or research funding, they probably could have discovered no better way to legitimate such goals than by announcing the pursuit of absolutely valid knowledge of nature – which itself is a prerequisite for its unrestricted utility.

The changes that the concept of science has undergone since the nineteenth century call for a very conscious effort to understand the previous self-image so widespread in natural research. Immersion in the historical material gives us a sense of how earnestly these scientists sought the truth, how little they questioned the notion itself and how bitter the gradual revelation must have been, that the goal they pursued might, in principle, perhaps not be attainable at all. In terms of claims to validity, historical reflection reveals the remoteness of a past that in other respects still seems immediately tied to the present.

Remoteness and proximity characterize my study of a contradictory chapter in the history of science. This work originated at the Institute of Philosophy at the Technical University of Darmstadt (Germany), funded by a doctoral grant from the *Studienstiftung des Deutschen Volkes*, for which I am grateful. I especially thank Gernot Böhme for supervising my work, supporting it wholeheartedly from the start and exercising an untiring willingness to discuss it. Reading work by Alwin Diemer and Gert König initially stimulated my interest in elaborating the basic idea of the changes the concept of science underwent during the nineteenth century. König was the first to examine the process of change reflected in Helmholtz's notion of science. I presented my theses in Gernot Böhme's postgraduate colloquium and discussed them there with other doctoral candidates. I also encountered critical debate at the International Helmholtz Congress at Ringberg Castle and in lectures at the Faculty for Philosophy at the Ruhr University in Bochum and the Institute for the History of Sciences at the Georg-August University in Göttingen. Timothy Lenoir, Jed Z. Buchwald, David Cahan, Helmut Pulte and Michael Heidelberger discussed separate aspects of my work with me in great detail. I thank them all for their suggestions and encouragement and also thank Sidonia Blättler and Helmut Pulte for carefully reading the manuscript and proposing numerous improvements.

Contents

Introduction

Searching for truth is still exciting in contrast to drab and dreary error; but the excitement is dwindling

(Friedrich Nietzsche, *Human, All Too Human*).

What is science? Today more substantiated, diverse answers to this question present themselves than ever before in the history of European culture and ideas. On the one hand, lingering, yet lively traditions in logical empiricism and critical rationalism still fundamentally and methodologically discern scientific knowledge from other, namely, aesthetic, kinds of knowledge. On the other hand we find equally convincing arguments, as propounded by Paul Feyerabend and Richard Rorty that no grounds can be found for distinguishing various kinds of knowledge from one another.[1] While some would characterize modern empirical science as simply a technically organized, basically inhumane mastery of nature, others have equally strong reasons for thinking that scientific knowledge is precisely what we need for dealing with nature rationally.[2] While some criteria for science are linked to universality, some notably sociological approaches reject all uniform concepts of science altogether and define science by a plurality of contingent, merely locally valid conditions.[3]

This confounding diversity of debatable definitions in the theory of science stands in notable contrast to the unanimity with which, in the theory of science, doubt about science's increasing cultural and social relevance is practically nonexistent. While science's growing significance remains uncontroversial, the sciences themselves, as a topic of reflection, continue to unravel into coexisting, partly diverging, partly converging concepts.

[1] Exemplary for one side are **Popper** (1935), **Carnap** (1936f.) and **Stegmüller** (1973ff.), Vol. II, Ch. IX; for the other see **Feyerabend** (1976) and **Rorty** (1991).

[2] Divergent positions have been taken on the scientific and technological command of nature. **Horkheimer** (1947) and **Heidegger** (1955) set the direction for the debate. Contemporary authors arguing that a rational relationship to nature will be judged by how science deals with ecological problems, are, among others, **Schäfer** (1993) and **Mittelstraß** (1992).

[3] **Popper** (1935), **Carnap** (1936f.), **Stegmüller** (1973ff.), **Horkheimer** (1947) and **Heidegger** (1955) can be contrasted with **Knorr Cetina's** (1984) sociological approach in the theory of science.

G. Schiemann, *Hermann von Helmholtz's Mechanism: The Loss of Certainty*,

A leading motive for my study is to respond to this highly unsatisfactory state of affairs with an elucidating historical analysis. It is based on the assumption that the prevailing complexity in the theory of science is an expression of a radical revolution within science at large and which has less to do with the criteria for *what* science investigates, than with formal criteria for science proper: the sciences, so the assumption goes, are in the process of abandoning the modern claim to absolute, valid knowledge, formerly known as the truth. The emphatic claim to truth is being substituted by modern insight as to the hypothetical character of theoretical knowledge. I express this as an *assumption* because it is not my intent to illustrate current forms of the alleged loss of truth in detail. Instead, my study explores the opening of an epochal transition, the end of which is not yet in sight. It is a reply to the clarion call to revisit the circumstances that shaped modern problems.

I am interested in those early stages in two respects. For one, by discerning more accurately the *early modern* from the *modern* concept of science,[4] I intend to approximately outline the general circumstances under which socially relevant origins of the epochal process developed. It is not necessary to examine every form of the scientific relativization of truth that has challenged European thought since the time of the ancient sophists. In fact, I concentrate on the second half of the nineteenth century. I view this period as pivotal for the nascent relativization of claims to validity, as it has remained constitutive until today. For another, I find it important to illustrate this emerging process of relativization with an exemplary case.

For this purpose I shall draw upon texts written by Hermann von Helmholtz. During the second half of the nineteenth century, Helmholtz was one of Germany's most eminent natural scientists. Due to the compass and the relevance of his work Helmholtz was highly esteemed, comparable by all means to Alexander von Humboldt. His work, though, is so diverse, that I must limit my study to one aspect of it. The mechanistic notion of nature that Helmholtz advocated throughout his life and that will be the subject of our case study is also representative in terms of content. For contemplating nature in terms of machines and attempting to reduce natural phenomena to matter in motion is historically inseparably linked to early modern scientific method.

At first it may seem a paradox to view the mechanistic notion of nature as central to the epochal transition that induced the modern concept of science.[5] "Modern philosophy of nature" is usually understood as precisely having gotten over the mechanistic concept of nature still virulent at the onset of the twentieth century. The anti-mechanistic concepts frequently mentioned in this context refer in part to physical theories of sub-microscopic systems, more recently particularly theories of chaotic systems and in part to biological theories of evolution and self-organization.[6] But as long as their critique clings to the notion that experimental research will

[4] Note on translation: For the purposes of this study, "early modern times" designates what in German is known of as "neuzeitlich", the period lasting approximately from the end of the Middle Ages up until Modern Times. In terms of the theses elaborated hereafter in this introduction, "modern times" begin approximately in the nineteenth century and continue into the present.

[5] In the following I use the terms "science" and "natural science" synonymously.

[6] Cf. **Kanitscheider** (1993) and a representative collection of papers in **Kanitscheider (Ed.)** (1984).

someday deliver a uniform concept of nature, whose validity will be consolidated at some future time, it can hardly be called "modern" in the meaning of the word developed here. Knowledge is only modern when it is marked by a relativeness of its claim to validity. The sign of modernity in the philosophy of nature is not the substitution of one concept of nature with an exclusive claim to truth for another; it is the formal acknowledgment of the narrowness and contingency of validity conditions and granting equal standing to alternative approaches. For mechanism, this means that it loses the entitlement to exclusive agency that it enjoyed from the onset of early modern times and becomes just one worldview among others. As hypotheses, its tenets can then also contribute to the plurality of approaches to nature, irrespective of any otherwise required empirical confirmation.

This case study consists in reconstructing the process of hypothetization that mechanism underwent within the work and writings of Hermann von Helmholtz. It shows how the relativeness of claims to validity so typical of present-day science already grew deep within mechanism. Within the framework of the tradition of mechanism, the onset of the epochal transition lies in the latter half of the nineteenth century, where the final decades of this concept of nature commence with a considerable renaissance around the middle of the century. In the concrete historical situation, Helmholtz's view of nature was confronted with growing criticism from about 1870 on. This suggests that he surrendered the claim to truth he had previously advocated for his view of nature, in order to further uphold that view as a hypothesis. Paradoxically, a characteristic feature of modern concepts of science might have contributed to the perpetuation of a concept of nature, which, at a first glance, looks hardly modern.

There exists some previous groundwork on the topic with which my exploration could be associated but it is astoundingly scarce. In recent times, it was Auguste Comte who noted epochal transitions in the history of science embedded in cultural contexts; he sees sciences first entering their own true epochs once they have first passed through a theological and then a metaphysical epoch.[7] In the twentieth century Gaston Bachelard was the first to advocate, with any long-term influence on the history of reception, that the sciences can only approximate objective knowledge by removing obstacles one stage at a time. According to Bachelard, in a fractured process, the sciences must overcome, one after another, successive traditional notions of substance, animist preconceptions and the idealistic pursuit of unbounded accuracy.[8] On his interpretation, scientific method could not have existed prior to the mid-nineteenth century. The "scientific mind" typical of scientific method is allegedly radically different from the spirit of philosophy prevailing up to that point and which was established on "absolute, unchanging reason".[9] "Scientific spirit" needs no justification based on timeless reason: it legitimates itself, works on the unknown and searches reality precisely for phenomena that contradict the knowledge it takes for granted.[10]

---

[7] **Comte** (1830ff.).

[8] **Bachelard** (1938).

[9] **Bachelard** (1940), p. 10 and 123.

[10] Loc. cit. p. 22ff.

Chronologically and thematically, Bachelard thus articulated a key conception in the historiography of science that significantly stimulated my study. Typical features of current science, to put the main idea very simply, originated in the nineteenth century in a period isolated from previous development by a phase of radical upheaval. After Bachelard, in France Michel Foucault and in Germany Wolf Lepenies described this transformation process and the features of modern scientific method that emerged from within it. Their accounts allow the conclusion that when seen from the viewpoint of historical genesis, the epistemic mark of modern science is the unconditional abandonment of its former certainty regarding truth. Foucault sees science prior to the threshold of the epoch, which he dates as being the years between 1775 and 1825, as taking its object of investigation to be representations of ideas that exist independently of physical objects. When science dismisses that truth-avouching representation relation, it begins to organize its objects of study in terms of inner principles.[11] "A profound historicity penetrates into the heart of things, isolates and defines them in their own coherence, imposes upon them the forms of order implied by the continuity of time".[12] Using the term "temporalization", Lepenies groups the elements characteristic of the transition in science at the dawn of modernity, which he dates similarly. By this he particularly means the breakthrough of the idea of evolution and thinking of time in terms of processes.[13] But to the extent that the historical transition itself becomes a topic of scientific study and scientific knowledge itself becomes historical, each one's claim to truth begins to relativize itself within its own historical context.

At this point, the historiography of science's interest in epochal transitions converges with discourse on the concept of modernity led predominantly by philosophers.[14] Inasmuch as that discourse refers to the sciences, it shows us more clearly than the historiography of science that it is necessary to make a conceptual distinction between early modern times that began in the sixteenth and seventeenth century and the modernity that defines the twentieth century. Wolfgang Welsch has suggested that in terms of content, we identify the cultural-historical onset of early modern times (excluding ensconced counter movements) as the program set forth by René Descartes and Francis Bacon for mastering the world through science and technology and in terms of form, identify it with the pathos of a radical new start and the claim to universality.[15] From these early modern times he then distinguishes culturally-historically defined modernity merely by the fact that it has given up the claim to universality. In early modern times, we could say, following Welsch, it would "not be possible for truth to appear in any other way than with an exclusive claim to validity"; for modernity, however, this is fundamentally different. In modernity "plurality and particularity are not only conceivable, they are dominant and [paradoxically – G.S.] binding".[16]

---

[11] **Foucault** (1966).

[12] Loc. cit. p. XXV.

[13] **Lepenies** (1976).

[14] **Lyotard** (1979), **Frank** (1987), **Spaemann** (1986).

[15] **Welsch** (1988), p. 66 ff.

[16] Loc. cit. p. 76 f.

Yet it is characteristic, not only of early modern times but also of all occidental history of science (except for counter movements like Greek sophistic philosophy, ancient skepticism and British empiricism) until the onset of modern times, that truth has always been advocated emphatically, marked by an absolute claim to validity. It exclusively denoted something used as a starting point for determining both the essence of a being qua being and the properties of an object considered congruent to those of another object. Regarding the nature of claims to truth, one is thus led to a dichotomy in the history of science. Following A. Diemer and H. Schnädelbach I shall call the conception of science characterized by an emphatic claim to truth "classical" and view it as having roots in Aristotle. This I contrast with the "modern" conception of science. The latter I define using the criteria of hypotheticity, i.e., the increasing openness of truth in theoretical knowledge (cf. Chapter 4: Contours of Modern Philosophy of Nature). On both the classical and the modern understanding of science, criteria for examining claims to truth are derived from validity conditions.

Now, for elaborating a case study from the second half of the nineteenth century, it is advisable to define the classical-modern conception of science more carefully. In order to better classify Helmholtz's mechanism formally, I explore the special features of the early modern notion of science and how it relates to our contemporary notion. Early modern and modern notions of science differ from earlier conceptions not only by the fact that the latter pursued no goal of mankind mastering nature, nor did it emphasize a radical new start. While sciences prior to early modernity legitimated their claims to truth primarily by reference to transcendent entities, early modern sciences developed two basically non-transcendent and mutually exclusive strategies for legitimating their equally absolute claims to truth: principles of scientific explanation were obtained and justified either metaphysically (Descartes, Kant) by pure contemplation, or scientistically (Galileo, Helmholtz) with reference to practical experimental research. What I call "metaphysical" legitimation abstains from any reference to experience and emanates solely from the intellectual power of the autonomous subject. In contrast, the experience that corroborates "scientistic" legitimation is generated rationally in specific scientific procedures, such that one could say it is one special, namely scientific, variety of empiricism. Not until the onset of the early modern epoch was experience as such either discovered or rejected in natural research, in terms of being an independent foundation for knowledge. Both strategies sought to establish truth as systematically ordered, universal and necessary, i.e., certain knowledge. Kant writes, "Whatever I . . . know", I reckon "*apodictically* certain, i.e., as universal and objectively necessary".[17]

Hence, it is crucial for my approach to truth's modern loss of certainty that the scientistic strategy can be found in both the early modern and our modern concept of science. While the relativization of truth claims is associated with the demise of all metaphysics, the transition from the classical to the modern concept of science

[17] **Kant** (1900 ff.), Vol. IX (*Logic*), p. 66 (emphasis in the original text).

happens within the framework of the scientistic conception of science.[18] This narrows down the topic of transition to one particular tradition of legitimating science. It also settles the issue that the positive reference to experience intrinsic to modern scientism did not – as Lepenies suggests – lead from the start to the deterioration of absolute claims to truth.[19] In addition, a potential array of causes for the modernization of truth claims can be addressed: if relativity of truth is not automatically intrinsic to modern scientism, perhaps circumstances *beyond* all scientistic rationale for science contributed to the transformation of truth claims. Such a combination could render discontinuities of the transformation process plausible.

Of course, within the historiography of European culture and science it is not novel to distinguish different strategies in modern foundations of science. It is a well-known historiographical commonplace that in terms of legitimation, Galileo's, Francis Bacon's and Isaac Newton's empiricism is the opposite of the metaphysical foundation for science propounded by René Descartes, Gottfried Wilhelm Leibniz and Immanuel Kant. Yet, in terms of their claims to truth, these legitimating strategies have been evaluated very differently. In particular, an attempt has often been made to prove that the origins of hypotheticity so typical of the modern conception of science are to be found in the writings of early modern authors.[20] Although some evidence of such harbingering is indisputable, to me it appears that generally early modern times understood the scientific claim to truth in a non-absolute way. But since there is, as of yet, little consensus on this, I shall expatiate both modern strategies, emphasizing the absolute claim to truth that they have in common (cf. Chapter 2: The Classical Conception of Science).

I regard Helmholtz as a representative advocate of the scientistic notion of science in nineteenth-century natural research. To me, what supports this classification is the relationship between Helmholtz and Kant, which in the past has been a source of frequent, contradictory and culturally charged examination. Far from desiring to globally assess the nature of validity for the possibility of objective knowledge that Kant establishes, I think that his metaphysical principles of mechanism provide a paradigmatic example of the classical and early modern foundation for science. In his early writings, Helmholtz also provides one such example. But in contrast to Kant,

---

[18] At this point I shall refrain from further explaining this statement, which would be worthy of a study in itself. Disagreement over the evaluation of German idealism's contribution to the modernization of the concept of science shows that how metaphysics is related to the relativity of claims to truth can be a matter of controversy. While **Diemer** (1968), for example, sees Kant as belonging entirely to classical science, **Foucault** (1966) puts him on the side of modern science. For present work on Helmholtz's theory of knowledge, it would also be interesting to elaborate J.G. Fichte's role in the transition process.

[19] Lepenies links the development of temporalization to the growing weight of experience in the 18th century (**Lepenies** (1976), p. 16ff.), making it seem as if the early modern direction itself led right up to the modern type of science he describes.

[20] See **Carrier and Mittelstraß** (1989) on Kepler, **Mittelstraß** (1970), p. 169, on Galileo, as well as **Krohn** (1990) on Bacon and **Loeck** (1986) on Descartes.

he does not set out with non-empirical principles; Helmholtz develops and argues his foundation for mechanism based on scientific experience. The difference between the two strategies is underscored by the entirely different structure of each one's concept of nature. While Kant sets no bounds to the scope of his mechanistic concept of nature and the world, Helmholtz divides the world of phenomena into those that are causal and those that are not. While Kant establishes a dynamic mechanism that traces all material properties back to forces and which is most easily derived from metaphysics, Helmholtz deduces his preference for an irreducible duality of force and matter from advances in chemistry research.

I make this comparison solely within the framework of the mechanistic concept of nature. Going by my definition of mechanism, I already, in a way, apply the standard from Helmholtz's program to the history of mechanism; that standard consisting of "the ultimate goal of the sciences" being to "merge into mechanism".[21] For this reason I understand mechanism as meaning not simply, as is often the case, a concept that views matter in motion as the primal cause of all natural phenomena and which I call mechanism in the wider sense. I also use the term mechanism in a narrower sense and apply it to those approaches in the philosophy of nature and the theory of science that establish their explanations on the principles of mechanics, as enunciated paradigmatically by Newton and Lagrange (cf. Chapter 1: The Conception of Mechanism).

While formulations for mechanics were available soon after early modern times began and their fit to nature became ever less controversial within the scientific community, various lines of tradition can be distinguished for the concepts of nature that, in terms of content, were guided by it. The relationship between Helmholtz's and Kant's mechanism shows that dissimilar combinations of concepts of matter and force can correspond to both of the legitimating strategies mentioned. In contrast to problems involved with foundations, to which the historiography of science has already devoted some attention, the possibility of distinguishing different traditional lines within the history of the mechanistic concept of nature has only rudimentarily been suggested.[22] It was necessary to remedy this desideratum, in order to classify Helmholtz's mechanism in terms of content.

As I see it, we can distinguish three different traditional lines of mechanism in the narrower sense: the first line is materialist. Its fundamental principle admits only matter in motion and rejects an irreducible concept of force (Descartes, Huygens). The second is dynamic. This line deduces all properties of matter from forces (Leibniz, Kant). The third line is dual and puts the concept of inert matter on equal standing with the concept of force (Newton) (see Chapter 3: Three Traditions in Mechanism). Once mechanism has been defined in the narrower sense as a scientifically oriented concept of nature – which naturally by no means

[21] **Helmholtz** (1869), p. 379.

[22] See Chapter 3, footnote 88.

suppresses its worldview implications – further definitions essentially arise out of the concept of science itself.[23]

Now, within the framework Helmholtz's lifelong held scientism and mechanism, his notion of science underwent a lively change from the start, the investigation of which constitutes the second half of this book. Its development led to such profound modifications that Helmholtz's view of science and nature can be described as uniform only to a very limited degree. I make use of the elements that did remain stable solely for the purpose of staking out the process of transition, the scope of which can hardly be overestimated: within the sciences, it provides one of the most impressive pieces of evidence for the erosion of a classical position once advocated with the greatest conviction. In terms of emphasis, universality and necessity, the claim to truth asserted by Helmholtz well into the 1860s falls barely short of Kant's. Then, in the course of the 1870s, his certainty concerning truth begins to falter. One of the most conspicuous outward signs was a new kind of significance that was awarded the concept of the hypothesis within the scientific vocabulary. The second half of my study intends to elucidate, above all, the extent to which the elements of modern hypothetization were already part of the modification undergone by Helmholtz's stand on science.

Although detected early, the change in Helmholtz's notion of science and nature was for a long time considered secondary for evaluating both his complete works and individual aspects of his research. This was perhaps due to the fact that until recently the reception of his notion of science and nature has only on rare occasions been made an isolated object of study. His notion of science attracted more interest in commemoration publications. There, more in-depth attention has been given to the elements of change in Helmholtz's thought, chiefly when investigating issues not exclusively or even explicitly tied to topics in theories of science and nature. Remarkably, these often have less to do with inclinations toward hypothesizing claims to scientific truth, than with reactions to

[23] What I have elaborated so far can be illustrated by a table. Note, however, that a scheme such as this does injustice to reciprocity. "Medieval times", "early modern times" and "modern times" stand for the concepts of science entertained during these epochs. The transition from the early modern to the modern concept of science is considered ongoing, making both types characteristic of the concept we have today.

<table>
<tr><th colspan="2"></th><th>Medieval Times</th><th>Early Modern Times</th><th>Modern Times</th></tr>
<tr><td rowspan="3">In terms of form</td><td>Claim to truth</td><td colspan="2">absolute</td><td>relative</td></tr>
<tr><td rowspan="2">Foundation</td><td rowspan="2">transcendent</td><td>metaphysical</td><td></td></tr>
<tr><td colspan="2">scientistic</td></tr>
<tr><td colspan="2" rowspan="3">In terms of content: conceptions of nature/traditions in mechanism</td><td colspan="3">dual</td></tr>
<tr><td colspan="3">dynamic</td></tr>
<tr><td colspan="3">materialist</td></tr>
</table>

that development, which themselves can also be found in Helmholtz's writing. This is all the more surprising because very early discussions of the transition already indicated that it included relativizing claims to validity.[24] Helmholtz's work is doubly topical for science and the theory of nature: it is useful not only for studying the onset of a modernization trend formative of the present but also for studying approaches of apparently still current attempts to compensate that loss of certainty.

The enormous public impact that Helmhlotz's writing had should not let us forget that evidence of hypothetization in mechanistic theory can most likely be found in work by other mathematicians and natural researchers as well, such as Carl Gustav Jacob Jacobi, Bernhard Riemann, Carl Gottfried Neumann, Emil du Bois-Reymond and Ludwig Boltzmann. But merely scattered research is available.[25] Also Pierre Duhem, Henri Poincaré and Ernst Mach – just to name those most important – tied their criticism of absolute claims to truth in science to a denial of the mechanistic program of explanation.[26]

Other than general overall accounts, a detailed isolated study like this is also always faced with the fact that the effects and the signs of the alleged epochal transition it seeks happen so subtly and are always shaped by the contingent circumstances of a person's life and research. Yet, even while limiting the subject to mechanism, the relevant spectrum of areas Helmholtz worked in is profound:

---

[24] First mentions of a change in Helmholtz's conception of science and nature can be found in **Koenigsberger** (1895), **Riehl** (1904), **Conrat** (1904), **Erdmann** (1921) and **König** (1968). **König** (1968), **Hörz and Wollgast** (1971), p. XXXIX ff., **Cahan** (1994b), **Cahan** (1994c), **Hatfield** (1994), **Heidelberger** (1994a), **Heidelberger** (1994b), **Krüger** (1994), **Schiemann** (1994), **Röseberg** (1994), **Leroux** (1995), **McDonald** (2001), Chap 1, and **Jurkowitz** (2002) published studies on Helmholtz's conception of science. His conception of science has also been discussed in context with his works on electrodynamics: **Buchwald** (1994a) and **Darrigol** (1994). Finally, studies on individual topics, such as acoustics (**Vogel** (1994)), the theory of perception (**Conrat** (1904) and **Hatfield** (1990)) or the conception of force (**Winters** (1985)) allow inferences to his views on the theory of science (see also Sections 6.3.1, footnote 403, and 7.1, footnote 625, for secondary literature on the change in Helmholtz's notion of science). Parts of his concept of science have been discussed in a variety of connections but fairly rarely made a topic of study themselves (cf. Section 6.3.2, footnote 540). **Riehl** (1904), **Erdmann** (1921), **König** (1968) and **Friedman** (1997) addressed the trend towards hypothesizing. **Schulz** (1995), **Pulte/König** (1998) and **Leiber** (2000) have adopted the hypothetization that I pointed out. Undoubtedly, I found important suggestions for my own study in König's groundbreaking work (see the preface). König as well sees the evidence he found for a change in Helmholtz's conception of science as part of the tension between a classical and a modern understanding of science (although he defines the period differently). He ultimately concludes that "the swing from the 'classical concept of science' to the 'modern concept of science'" "is obvious in Helmholtz's development" (loc. cit. p. 100). His brief expositions, however, unfortunately provide only random samples of evidence for this. Some of his assessments of individual texts depart from mine, for example, how he sees the empirical principles of geometry as being modern (loc. cit. p. 100; cf. also Section 6.2 of this book) and how he lists the mechanistic program that Helmholtz proclaimed at the conference for natural research as an example of the modern tendency to make theories conditional (loc. Cit. p. 99 f. and Section 6.3.2, footnote 555, in this book).

[25] Namely for C.G.J. Jacobi: **Pulte** (1993), **Pulte** (1994), **Pulte** (1996) and **Knobloch et al.** (1995), for C.G.J. Jacobi, B. Riemann and C.G. Neumann **Pulte** (2005) and for L. Boltzmann: **Curd** (1978).

[26] **Duhem** (1904 f.), **Poincaré** (1902) and **Mach** (1872) and (1883).

we shall be dealing with his theory of matter, his investigations into energy, his work on the empirical foundation of geometry, certain aspects of his studies in electrodynamics and thermodynamics, fundamental elements of his theory of perception and his conception of science and nature.

While focusing on an exemplary case allows us to unfold a text's micro-structure complexity as carefully as possible, it may also neglect cultural-historical surroundings, factors within and outside of the scientific scene and parallel, as well as counter-developments. In my opinion this tendency need not prove to be disadvantageous, as long as it does not prevail. Omitting excessive reference to peripheral contexts may counteract the misconception that we may derive conclusions about the existence of a general process of transformation by scrutinizing one case. On the other hand, however, a fundamental obstacle to understanding individual phenomena can arise from elaborating an isolated case, if its manifestations and modifications cannot be understood without knowledge of the respective contexts. Thus we cannot ignore the fact that the onset of transition in Helmholtz's notion of science happens at a time marked by crucial new biographical circumstances, such as the move from holding a professorship for physiology in dreamy provincial Heidelberg to obtaining a university chair for physics in metropolitan Berlin. Of course, not all conditions and causes of the transition were external, some were intrinsic to his work, like the interrelatedness of all the projects he pursued successively or simultaneously across various scientific disciplines. It is wrong, however, to think that we cannot recognize a phenomenon as such until it has been traced back to its alleged causes or until we realize its embedment in a complex context. All too often that makes it vanish altogether.

To ward off that danger, I shall initially refrain from all causal analysis of the transition process and reconstruct following the transition's own pattern of development. My reconstruction will follow the course taken by the development itself. Indeed, since this is about an open process, a causal analysis cannot be the point of reconstruction. Novel terms and concepts emerge and structural features evolve unpredictably and we cannot classify or rudimentarily explain them before the process has come to an end. For this reason, I confine myself to text-immanent investigation and shall discuss possible conditions and causes of the transition process later. The textual basis I use is a selection of Helmholtz's published writings in which he mentions his concept of science and nature in general. These pieces sufficiently document the individual steps of the development and serve as a basis for a preliminary understanding of the transition.

Finally, my procedure can be called "rational reconstruction" (I. Lakatos). The classical or early modern and the modern concepts of science provide two distinctly different sets of criteria for scientific rationality that we must use as a basis, not only for a later evaluation of Helmholtz's mechanism but from the start, when describing how his concept of mechanism developed. Every study in the history of science assumes standards of rationality when selecting and evaluating its material. Instead of a futile attempt to write an objective report, I – with slight exaggeration – intend to fit, as well as possible, Helmholtz's mechanism into the framework I lay out but without coercion. Of course, the framework has already been erected in terms of the object of study but the particular criteria for scientific rationality raise an epochal claim and are justified without recourse to Helmholtz.

Other classifications are conceivable and other authors have tried them. I would like to suggest, however, that my classification manages with a comparative minimum of argumentative effort. Associating Helmholtz's conception of science with, for instance, the Neo-Kantian schema or Fichte's type of idealism takes considerably more effort than I am willing to exert. The fact that in the following I shall not repeatedly mention reconstruction and that occasionally the impression arises that this is a debate about historically indisputable facts, is the price paid for readability.

The first part of my study (Mechanism between the Classical and the Modern Conception of Science) begins with the conceptual definition of mechanism. Since I apply the concept in the narrower sense to early modern mechanics, this also includes the elementary characteristics of that discipline (Chapter 1: The Conception of Mechanism). Having thus determined the content framework for further investigation and the historical framework for the prehistory to Helmholtz's mechanism, I introduce the concept of the classical notion of science as the first formal criterion. Within classical mechanism I distinguish the metaphysical from the scientistic strategy of legitimation. (Chapter 2: The Classical Conception of Science). I then turn to the subsequent historical development in mechanism, distinguishing three separate lines from one another (Chapter 3: Three Traditions in Mechanism). That overview provides all the relevant definitions, in terms of content and form, for the prehistory to Helmholtz's mechanism, which lasts up to the nineteenth century. The final section leaps to the present and elaborates hypotheticity as the mark of our contemporary concept of science (Chapter 4: Contours of Modern Philosophy of Nature).

Part I
Mechanism Between the Classical and the Modern Conception of Science

Chapter 1
The Conception of Mechanism

We cannot triumph over the machinery of matter by ignoring it; we can triumph over it only by subordinating it to the aims of our moral intelligence. We must familiarize ourselves with its levers and pulleys [. . .] in order to be able to govern them after our own will and therein lies the complete justification of physical investigation and its vast importance for the advance of human civilization
(Hermann von Helmholtz, *On Goethe's Scientific Researches*).

1.1 What is Mechanism?

Hermann von Helmholtz not only rarely made a major issue out of the idea of nature that he advanced most notably in popular writing and which was linked to his scientific work in many ways; he also left it up to others to coin a term for it.[27] What I would like to call Helmholtz's mechanism was one of the fundamental convictions shared by most physicists and large circles of other scientists throughout the second half of the nineteenth century.[28] In two passages, Helmholtz very clearly exemplifies his idea of nature. One is found in the introduction to his most famous treatise *The Conservation of Force: A Physical Memoir* [*Ueber die Erhaltung der Kraft*], published in 1847, where he writes programmatically: "Natural phenomena should be traced back to the movements of material objects which possess inalterable motive forces that are dependent only on spatial relations".[29] He defines these forces more closely as "inalterable forces of attraction and repulsion, the intensity of the forces depending on the distance".[30] Twenty-two years later, basking in twofold fame as a physiologist and physicist,

[27] **Mach** (1872), p. 18 ff., **Schwertschlager** (1883), p. 81, **Helmholtz** (1853), p. 45, **Helmholtz** (1869), p. 396.

[28] **Jungnickel and McCormmach** (1986), Vol. II, p. 211 ff., **Merz** (1907 ff.), Vol. III, p. 399 and 564 ff.

[29] **Helmholtz** (1847a), p. 5.

[30] **Helmholtz** (1847a), p. 6.

he says in his opening address to the Association of German Natural Researchers and Doctors [*Versammlung Deutscher Naturforscher und Ärzte*] in Innsbruck:

> Chemistry shows by actual investigation that all matter is made up of the elements which have been already isolated. [...] In their *properties* [...] they are altogether unchangeable; [...] If, then, all elementary substances are unchangeable in respect to their properties and only changeable as regards their mixture and their states of aggregation – that is, in respect to their distribution in space – it follows that all changes in the world are changes in the local distribution of elementary matter and are eventually brought about through *motion*.
>
> If, however, motion is the primordial change which lies at the root of all the other changes occurring in the world, every elementary force is a force of motion and the ultimate aim of physical science must be to determine the movements which are the real causes of all other phenomena and discover the motive powers on which they depend; in other words, to merge itself into mechanics.[31]

What these two expositions have in common is obvious: material elements ("material objects" or chemical "elements"), change of position and forces of motion are the only fundamental concepts Helmholtz permits. The elements, discretely distributed in space, are incapable of any kind of change other than a continual change of position. To this process, upon which presumably all phenomena are founded, are linked forces of motion whose definitions can be derived from the principles of physical mechanics. The first quotation characterizes them as so-called central forces, as Isaac Newton had defined them in his mechanics and which were widely considered paradigmatic in physics until well into the nineteenth century. For the second quotation we can also assume that by explicitly mentioning mechanics, Helmholtz also identifies the kind of simple forces he had in mind.

The concept expressed in both excerpts can be called mechanistic in two respects: it is mechanistic in the broader sense in that it views matter and motion very generally as the primordial and exclusive causes of all natural phenomena. It is mechanistic in the narrower sense in that the kinds of motion it allows are fixed by the principles of a specific discipline, namely mechanics. I would like to more accurately elucidate the concept of mechanism in the narrower sense by first explaining the distinction between the narrower and the broader meaning and then by outlining the philosophy of nature aspect inherent in the narrower meaning (the focus of my study) as distinguished from the scientific aspect of mechanics. Finally, I shall contrast both meanings with other ways of understanding the concept.

In its broader sense the concept of mechanism includes fundamental definitions of matter and motion. These presuppose the "logical primacy" (Cassirer) of the traditional notion of substance: "Only in given, existing substances are the various determinations of being thinkable".[32] In order to understand all phenomena as expressions of matter in motion, or to explain them in any way using such terms, we discretely or continually split matter into segments that are not necessarily soul- or spiritless but in themselves *are* unchanging and differ from one another at the most

[31] **Helmholtz** (1869), p. 378 f. (Germ.)/211 (Engl.) (italics in original).

[32] **Cassirer** (1910), p. 10.

in terms of purely quantitative properties of form, such as size and shape; thus the concept of motion reduces natural change to the spatial movement of particles. Not only does this broader sense of the term set no limit to the diversity of forms, it also reveals nothing about its position on the elementary forces that in mechanistic systems might be responsible for cohesion among the various material particles, for their gravity and for the movements, or kinds of movements, they make.

A glance at the history of mechanism gives one the feeling that we could hardly grasp the compass and complexity of it, were it founded on the concept of mechanism in the broadly construed sense.[33] Restricting the definition of mechanism to its narrower sense is thus in and of itself indispensable for reasonably limiting the wealth of material relevant to the prehistory of Helmholtz's mechanism. In doing so, however, I shall not revert to science's internal debates on the principles of mechanics and how these are interpreted by the philosophy of nature. Choosing the science of mechanics as our bearings provides a criterion-directed guideline immanent to philosophy of nature discourse that sends those concepts to the sidelines, which themselves (usually of their own accord) cannot be measured by the canon of mechanics.

In the narrower sense, the contents of the concept of mechanism are specified by the principles and historical development of physical mechanics. Assuming that these principles hold as enunciated by Newton and Lagrange, which I shall expound below, the mechanistic concept of force can be considered on equal standing with the concepts of matter and motion and all three can be defined and operationalized using dimensions of time, space and mass. Historically, alignment with mechanics was seen as a way of isolating mechanism from fundamental atomist positions. While mechanics must include substances, substances must not necessarily be distributed discretely in space. They may be continuously distributed media, like those suggested by nineteenth century ether theories favored by mechanistic thinking at the time.

Mechanism extends the special context of application for mechanics in two directions: for one, mechanism spans the entire system of natural science. It becomes the discipline into which – as Helmholtz puts it – all other sciences must "merge". This scientific task must be distinguished from the philosophy of nature position that the object of mechanics are – to use Helmholtz's expression – "primordial changes".[34]

By the scientific aspect of mechanism I mean that scientific research was internally guided by mechanics. The purposes of that end – an end pursued generally by all disciplines of early modern natural research – were typically served by mechanistic heuristics and by partially breaking down non-mechanical concepts and theorems into mechanical ones. As heuristics, i.e., as a regulative orientation for experimental and theoretical work on as yet hardly researched phenomena, mechanics mattered from early modern times on and well into the twentieth century, particularly because of the exemplary mathematical structure of its equations and the clarity of its models that correlated to physical

---

[33] A brief survey of mechanism's history in the wider sense is given in **Schiemann** (1997), p. 22f.

[34] Whenever unnecessary to distinguish the scientific aspect from the philosophy-of-nature aspect, I shall simply call mechanism a theory of nature.

phenomena.[35] The mechanistic attempts at reduction also undertaken from the onset of early modern times form the prelude to a story of difficulties which persists into contemporary philosophy and that after the close of the nineteenth century no longer referred unilaterally to mechanics but likewise to the relation of physics to other sciences, particularly chemistry and biology.[36] The mechanistic part of that story includes attempts to deduce laws for thermodynamics and electrodynamics from the mechanical movements of elementary particles[37] and the effort to replace nonphysical methods of research in biological disciplines with methods from physics ("physicalism").

In terms of the theory of science, the scientific aspect of mechanism is also expressed by the impulse to ordain a particular idea of explanation as the norm. From modern times onward dominant causal explanation of nature initially understood causes as mechanically moved substances. These were later gradually replaced by mechanical laws of force. Through this the concept of explanation within the tradition of mechanism begins to disengage itself from notions of substance, such that explanation comes to be understood as the logical inference of a scientifically comprehended phenomenon from laws under certain constraints. A scientific explanation is therefore mechanistic (in the narrower sense), when it avails itself exclusively of mechanistic concepts and laws.[38]

I count as philosophy of nature every text that focuses on nature as a whole and asks what its basic traits are.[39] The holistic aspect can be accommodated in very different ways: one task of philosophy of nature is to summarily describe life's experience and scientific insight. Faced with profoundly developed specialization, that has already become difficult without synthetic efforts on one's own. Some of the special kinds of access to nature thus far for the most part pursued by philosophy alone can also be called holistic. Of these, of importance for the philosophy of nature in the early nineteenth century were approaches categorically distinguished from the scientific approach to knowledge. These approaches were based on conceptual contrasts, like the distinction that Schelling adopted from Spinoza: natura naturata versus natura naturans. In contrast to natura naturata as studied by science, natura naturans is the object of reflection for the philosophy of nature.[40] The critical potential of this approach,

---

[35] Cf. **Cardwell** (1971), p. 118, and **Klein** (1972), p. 67 ff.

[36] Secondary literature on the challenge induced by the notion of reduction is legion. **Nagel** (1961) provides a standard definition of the problem, **Cohen et al.** (1976) provide an overview.

[37] **Descartes** (1644), III, 55 ff., **Newton** (1872), p. 282 ff. Overviews of nineteenth century mechanistic attempts at reduction can be found in **Winckelmann** (Ed.) (1894 ff.), Vol. 2.1. (1894), p. 623 ff. (*Die Natur des Lichtes*), Vol. 2.2. (1896), p. 416 ff. (*Mechanische Wärmetheorie*), and Vol. 3.2. (1895), p. 550 ff. (*Erklärungsversuche für die elektrischen Erscheinungen*).

[38] **Nagel** (1961), p. 79 ff., and **Scheibe** (1971 ff.).

[39] Philosophy of nature (in contrast to romantic *Naturphilosophie*) could be understood generally as the philosophical search for a concept of nature, something that can only be seen as a legitimate undertaking to the extent that it separates its study from other philosophical and scientific endeavors. To this day this demarcation has traditionally been accomplished by the contrasts typical of European cultural history, which make a distinction between what is nature and what is convention, technology, or man-made and so on. Cf. **Böhme** (1992), p. 11 ff.

[40] **Siebeck** (1890) and **Schelling** (1799), p. 33 f.

as compared to mechanism guided by science, is still quite observable in writings by Helmholtz's teacher, Johannes Müller. Conversely, Helmholtz makes this itself a topic of mechanistic critique.

The theory of mechanism can only be taken as philosophy of nature if it satisfies the requirement of inquiry into the basic traits of nature, i.e., if it makes essential definitions of nature its central issue, in search of an answer to the question: "What is nature?" If those traits are physical, that search may be called ontological philosophy of nature. Statements in the mechanistic philosophy of nature are physical when they define the contents of what they refer to, namely substance in motion. These include, for instance, specifications regarding the quantitative properties of atomic elements (size, shape, number) and the types of motion realized in nature.

Nevertheless, the basic traits of nature defined by the philosophy of nature need not necessarily be physical. Philosophy of nature can also be done as epistemology – whether by transcendentally analyzing on the formal structures of space and time and attained synthetic knowledge or by analytically scrutinizing the logics and pragmatics of scientific *linguae francae*. These configurations, to which statements about content, respectively meanings, are subject, may also be rightly called "basic traits". They become mechanistic when they designate formal traits characteristic of physical mechanics such as the asymmetry of space and time, to which I shall return in Section 1.2.

Defining the philosophy of nature as thought that designates as its central theme the whole of nature viewed in terms of nature's own fundamental traits allows for including not only nature-related discourse within academic philosophy but also for embracing efforts made outside of academic philosophy. Some of these were debates among natural scientists and particularly important for mechanism. Unquestionably, both of Helmholtz's passages quoted at the outset of this chapter can be counted as philosophy of nature: he speaks of "natural phenomena" and very generally of "all changes in the world", where "world" means something that excludes the mind and includes life only in a constrained sense. He believes to be naming the basic traits of the totality of all phenomena by reducing them to inert matter in motion and movement-altering forces and by defining their essence or foundation, as substance-related spatial motion, also called "primordial change".

Inasmuch as the main feature of the thus introduced broader and narrower sense of the concept of mechanism is its direct reference to substance in motion, it can be sundered from other significations of the term advocated both throughout the prehistory of mechanism (which we shall discuss below) and within studies of the historiography of science and which I divide into two groups.

For one, the term mechanism is often used when nature is compared to mechanical apparatuses, or the manufacturing of such, for the purpose of typifying the entire course of nature allegorically, or individual natural phenomena metaphorically.[41] Analogies of this kind, however, are not based on the knowledge that mechanical equipment itself is subject to the laws of science and thus may be much older than

---

[41] **Hesse** (1961), p. 51 ff., Art.: *Organ, Organismus, Organisation, politischer Körper,* in: **Brunner et al. (Ed.)** (1977), **Kirchner** (1833), p. 213, and **Brugger (Ed.)** (1950), p. 213.

that knowledge. Thus the ancient and medieval notion of the cosmos being a "machina mundi", which can be traced back to Plato's *Timaeus* and Chalcidius's comments on it, does imply a mechanical order. But while for Plato mechanics were part of lifeless technology, the cosmos was a "visible living thing".[42]

For another, not only were scientific generalizations of the principles of mechanics and those found in the philosophy of nature called mechanical or mechanistic but also the elements of their epistemological implications were so designated. In this sense, the causal method was primarily called mechanistic and contrasted with the teleological manner of viewing nature and the deterministic view of nature was associated with the mathematical form of mechanistic equations.[43] The term thus becomes the embodiment of all ways of gaining knowledge that proceed by using methods like those of mechanics. This stretches the term almost beyond manageability. The causal principle ("if, then always"), historically speaking first scientifically applied successfully in mechanics, has meanwhile throughout science gained recognition as a "comprehensive hypothesis of great heuristic value".[44] In physics the type of differential equation originally used only in mechanics (second-order linear equation where time is the variable) has enjoyed so many applications (generally for non-deterministic problems) that one can find neither the formal structure of the equation nor the theoretical determinism it presupposes particularly characteristic of mechanism.

Helmholtz's mechanism, as we shall see, contains in its entirety only a few elements that would match the latter two meanings given for mechanism. His concept of nature transcends on the one hand so thoroughly all questions posed by individual sciences and on the other is so immediately directed by physical mechanics that it seems appropriate to arrange the examination of its prehistory along the lines of the philosophy of nature's concept of mechanism in the narrower sense. This does not, of course, imply that different alignments with mechanics that are part of other meanings of the concept of mechanism are irrelevant. Rarely an advocate of mechanism, including Helmholtz himself, gets by without making use of analogies or did not participate in disseminating ways of knowledge stemming originally from mechanics. They are only irrelevant for portraying the prehistory of mechanism inasmuch as they provide no criterion for selecting material.

1.2 The Concept of Classical Mechanics

As teachings on the equilibrium conditions and motion of physical bodies, mechanics has been part of the cultural history of human civilization since ancient times.[45] It was always both a technical activity, manifesting itself in specific social contexts,

[42] Tim. 30d 3, in **Plato** (1959), Vol. V, p. 156, **Chalcidius** (1962), 25.7.

[43] **Eisler** (1899 ff.), 4th edition, p. 107, **Weizsäcker** (1974), p. 136 ff.

[44] **Bunge** (1987), p. 372.

[45] **Dijksterhuis** (1956) and **Dugas** (1957) can be considered standard portrayals of the history of mechanics. See also **Dühring** (1873), **Mach** (1883), **Duhem** (1912) and **Haas** (1914).

as well as a related, yet relatively independent elaboration of theory that generated a tradition transcending the historical horizon of a time. My portrayal of mechanics begins with the history of how mechanistic theories were received and assumes that this history can be understood even when disengaged from the practical matters with which it was more or less interwoven.[46] To that effect I shall identify the concept of mechanics using structural features of theories that had a major impact on history.

In contrast to the foreshortening achieved by Thomas S. Kuhn's and Imre Lakatos' post-positivist theories of the history of science, in recent decades the historiography of science, concentrating primarily on the history of ideas, has drawn such an impressively multifaceted mural of the development of early modern mechanics that doubts inevitably arise as to whether a single concept is still appropriate for capturing the entire spectrum of the historically effective conceptions it contains. While Kuhn and Lakatos still spoke globally of the Newtonian paradigm[47] or research program[48] constitutive of the reconstruction of crucial developmental phases in early modern mechanics, that historical period is meanwhile seen as having contained various programs,[49] opposing traditions,[50] an alteration of scientific concepts[51] and a paradigm switch within mechanics itself.[52] Also, following up Pierre Duhem and Anneliese Maier's studies on the medieval origins of early modern mechanics, the difference between scholastic and early modern mechanics has been effectively relativized.[53]

In spite of reservations about uniform terminology, which in light of the complexity of recent research suggest themselves, I have chosen, as a major thread for the prehistory to Helmholtz's mechanism, a portrayal of classical mechanics that proceeds from a coherent and historically extensive enunciation. I would like to provide and explain three reasons for this choice. These may not suffice for a well-founded position on the state-of-the-art for the history of mechanics within the history of science but may do so for the purpose of the present study in the philosophy of nature.[54]

First: the subject matter of early modern mechanics is difficult to isolate from that of pre-modern mechanics. For instance, the early modern concept of force, as we shall see, was still closely tied to the scholastic notion. But the idea of science

---

[46] The relation between mechanical theory and mechanical practice is particularly relevant for the modern origin of scientific mechanics (see, for instance, **Grossmann** (1935), **Zilsel** (1976) and **Freudenthal** (1982)) and was a topic of controversy (cf. **Wolff** (1978), p. 60ff.).

[47] **Kuhn** (1962), p. 44ff.

[48] **Lakatos** (1970), p. 132ff.

[49] **Pulte** (1989), p. 23.

[50] **Truesdell** (1968), p. 133f.

[51] **Mittelstraß** (1974), p. 119, **Mittelstraß** (1970), p. 302f.

[52] **Dellian** (1988), p. IX ff.

[53] **Duhem** (1906ff.), Vol. III, **Maier** (1949), **Feyerabend** (1976) and **Wolff** (1978).

[54] These three reasons are explained in greater detail in **Schiemann** (1997), p. 36ff.

entertained by these two periods differed radically. What was new about early modern mechanics – as Galileo Galilei put it in his critique of the Aristotelian concept of mechanics – was that it did not work against but instead worked according to nature. This indicates where the story begins and already addresses a criterion of proper science that will be shared in subsequent history.

Second: in terms of content, the theory of mechanics that followed was heavily synthetic. Its important statements claim to combine various approaches and most of the scientific community considered them successful at doing so. Isaac Newton brought Galilean mechanics and Kepler's celestial mechanics together in one uniform system of mechanical forces of motion. Newton's system was ferment for mechanical theory. Although his formulations underwent radical modification, they continued to guide the development of the theory. Joseph Louis Lagrange combined the forces introduced by Newton with theorems of conservation that can be traced back to Christiaan Huygens, Leonhard Euler and Daniel Bernoulli and with the principle of minimal action worked out by Pierre Louis Moreau de Maupertuis and Euler to create a closed analytic formula.[55] This was followed by a gradual conviction in physics that Newton's and Lagrange's formulations were mathematically equivalent and physically equal.[56] In retrospect, at least those values that came to be structurally significant in important mechanical philosophies of nature can be considered mechanistic precisely because they were integral to mechanics as described by Newton and Lagrange.

Third: this way of viewing the topic masks neither the discrepancies among various mechanistic approaches, nor the controversies that persisted despite all eighteenth-century efforts to unify mechanics' conceptualization. Indeed, if we are to appraise them, it is essential to take these common underpinnings and the evolving of a tradition in mechanics into account. We cannot discuss, for instance, the differences between Newton's and Lagrange's formulations meaningfully until we have established the mathematical and physical consistency of their foundations. We can only do justice to the conceptual disputes over the systematic design of mechanics by realizing that far-reaching consensus existed regarding the empirical content of Newton's laws. Debate on conceptualization can be understood as one way of clarifying basic positions within a science. It normally occurs within a single discipline and is thus subject to that field's specific type of specialization, usually not transcending the framework of scientific practice and the theory constitutive of that discipline.

---

[55] **Lagrange** (1887), p. 198 ff.

[56] See contemporary pertinent portrayals of classical mechanics, such as **Sommerfeld** (1950), **Goldstein** (1963) and **Mittelstaedt** (1970).

Chapter 2
The Classical Conception of Science

It is right to call philosophy the science of truth. For the object of observational science is truth (ἀλήθεια), the object of operational science is works (ἔργον)

(Aristotle, *Metaphysics*).

Physical Mechanics and the theory of mechanism constructed upon it can be considered prototypes of early modern science and philosophy of nature. Since from the moment it was established this new type of mechanical science exemplified the redefinition of the relation between science and nature, the principles of mechanics continued to constitute the overall ideal of scientific knowledge of nature right up to the twentieth century. Yet not only in terms of content but also in terms of the formal conditions for the validity of knowledge, a far-reaching change occurred as mechanics became the predominant science. Compared to medieval scholastic ideas, on the one hand empirical experience was upgraded and redefined; on the other hand the rational principles of knowledge founded on reason became central for grounding and deducing scientific knowledge. These two clearly distinct and yet closely related trends, both sharing the goal of the complete command of nature, can be called the empiricizing and rationalizing of the notion of science. If today we were still inclined to consider these features the characteristic mark of science, their origin would fall in the period of early modern times.[57]

But for analyzing Helmholtz's mechanism, I find it more appropriate to arrange European conceptualizations of science into different periods. This arrangement of periods does not take the empiricizing and rationalizing of science that occurred in early modern times as the major criterion but instead, sees the major criterion as being science's claim to truth – advanced from ancient times onward. Measured against that claim, the changes involved in establishing early modern mechanics were still basically

[57] Hans Blumenberg has impressively described the highly complex interwovenness characteristic of the change in the concept of science at the turn from the Middle Ages to modern times; see **Blumenberg** (1973) and **Blumenberg** (1981). On the origin of modern science see generally **Cohen** (1985) and **Bernal** (1961).

part of the framework of a superior, "classical" concept of science[58] that was just as compulsory in the Middle Ages as it was throughout the entire modern age until well into the nineteenth century. Classical science has been characterized by its emphatic commitment to truth ever since Aristotle first systematically argued the methodological foundation of rational knowledge for occidental philosophy: knowledge was only acknowledged as scientific if it could be claimed to be true and to serve the purpose of achieving truth in the comprehensive meaning of complete knowledge. The early modern traditions of mechanics and mechanism pursued that goal in very different ways but they were both thoroughly obligated to its pursuit.[59]

While I adopt the term "classical mechanics" in the sense commonly used today, as a terminus technicus in physics designating a specific theoretical schema, I would like to introduce the term "classical mechanism" to indicate that mechanism was part of the classical system of the sciences. In contrast to mechanics, which is called classical because for a limited range of validity it can still be and is used in science today, the classical concept of science itself has since the nineteenth century to a great extent lost much of its former relevance. German idealism can be considered the last historically significant expression of the classical idea; with the demise of systematic thinking modeled after Hegel it increasingly lost its function as a leading concept.

In terms of claims to truth, the onset of the transitional period headed towards the "modern" concept of science did not begin with early modernity but later, namely in the nineteenth century. The concept is modern, meaning both new and still definitive today, not because it is guided by empiricism and rationality as laid down in early modernity but because it makes the absolute validity of scientific knowledge a central problem and grants its own statements in principle only a hypothetical status. The way the nineteenth century increasingly made the classical conception of validity a problem, relativized it and even rejected it, has continued to this day and upgraded a concept of science that bestowed the concept of the hypothesis with a meaning no longer compatible with its traditional one.

Without denying that they have elements in common, I now want to make a distinction between two concepts of science found in European cultural history that disagree on how to evaluate claims to truth and whose opposition has become increasingly clear since the nineteenth century. Previous studies on the change in the concept of science that took place during the nineteenth century have similarly distinguished two conceptions of science, one of which can be traced back to

[58] I have taken the term "classical" from Alwin Diemer (**Diemer** (1964), (1968) and (1970)) and Herbert Schnädelbach (**Schnädelbach** (1983)) to designate the concept of science entertained from Aristotle up to German idealism.

[59] A significant counter-trend was the influence that British empiricism had on the conception of science in mechanism. In critical remarks on science, David Hume denied both the absolute validity of natural causal regularity (see his expositions on causality in **Hume** (1748) and mathematical truths (loc. cit. and **Hume** (1739 f.) I), thus shaking the fundament of the classical understanding of science. Kant saw this very clearly (see **Kant** (1788), A 88–93). Nevertheless British empiricism, as advocated by Hume, was not particularly influential in the German-speaking scientific community until the nineteenth century (cf. **Poggi and Röd** (1989), p. 84 ff., and **Merz** (1907 ff.), Vol. 1, p. 211 ff.).

ancient times and the other which did not become perspicuous until contemporary times. If those studies do not always explicitly focus on raising or negating emphatic claims to truth, this is because they are more concerned with revealing the transition in conceptions of science in terms of its entire historical intellectual and cultural-sociological significance and such a dimension cannot be accessed solely through an epoch's respective concept of truth.[60]

If we limit our study to aspects of the theory of science and the philosophy of nature, in other words: to one topic of theoretical philosophy, it makes sense to concentrate on the kind and conditions of validity considered prerequisite for knowledge. Even narrowing it down like this, still leaves a wide frame to which the analytical net for studying the formal aspects of Helmholtz's mechanism must be pinned. Some of the supports – to remain within the metaphor – belong to ancient times, others are contemporary. The net of early modern times is also knotted differently than that of the Middle Ages. But we can keep it taut in the right place by deciding upon comparatively few, yet particularly salient fasteners and knots.

This will be done in three steps: I begin with a brief description of the classical claim to truth (a). Then I separate into two main classes the early modern strategies that justified that claim – one metaphysical, the other scientistic. Thus without leaving the framework of the classical concept of science, this classification does justice to

---

[60] Here we must first mention Alwin Diemer, who has pointed out in a series of publications on the concept of science, that such polarity exists between the understanding of science in the "modern" sense definitive of contemporary science and the in "classical" sense originating in early modern times (classical science in the narrower sense) and Aristotle (in the wider sense) respectively, that it would justify giving each of these ways of understanding what science is a name of its own, despite the distinctions that each of them makes. He locates the onset of the transition in the theory of science, from the classical to the modern understanding, in the nineteenth century (cf. **Diemer** (1964), p. 22) and lists four elements constitutive of classical science: the thesis of absoluteness, the assumption of truth, a postulate of universality and a method of attaining evidence by logical deduction, which is supposed to designate the systematic structure of classical science (**Diemer** (1968), p. 24 ff.). See also **Diemer** (1970), **Redner** (1987), **Diemer and Seiffert** (1989), **Diemer and König** (1991) and **Pulte** (2005).

Like Diemer, Herbert Schnädelbach lets a "classical" period of science that began in ancient times, also end in the nineteenth century, when "modern" notions started becoming apparent. He lists the following elements as basic to the classical conception of science: universality (in terms and statements), necessity (of the systematic context) and truth (**Schnädelbach** (1983), p. 106). Schnädelbach makes reference to Helmuth Plessner, whom I would also like to mention here, because the problem of truth becomes explicit in his remarks (covering more than nineteenth-century science) on the change of the concept of science. **Plessner** (1924) found "*law-like relationships of meaning* between types of society and types of knowledge" and distinguished "three 'evolutionary' phases" in occidental science: The phase of the hierarchical-feudal world of the Middle Ages, the phase of natural justice and absolutism in the seventeenth and eighteenth centuries and the phase of the evolutionist-democratic world of the nineteenth and twentieth centuries (loc. cit. p. 7 f. – emphasis in original). Only the two first phases, Plessner says, are marked by the fact that "in both systems truth is fixed materially, a treasure of supernatural revelation or reason-immanent laws of essence, whose successive clarification in the course of research is part of a finite order that was given once and for all" (loc. cit. p. 8).

In Section 4.1, I shall return to what Diemer and Schnädelbach considered the characteristics of modern science.

that specifically early modern type of science that was in part very rationalistic and in part very empirical. That is possible, of course, only to the extent that the relevant forms of rationalism and empiricism did not question the classical claim to truth (b). Not until the final section of the first part of the book (Chapter 4: Contours of Modern Philosophy of Nature) shall I, in a third step, roughly sketch the modern concept of science and the scientific worldviews erected on it.

(a) Historically, the concept of truth, which in classical science designated both individually sought knowledge as well as the ultimate goal of total knowledge, was not only defined in many different ways but using it could also be so taken for granted that generally no need arose to question it. The lack of a unifying definition for truth and the fact that no one demanded one while the concept was powerful and effective, indicate in my opinion not only that the problem of truth was almost non-existent in the classical concept of science but also that it was part of the scope of the classical concept of science that the matter of truth could not be settled by a definition.[61] In order to clarify the formal aspects of classical mechanism at this point I can only roughly disclose that idea of truth to the extent that it is relevant for the historical contexts of Helmholtz's notion of nature.

One distinction sufficient for this task is to distinguish between using the concept of truth substantially and using it as an attribute.[62] Substantive truths are real or ideal contents, which can be expressed in any way – be it by demonstration or by statement.[63] The outward sign of this kind of truth is the option of using it in the plural.[64] Thus historically, European philosophy knows a number of truths and ways to classify them, the most familiar one being the division of truths into truths of divine revelation, truths of reasoning and empirical truths gained by experience. Within that order, early modern times are marked by a loss of significance attached to truths revealed directly by God. Thus for the modern substantive conception of truth, the focus shifted to the relation between reason and empirical experience.

A paradigmatic example is Gottfried Wilhelm Leibniz's distinction between timeless truths of reasoning and time-bound truths of fact, which picks up Descartes' distinction between "eternal truths" and things and is continued by Kant in his distinction between a priori and a posteriori knowledge:

> There are[. . .] two kinds of truths, those of reasoning and those of fact: truths of reasoning are necessary and their opposite is impossible; truths of fact are contingent and their opposite is possible. When a truth is necessary, its reason can be found analytically by breaking it down into simple ideas and truths, until one finally arrives at elementary basic truths.[65]

Leibniz's concept of truths of reason is modeled after the identical propositions of logics and mathematics that are necessarily true when their opposite contains a contradiction. For the entire tradition of classical mechanism, the truth of mathematical propositions and logical reasoning were presupposed as necessary in this way.

---

[61] Cf. **Gawlick** (1962).

[62] I take this distinction from **Diemer** (1964), p. 26f. and 53ff., and **Diemer** (1967ff.).

[63] **Diemer** (1964), p. 53.

[64] **Diemer** (1964), p. 26.

[65] **Leibniz** (1714), No. 33.

Until Kant, who subjects the issues of traditional ontology to a critique of knowledge, the use of the substantive concept of truth was normally linked to the idea that the meaning of truth is ontological: both truths of reasoning and truth of fact gained from experience tell us something about worldly being, its origin and essence. What is more, in the tradition of conceptual realism modeled after Plato, logical mathematical statements even have their own "ideal being", raising them to the level of a truth that coincides with itself and is valid in and of itself.

The attributive meaning of the concept of truth can be understood, ontologically speaking, as that "trait of a given" that "makes it correspond to something else".[66] For the classical concept of science this meaning is above all relevant for theories that try to enunciate a correspondence theory of truth. These use the adjective "true" to distinguish judgments that can be correlated to circumstances in the world of objects. The writings of Thomas Aquinas ("veritas est adaequatio rei et intellectus")[67] and Kant ("explaining the word truth, as meaning the correspondence of knowledge with the object of that knowledge")[68] contain prime formulations for the tendency of correspondence theory, which fans out in a number of ways to accommodate both extremely diverse notions of the correspondence relation itself (equivalence, assimilation, etc.), as well as the very diverse *relata* (thinking – being, subject – object, language – world, etc.) involved.[69]

As the mark of the classical conception of science, the concept of truth needs to be made more precise in two respects.[70] First, the formal condition of universality must define the range of validity for a truth or a true judgment. No matter what the truth is about, it is universally true, i.e., true "for 'everything', 'always' and 'everywhere' ".[71] This condition does not exclude from the concept of science that which is particular, including what is singular and individual. The particular remains an object of classical science inasmuch as information about it can be used to discuss something general, or a form, or inasmuch as it is considered a specimen of a general trait. Leibniz's conception of science, for example, views individuality itself as an ubiquitous essential trait of nature.[72] Second, any single scientific finding in the classical sense can only be called true if it fits into a system of general knowledge guided by the ideal of completeness.[73] All statements in the scientific system must be logically deducible from absolutely valid principles. Only the entirety of such a "categorically deductive system" (as A. Diemer calls it)[74] can, in the full meaning of the word, claim to be true.

---

[66] **Diemer** (1964), p. 55 (emphasis in original).

[67] **Thomas Aquinas** (1256 ff.), p. 317. Cf. **Heckmann** (1981), p. 102, and **Hoven** (1989), p. 63.

[68] **Kant** (1781), B 82.

[69] **Puntel** (1978), p. 28.

[70] With these specifications I incorporate into the explication of the classical conception of truth some individual definitions of the classical concept of science that Diemer and Schnädelbach mention, along with the claim to truth.

[71] **Diemer** (1968), p. 29.

[72] **Leibniz** (1904 ff.), Vol. 2, p. 191.

[73] Cf. **Diemer** (1968), p. 29 ff., **Schnädelbach** (1983), p. 106.

[74] **Diemer** (1968), p. 28.

Finally, the way the classical concept of science understood truth is also reflected in its abatement of the importance of scientific hypotheses.[75]

(b) Within the framework of traditions in mechanism, the early modern transition in the idea of science did not question the classical claim to truth but – as far as the problem of the validity of knowledge is concerned – had more to do with grounds for legitimacy. Theological authority, so crucial in the Middle Ages for legitimating scientific truth, became substituted in early modern times by reasons appealing to certain kinds of human experience and human thought. A closer look at these specifically early modern types of legitimating now permits making a distinction within the classical concept of science, which is cogently relevant for classifying Helmholtz's mechanism. This distinction allows us to formally classify Helmholtz's concept of nature, to the extent that it is still part of the traditional line of classical mechanism, as belonging to a smaller group of mechanistic positions and to separate it from others. For this purpose we distinguish scientistic from metaphysical legitimating.

Inasmuch as the claim to truth was founded on experience, it had to do primarily with the experience of practical scientific research, which can be considered a genuine product of early modern times. Its general characteristics can be traced back to Francis Bacon, while Galileo's work is particularly suitable for studying features that were specific to mechanism. In order to accentuate the special scientific kind of experiential reference involved, I call the legitimating of claims to truth founded on that practical research "scientistic".[76]

The concept of experience is disproportionately broadened by the second strategy, which – in contrast to scientistic legitimacy – justifies its claim to truth not by appealing to a special kind of experience but by abandoning all experience. It defines experience as the opposite of the autonomous subject's pure thought, as what is merely receptively perceptible, the spatio-temporally given and what is merely contingent and changing. This term thus covers both quotidian experience, as well as the different, rationally guided practice of the sciences. It views pure thought alone as the fundament of any claim to truth, being "pure" only because it knows itself to be entirely independent of what it itself defines as "the given". Chiefly because of its independence of all experience, I call this strategy for legitimating claims to truth "metaphysical". By metaphysics I thus basically understand this separation from all experience undertaken from the perspective of pure thought. Nevertheless, the activity of the mind done in awareness of the autonomy of thought is also called metaphysical. On this interpretation both the metaphysical legitimacy of science's claim to truth and the scientistic one are inventions of early modern times. Descartes' establishment of metaphysical legitimacy also marks the

---

[75] In **Schiemann** (1997), p. 85 ff. I explain this aspect of the greatest importance for the transition to the modern concept of science in a discussion of Descartes' concept of hypotheses.

[76] **Apel** (1970), p. 110, **Lenk** (1989), p. 353, **Lenk** (1979), p. 80 ff. In this study, "scientism" means scientifically-oriented empiricism as a part of philosophy of nature. **Sorell** (1991) defines it similarly.

commencement of efforts to deduce and justify the principles of mechanistic philosophy of nature by pure thought.[77] Subsequently, within the traditions of mechanism, Leibniz and Kant were representatives of this strategy.

The properties of being universal and systematic characteristic of scientific truth can be traced metaphysically back to subjective certainty and scientistically back to objective certainty. At the same time, the scientistic strategy need not entirely reject legitimating elements that are not based on experience and conversely, the metaphysical strategy must not forego every reference to scientific practice. Not all elements brought together for a justification are important; in the end, the decisive reasons provided for justifying validity are crucial. We have here a typology that distinguishes two extreme positions, between which historically a number of transitional and mixed forms have existed. Only when stylized as antithetical variants do they mutually exclude one another. When interpreted as opposites, scientism denies all possibility of pure thought, while the metaphysical standpoint rejects any foundation for legitimacy that is even slightly independent of pure thought.

The difference between these strategies becomes perspicuous in their different evaluation of the practice of scientific research: while metaphysically the truth of mechanism must exist prior to all experience gained through scientific research, scientistically it may originate and be founded only through such experience. The difference between the two strategies lies less in their definitions of the concept of experience than in their relation to a basically undeniable scientific practice.

Of the features of scientific experience that are characteristic of early modern mechanics, I would like to mention three that already marked Bacon's and Galileo's work: first, the experimental context of action, which for terrestrial mechanics is technical; second, the quantifiable form of empirical data that corresponds to the mathematical character of mechanic theory and third, the notion of progress typical of practical research, an aspect which mechanics very generally shares with all scientific experience.

I will start with elements that can be traced back to Bacon. According to Bacon's dictum, true knowledge about nature can be acquired and legitimated solely through purposefully executing intervention with nature.[78] In other words: knowledge about nature results from the collaboration of free action and natural causality within the context of the experiment, to which the claim to truth owes its legitimacy. Man changes natural bodies by experimenting, in order to draw from the resultant phenomena conclusions about the causes effectual in nature.[79] Causes are neither generated nor altered by experiments; they are discovered for what they are. But they can only be identified when experiments are performed methodically in a test series and are evaluated as part of the inductive method,[80] where Bacon understands

[77] On Descartes' metaphysical reasons for mechanism, see **Schiemann** (1997), p. 76ff.

[78] **Bacon** (1620), I, 50 and I, 70.

[79] **Bacon** (1620), I, 4 and II, 4.

[80] **Bacon** (1620), I, 70, 8 and 100, mentions these kinds of test series and describes them in more detail in **Bacon** (1858ff.), 1, 622ff. On applying the inductive method to experiments see **Bacon** (1620), I, 18f.

induction as the generalizing of individual facts about nature to create general theorems or laws of nature.[81] His experimental method also already indicates another meaning of causality: the triggering act on the part of the researcher can also be interpreted as a cause of the resultant effect.[82]

Also, for Bacon the practice of research is also characterized by temporality, underscoring the contrast to all timeless and non-spatial metaphysical foundations. He thinks of scientific experience as a process of progress in which truth is a "daughter of time".[83] Advance for Bacon means that each new bit of knowledge accretes the body of knowledge achieved thus far and part of knowledge's essence and its technical application is its contribution to humanizing mankind.[84] Crucial for sustaining the classical claim to truth is that here progress is not thought of as an endless endeavor but is something guided by attainable goals.[85] If Bacon's goals become no longer achievable for an individual subject and all of mankind is required to accomplish them sometime in the future, then truth exits the horizon of that period's research. From today's vantage point it appears that here the modern absence of truth already announced itself in classical early modern science's idea of progress. Obliterating truth this way would have been unthinkable in contexts of metaphysical truth legitimation, where at least the truth of knowledge (no matter how knowledge is understood) is unrestrictedly accessible to every individual.

In order to more carefully characterize the scientific practice of mechanics and mechanism, it is essential to include Galileo's work.[86] The technical practice central to his research accentuates the particular kind of experimental method especially typical of mechanics: technical equipment is used or constructed for the purpose of measuring specific quantities. The experiment is subordinate to the purpose of measurement and thus subordinate to a mathematical theory in which the quantities to be measured can be defined and calculated as numerical values. The theory thus shapes the experiment but is not itself an idea derived independently of experience; theory is rather itself determined by experimental conditions.

Moreover, with Galileo it becomes evident that the scientific practice of both mechanics and mechanism is not totally determined by experimental method alone. Galileo's observations in astronomy are also part of that practice. One feature

---

[81] See **Bacon** (1620), I, p. 45, and I, 19 and 104 f., on induction; in II, 15 f. he develops an approach for "induction by exclusion". For the concept of a law ("servatis legibus materiae") cf. II, 4 and I, 51 and 75 and II, 2 and 4.

[82] Today this type of causality is known as "experimentalist causality" (**Wright** (1974), p. 44 and 72) or "interventionist causality" (**Stegmüller** (1978 ff.), Vol. II, p. 108). I think Bacon's broad definition of the experiment contains what has remained essential for experimental method in natural science.

[83] **Bacon** (1620), I, 44. On the metaphor "truth is the daughter of time" see **Blumenberg** (1986), p. 153 ff.

[84] On the concept of progress, see **Ritter** (1971 ff.) and **Pot** (1985).

[85] Blumenberg calls the plan to recover the "lost paradise" on earth Bacon's "regulative idea" (**Blumenberg** (1973), p. 198).

[86] On Galileo's scientism see **Schiemann** (1997), p. 62 ff.

astronomy shares with terrestrial mechanics is that it prepares empirical experience to become quantitative data, making it available in dimensions having the property of being equal or unequal to others, which, in turn, enables mathematical evaluation. This then, is the component that the practice of mechanics shares with the experience of all mathematical scientific theories and for which metaphysical legitimation has no equivalent. For the metaphysical legitimation of mechanism is conceptual and involves absolutely no mathematical formulations.

Finally, Galileo's work in natural science shows that the practice of mechanics must not necessarily include the inductive evaluation of series of data. Not only did he *not* – as a rule – arrive at his research results by induction, he also expressed explicit skepticism with respect to it.[87]

In conclusion, we can say that these three features of scientific practice do, on the one hand, clearly distinguish scientism from naive empiricism of the kind that limits itself to finding, collecting and cataloging things. Scientific research is normally a profoundly artificial, rationally guided practice. On the other hand, the rationality predominant here differs fundamentally from the rationality underlying metaphysical legitimation. From a metaphysical perspective, experimental experience cannot provide reasons for knowledge. Moreover, the conceptual work of metaphysics knows no mathematical application and is independent of the temporal dimension intrinsic to the process of progress.

[87] **Galileo** (1890a ff.), Vol. IV, p. 701.

Chapter 3
Three Traditions in Mechanism

Within the traditions of mechanism, both strategies for legitimating claims to truth – the scientistic and the metaphysical strategy – appear to have been largely indifferent to content. Thus when I now distinguish, in terms of content, three different lines within mechanism, this does not suggest which legitimating strategy was involved in which line.

The defining criterion, which will be further differentiated later on, is the position held by each mechanistic concept of nature (mostly understood in the narrower sense) on the notions of matter and force: concepts that reject an independent notion of force, I call "materialist" and those concepts that conversely explain all properties of matter from effects of force, I call "dynamic". An intermediate position between these two extremes is held by the "dual" mechanism, which views matter and force as irreducible fundamental ideas. This classification ideally typifies the lines and was historically effective for distinguishing the various theoretical strands. At least from the time of Newton's mechanics onward, forces were the central issue of debate in mechanistic philosophy of nature. The designations used for the various drifts, however, differ from mine. While the "materialist" line was normally considered the epitome of the "mechanistic" concept of nature, no unique name was associated with the dual line, although the literature rarely doubts its mechanistic character (in the broader sense). Only with respect to dynamic mechanism does my designation largely coincide with the term used historically, being the term introduced by Leibniz and taken up particularly by Kant and contrasted with the "mechanistic" concept of nature. While these kinds of dynamism adhered fairly closely to the notion of force used by mechanics, the ideas of force prevalent in romantic philosophy of nature following Schelling, likewise called "dynamic", defined the essence of natural events as an immaterial and universal agent.

Although mechanism crucially shaped the development of the natural sciences from early modern times until the end of the nineteenth century, remarkably no coherent historical account of it exists.[88] None of the numerous works that elaborate

[88] To date the most thorough account of early modern mechanism is by **Lange** (1866), covering authors of both materialist and other types of mechanism. For the mechanism of nascent early modern times and its origins in ancient times see **Koyré** (1957), **Dijksterhuis** (1956) and **Lasswitz** (1890). Some aspects of mechanist philosophy of nature have been discussed in the history of science's treatment of

individual aspects of mechanism to varying degrees contains a classification similar to the one I propose and that is itself suggested by the historical material.[89] Thus in the sections that follow, I will be treading new territory, the preliminary exploration of which may prove useful for further research. Nevertheless, the overriding purpose of my study is to develop first criteria for evaluating Helmholtz's foundation for mechanism. While the purpose of discerning scientistic from metaphysical legitimation was to gain formal criteria for the characterization of classical early modern mechanism, separating mechanistic thought into three distinct lines will provide a system that enables us to carefully classify it in terms of content.

3.1 Materialist Mechanism

> In fact Materialism [. . .] is not complete until matter is *understood* purely *materially*, i.e., when its components are understood as bodies that move according to purely physical principles
>
> (F.A. Lange, *History of Materialism*).

As "materialist" mechanism I count all concepts of nature that view pressure and impact processes between elementary bodies as the essence of all phenomena.[90] These reject immaterial forces as explanatory entities and consider the mere contact of purely mechanically moved matter the sole possible form of natural interaction. The first historically effective important articulation of this line of mechanism was provided by Robert Boyle's concept of nature, which, however, can only be called mechanistic in the broader sense of the term. Not until Christiaan Huygens, the second renowned advocate of materialist mechanism, did mechanics become the structuralizing principle.

Boyles's mechanism combines elements of both Cartesian corpuscle philosophy and early modern atomism, which was propagated in the seventeenth century particularly by Pierre Gassendi and differed little from Democritus's ancient notion.[91]

---

mechanics, particularly **Wolff** (1978), **Duhem** (1912) and **Mach** (1883). Due to the predominance of mechanist thought throughout all early modern times, we also find discussion on individual topics among the wealth of literature dedicated to the history of physics (e.g., in general: **Kuznecov** (1970); on ether theories: **Whittaker** (1958); for the nineteenth century: **Jungnickel and McCormmach** (1986)), the history of atomism (e.g., **Melson** (1957), **Toulmin and Goodfield** (1966) and **Scott** (1970)) and the history of science (e.g., **Mittelstraß** (1970), **Carrier** (1984), **Schofield** (1970) and **Buchdahl** (1969)) or cultural history (for the nineteenth century: e.g., **Merz** (1907 ff.)).

[89] On the history of mechanics, however, see also the reference to **Pulte** (1989) in footnote 49. Pulte writes that "18th century rational mechanics [. . .] can essentially be described as a struggle between the three major programs of Descartes, Newton and Leibniz" (loc. cit. p. 23).

[90] Cf. particularly definitions by **Kirchner** (1833), p. 212 f., and **Brugger (Ed.)** (1950), p. 213, previously mentioned in Section 1.1, footnote 37.

[91] On Boyle's mechanism see **Lasswitz** (1890), Vol. 2, p. 126 ff., **Hall** (1965), p. 57 ff., and **Shapin and Schaffer** (1985).

Like Democritus, Gassendi also assumes that nature consists only of void and atoms and that no external forces act upon the latter. Atoms possess an intrinsic principle of motion, consist of one identical (solid) substance and differ from one another solely in terms of the quantitative properties of size and shape.[92] Boyle particularly takes up those elements of the theory that are contiguous to corpuscular philosophy.[93]

Huygens, like Boyle (and Descartes) categorically rejects any presupposition of immaterial forces for explaining natural events.[94] Attributing gravity to atoms, then, was not explaining causes but rather simply claiming "*principes obscurs et non entendus*".[95] Besides explaining the phenomena of light, Huygens considers it the main challenge of mechanism to explain gravitation[96], which is subject to definitions fixed by the theory of matter.

Overall, for this line the notion of force has no validity of its own in mechanics. Phenomenal forces – whose real existence is not denied – must be derived, like all other natural events, from mechanical laws of motion. The bearers of this law-determined motion are invisible tiny material particles composed of an inert substance, which differ solely in size and shape and interact exclusively by pressure and impact. Issues of the infinite divisibility of matter and whether space can be devoid of matter remain undecided and are thus unsuitable criteria for evaluating this line of mechanism.

3.2 Dual Mechanism

> [We] have the authority of those the oldest and most celebrated philosophers of Greece and Phœnicia, who made a vacuum and atoms and the gravity of atoms, the first principles of their philosophy; tacitly attributing gravity to some other cause than dense matter
>
> (Isaac Newton, *Opticks*).

In "dual" mechanism the mechanical notion of force is on an equal standing with the idea of matter in motion. Fundamental for the early modern understanding of mechanical forces is its distinction between the motion of inert material bodies, which occurs without force and is constant on the one hand and on the other, a change of this motion caused exclusively by forces. Once matter has been initially set in motion, every subsequent change of motion for those material bodies can be achieved only by a measurable mechanical force, which in dual mechanism

[92] **Gassendi** (1727), Vol. I, p. 236 ff.

[93] On Gassendi and Boyle see also **Schiemann** (1997), p. 96 ff.

[94] On Huygens's mechanist philosophy of nature see **Lasswitz** (1890), Vol. 2, p. 341 ff., **Westman** (1980), **Snelders** (1980), **Gabbey** (1980) and **Schiemann** (1997), p. 98 ff.

[95] **Huygens** (1690a), p. 445.

[96] To solve both problems, Huygens published twice in 1690: **Huygens** (1690a) and **Huygens** (1690b).

(excluding effects caused by the shaping of atoms (cohesion, linkage, etc.)) is ranked as the sole permissible cause of changes in nature.

This definition of force immediately reveals where dual mechanism is contiguous with materialist mechanism: inasmuch as mechanical forces are clearly correlated with changes of motion, degrees of force can be substituted by the gradient (i.e., by direction-dependent differential change) in acceleration (for immutable mass) and reduced in form to matter in motion, which is the underlying concept of materialist mechanisms. Ontologically speaking, however, this formal reduction does injustice to materialist mechanism's claim to explanatory power: for the changes in movement caused by mechanical forces can, in contrast to those caused by impact, occur without any contact at all. But in materialist mechanism such changes, apparently caused invisibly, must be explained as phenomena that in reality are erected upon processes of impact.

Dual mechanism is not the reverse of this materialist doctrine. Impact events can but need not be explained as an expression of forces. The reason for this ambiguity is the independent concept of matter not reduced to forces, which does not exclude contact with negligible force effects. On the other hand, the occurrence of force effects is not bound to contact among matter, because the forces that in dual mechanism are at work solely among bits of matter continue to be effective when that matter is spatially separated. It follows from the duality of matter and force that the advocates of this line hold an essentially firmer position on issues of divisibility and the vacuum than advocates of the materialist line. Dual mechanism's theory of matter is atomistic and the vacuum is where forces are at work. I shall illustrate this line of mechanism by way of Isaac Newton, the founder of the dual concept.[97]

Newton's mechanistic theory of nature[98] is modeled on his theory of gravitation. Gravitation is the attractive force postulated by Newton, which in a vacuum operates along the connecting line between the centers of gravity of two macroscopic bodies and whose absolute value is proportional to the mass and the reciprocal value of the square of the distance.[99] Since nature is always "conformable to herself and very simple"[100] similar forces should also be at work among the submicroscopic particles, of which, as Newton assumes, all material things are composed.[101] He establishes the invisible existence of these elementary components of matter scientistically by referring to natural phenomena that (presumably) can only be discovered by experiment.

---

[97] Roger Boscovich, whose comparatively uninfluential natural philosophy carries Newton's theory to its full consequence, provides another specimen of the dual line of mechanism, which I describe in **Schiemann** (1997), p. 107 ff.

[98] On Newton's mechanism see especially **Thackray** (1970), **Westfall** (1971) and **Freudenthal** (1982); on the alchemist background (not dealt with here) of Newton's natural philosophy, see **Dobbs** (1975). A survey of the most important secondary literature is given in **Schneider** (1988).

[99] **Newton** (1872), p. 69, and **Newton** (1726), p. 54.

[100] **Newton** (1898), p. 263.

[101] **Newton** (1704), p. 261, and **Newton** (1898), p. 267.

Both of his decisive arguments for this deal with the linear diffraction of light, which would seem to correspond to force-free motion of inert matter and thus suggest that one might think of light as being composed of particles[102] and to the reciprocal transformation of corporeal matter and light that demonstrates that bodies also possess a corpuscular structure.[103] Just as the entire universe is governed by gravity, so too are forces at work in the world of minuscule things, among the particles of light and matter. Question 31 in the third book of Newton's *Opticks* reads:

> And thus nature will be very conformable to herself and very simple; performing all the great motions of the heavenly bodies by the attraction of gravity, which intercedes those bodies; and almost all the small ones of their particles, by some other attractive and repelling powers, which intercede the particles.[104]

If one dismisses the limitation "almost all", alluding to contact among particles, what remains as the difference between the force of gravity and submicroscopically effective forces is the lack of a quantitative definition for the small-scale attraction effect and the repulsion accompanying it. Newton again finds evidence of a relation between attraction and repulsion on the minute scale in experimental situations: "Since metals dissolved in acids attract but a small quantity of the acid, their attractive force can reach but to a small distance from them"; on the other hand, it seems "to follow from the reflexions and inflexions of the rays of light" that "where attraction ceases, there a repulsive virtue ought to succeed".[105] Attraction and repulsion effects among material components can thus be understood as properties of one single force, whose characteristics depend entirely on distance. The question remains, however, whether submicroscopic attraction is governed by the same law as macroscopic gravitation. Newton mentions "powers, virtues or forces" in the plural[106] and as "attractions" he mentions not only gravitation but magnetism and electricity as well.[107]

---

[102] **Newton** (1704), p. 238 f., and **Newton** (1898), p. 244 f. In this passage Newton appropriately phrases his theory of matter suggestively: "Are not the rays of light very small bodies emitted from shining substances? For such bodies will pass through uniform mediums in right lines, without bending into the shadow" (**Newton** (1704), p. 238). In other words, he is not claiming that this *is* the truth but that, based on scientific experience, it could be. This conforms with the fact that while his theory of matter is not (yet) derivable from empirical data, it is nonetheless not to be considered purely hypothetical: the results are "probable" and true, inasmuch as "their truth appear[s] to us by phenomena, though their causes be not yet discovered" (**Newton** (1898), p. 266, and **Newton** (1704), p. 261). This problem of validity is a theme from the philosophy of nature that Newton clearly demarcates from the issue of legitimating theories in physics. Thus a physical theory of light required no presuppositions about the nature of optical processes: "To avoid dispute and make this hypothesis general, let every man here take his fancy; only whatever light be I suppose it consists of rays differing from one another in contingent circumstances, as bigness, form or vigour" (Royal Soc. 9.12.1675, cited in T. Birch, *History of the Royal Society of London*, 1756–57, Vol. 3, p. 255).

[103] **Newton** (1704), p. 241 f., and **Newton** (1898), p. 247 f.

[104] **Newton** (1704), p. 258, and **Newton** (1898), p. 263 f.

[105] **Newton** (1898), p. 262, and **Newton** (1704), p. 256.

[106] **Newton** (1704), p. 242.

[107] **Newton** (1704), p. 243.

Just as for forces, nature is "always conformable to herself", so she is also concerning matter. The essential properties of the tiniest particles are precisely the same as those of objects "that we handle" ("quae tractamus") and which can be investigated in simple experiments ("experimenta").[108] Newton names: "extension, hardness, impenetrability, mobility" and finally the "force of inertia" ("vis intertiae").[109] The latter he also calls "the inertia of matter" ("intertia materiae")[110] indicating the tendency of bodies to remain "in its state of being at rest or of moving uniformly straight forward" ("in statu suo quiescendi vel movendi uniformiter in directum")[111], unless disturbed by mechanical forces. In contrast to gravitation, where strength depends on distance, inertia is an immutable quality.[112] It is a "passive principle", through which "there never could have been any motion in the world".[113] According to Newton's theory, motion requires both an initial impulse and conservation against loss of motion due to friction. Inert matter itself can neither generate its own momentum nor change the direction it has been given. This can only be done by forces that work both externally on matter and generate interaction among bits of matter. For Newton, the effects of force are a "different"[114], "active principle"[115] that with the passive principle collectively constitute a dualism upon which all natural phenomena rest.

3.3 Dynamic Mechanism

> [All] philosophy of nature (consists) of reducing given, apparently dissimilar forces to a smaller number of forces and virtues that are sufficient for explaining the effects of the first but this reduction only reaches as far as the fundamental forces, beyond which human reason cannot get
>
> (I. Kant, *Metaphysical Principles of Natural Science*).

Regarding dynamic mechanism, it must be emphasized in general that of the three lines it distinguishes itself from the others by being the most consistent on the infinite divisibility of matter and being in least need of presupposing the existence of a vacuum. It uses metaphysical deduction to derive the properties of matter from forces that entirely fill space. Nevertheless, these metaphysical forces are not only the origin of matter and its modifications; they also determine the general structure of the forces used to explain natural phenomena and that in part correspond to physical quantities. The most influential conceptions of dynamic mechanism are considered

---

[108] **Newton** (1872), p. 38, and **Newton** (1726), p. 388.

[109] Ibid.

[110] **Newton** (1872), p. 2, and **Newton** (1726), p. 2.

[111] **Newton** (1872), p. 3, and **Newton** (1726), p. 13.

[112] **Newton** (1726), p. 38, and **Newton** (1872), p. 381.

[113] **Newton** (1898), p. 26, and **Newton** (1704), p. 258.

[114] Ibid.

[115] **Newton** (1898), p. 26, and **Newton** (1704), p. 258: "active principles".

to be those of Kant and Leibniz. Both claim to offer a foundation for a materialistic-mechanistic explanation of nature. However their theories disagree on the form and number of physical quantities. Leibniz's paradigm, the vital force, has no place in Kant's absolute polarity of fundamental forces.

3.3.1 Leibniz's Rationale for Dynamism

Gottfried Wilhelm Leibniz was the one to introduce the concept of dynamics into the tradition of mechanism. He uses this term to designate a science that deals with the forces that constitute the essence of natural bodies and whose principles, in physics, are on equal standing with the principles of geometry.

> We claim then, that its [the body's–G.S] essence is singularly ἐν τῶ δυναμικῶ, i.e., can consist of a primal, inner principle of change and persistence. Physics [. . .] is [. . .] subordinate to both *geometry* and *dynamics* [. . .].[116]

The inner principle of change and persistence follows the Aristotelian-scholastic notion of δύναμις. In ontology, Aristotle used this term to designate that manner of material being which "possess the principle of origin within itself" and therefore, "has the virtue – in the absence of external impediment – to come into being by itself".[117] Leibniz sees this power as composed of forces, which, "unless it is impeded by some contrary striving" (or tendency – G.S.), "attain its full effect. This effort" Leibniz continues, "often makes itself felt by the senses but in my view reason shows that it is everywhere in matter, even when it is not apparent to the senses".[118] Characteristic of Leibniz's force is its striving to be effective not in the past or future but momentarily, in the continuous, ubiquitous present.[119] Only by other, essentially similar tendencies striving equally to be active can forces be withheld from unfurling their efficacy. This restriction can be seen as the reason why forces must not necessarily be "apparent to the senses". Forces are potencies

---

[116] **Leibniz** (1702), p. 332 (emphasis in original). Cf. also **Leibniz** (1695b), p. 260, and **Leibniz** (1849 ff.), Vol. 6, p. 287 (*Dynamica*). Leibniz uses the term *dynamics* for the first time in 1694 in *De primae philosophiae emendatione, et de notione substantiae* (**Leibniz** (1875 ff.), Vol. 4, p. 468 ff.), later in the title of his programmatic treatise *Specimen dynamicum* (**Leibniz** (1695a). On Leibniz's philosophy of nature see: **Okruhlik and Brown (Ed.)** (1985), **Moll** (1978 ff.) and **Allen** (1983).

[117] **Aristotele** Met. IX, 1049 a 11. A "kinetic" meaning for the concept of dynamism can be distinguished from the ontological meaning given here. The kinetic interpretation also designates possibility "but precisely not the possibility to be something but rather, the possibility of doing something or of suffering something" (**Wieland** (1970), p. 295, cf. also **Düring** (1966), p. 671, and other places).

[118] **Leibniz** (1695a), p. 256. Cf. also **Leibniz** (1875 ff.), Vol. 4, p. 472 (*Système nouveau pour expliquer la nature* ... (First draft)).

[119] Furthermore, as Cassirer stresses, Leibniz generally defines what is real by using the concept of force: "for Leibniz, it is generally understood, force is synonymous with reality" (**Cassirer** (1902), p. 288).

and in collaboration they constitute not only the principle of spatio-temporal change of motion and activity in general but also the principle of resistance to any kind of motion, the sometimes covert principle of inertia in an entity.

Compared to the causally-effective concept of nature intrinsic to dual and materialist mechanism, in Leibniz's dynamism elements of a metaphysic are at work, which, as his adoption of Aristotelian terminology in itself suggests, are characterized by a teleological and organic approach. These elements are ensconced in the central concept, the concept of force. Nonetheless, one may reasonably call Leibniz's dynamism mechanism, not only in the broader but also in the narrower sense. For with his metaphysics of forces, Leibniz establishes not only a concept of nature that is meant to explain the entirety of natural events, including living organisms, based solely on the three basic notions of "magnitude, figure and motion, that is, through mechanism".[120] By using a mechanical quantity (which is proportional to contemporary kinetic energy) as a standard for measuring the forces at work in nature, Leibniz also establishes the foundation for a kind of mechanism in the narrower sense, which stands out clearly from the other two lines.[121]

3.3.2 Kant's Dynamic Theory of Matter

The success of Newtonian mechanics in the eighteenth century is what makes the changes distinguishing Kant's dynamic mechanism from Leibniz's theory easier to understand. The notion of action at a distance, rejected categorically by Leibniz, subsequently became extraordinarily successful, while materialist mechanism's efforts to explain phenomenal effects of force as the motion of elementary matter failed. For physical mechanics, then, Kant was compelled to assume that "Newton's system of universal gravities is established, although it involves the difficulty of explaining how attraction at a distance is possible".[122] Kant lets the issue rest, at least until writing his posthumously published treatises on the philosophy of nature[123] ("difficulties are not doubts")[124] and makes action at a distance the fundamental component of his dynamic theory of matter. His dynamism also reflects the fact that in contrast to Leibniz's lifetime, throughout the waning eighteenth century, approaches in natural research couched in terms of cause and universality had gained widespread acceptance.

[120] "ut explicentur per magnitudinem, figuram et motum, id est per Machinam" (**Leibniz** (1875 ff.), Vol. 7, p. 265 – emphasis in original). Leibniz stated this in several places: for instance, in **Leibniz** (1875 ff.), Vol. 7, p. 117 f., **Leibniz** (1692), p. 326, and in **Leibniz** (1875 ff.), Vol. 5, p. 429 ff. (*Nouveaux Essais*, Livre IV, Chap. 12 and 13).

[121] For more on the meaning of Leibniz's concept of force and the function it fulfills as a foundation for philosophy of nature, see **Schiemann** (1997), p. 115 ff.

[122] **Kant** (1786), p. 474.

[123] In these writings, not further accounted for here, Kant seeks to explain effects of force by presupposing a universal ether stuff. See **Hoppe** (1969) and **Mathieu** (1989).

[124] **Kant** (1786), p. 474 (emphasized in original).

Thus Kant eliminated from the concept of force all teleological and individualizing elements still found in Leibniz; they no longer fulfill a legitimating function but serve only purposes of admittedly indispensable heuristics for the theory of nature.[125]

Being more geared to the structure of mechanical forces and by accepting the philosophy of nature's tenet that those forces attain unrestricted efficacy, the basic position underlying the mechanistic conception of dynamism in the narrower sense emerges more distinctly in Kant: the essence of matter must be concluded from immaterial effects of force that structurally fit the laws of physical mechanics. To this end, matter must be viewed as infinitely divisible and no space devoid of matter may exist. Despite the incongruity of these tenets and those of (either materialist or dually conceived) atomism, Kant mitigates the contrast between dynamism and atomism, which seemed totally insurmountable for Leibniz, when it touched upon the fundament of the theory of nature. In contrast to Leibniz, Kant wants dynamic comprehension of, if not atomic, then at least elementary and spatially, definite particles, which is why he focuses systematically on a dynamic reconstruction of the concept of the extended body. Accordingly, his theory implies using an approach from which it should be possible to deduct the finiteness of the surfaces of bodies.[126]

3.4 Concluding Remark

Thus, in completing my reconstruction of these three traditional lines of mechanism it appears that in one respect dynamic mechanism is closer to materialist mechanism than to the dualist line. Each of the two approaches claims to derive the principle used in the other one from its own principle. Dynamic and materialist mechanism could also be called "monistic" versions of mechanism, each of which will allow, for the purpose of explaining nature, only one principle: either matter in motion (Descartes, Boyle, Huygens) or forces at work (Leibniz, Kant). Nonetheless, it should have become clear that the system-immanent scope of each of these principles is limited in its own way. While materialist mechanism must postulate an immaterial origin for primordial motion (Descartes, Boyle), the dynamic version cannot accomplish, from the presupposed principle of the infinite divisibility of matter, the deduction of either distinctly defined trajectories (Leibniz) or the definiteness of bodies (Leibniz, Kant), adequate for applying physical mechanics. Such difficulties, immanent to monist explanations of nature, can be considered an argument against attempts at complete reduction to one of the two principles.

It follows that dual mechanism could be viewed as not the mere combination of the materialist and dynamic approaches but as a concept in its own right, distinguished by its acknowledgement of two reciprocally irreducible yet mutually referential principles of mechanical passivity and activity.

---

[125] **Kant** (1788), p. 181, **Kant** (1790), p. 410ff.

[126] For more on Kant's dynamic theory of matter found in *Metaphysical Foundations of Natural Science* [*Metaphysische Anfangsgründe der Naturwissenschaft*] see **Schiemann** (1997), p. 124ff.

Chapter 4
Contours of Modern Philosophy of Nature

The philosophy one chooses [...] *depends on the kind of person one is*
(Johann Gottlieb Fichte, *First Introduction to the Science of Knowledge*).

No science is, strictly speaking, 'without presuppositions'
(Friedrich Nietzsche, *On the Genealogy of Morals*).

Any statement can be held true come what may, if we make drastic enough adjustments elsewhere in the system
(Willard Van Orman Quine, *Two Dogmas of Empiricism*).

The previous section itself no longer served the sole purpose of defining the concept of mechanism as a classical philosophy of nature. Historically, all three lines persisted well into the nineteenth century, predominantly accompanied by an absolute claim to validity, which in itself is sufficient for associating them with the classical notion of science. But at no time were they exclusively valid. By contesting, from the dawn of early modern times onward, the validity of each other's claim to comprehensive knowledge of nature and by coexisting, albeit to varying degrees of acceptance, as alternate approaches, they in fact constituted a variety of ideas of the philosophy of nature, which first came to be legitimized by modern conceptions of science. From the standpoint of mechanism's traditions, the renouncement of the classical claim to truth in a way does justice to the factual situation. Modernity unleashes a plurality of theories for which the scientific community already had a disposition.

On the other hand, in terms of the formal aspects of the legitimating strategies, the range of possibility gets narrowed down. Historically, modern conceptions of science dominate in their criticism of metaphysical foundations, which, advocated paradigmatically by Descartes, Leibniz and Kant, establish science's claim to truth a priori. Except for documenting it by way of Helmholtz's relationship to Kant, I shall not elaborate the demise of the metaphysical strategy.

I understand "modern philosophy of nature" as a scientistically legitimated philosophy of nature that relies on statements of modern science to justify its claim to validity. I distinguish the modernity of that science by the specific hypothetical character of the validity attached to its propositional contents, which transfers to the philosophy of nature statements inasmuch as these refer to natural science for legitimacy. The claim to truth raised by the classical concept of science is no longer constitutive of the modern understanding of science. The difference between the modern and the classical version of scientistic philosophy of nature can therefore be traced back to changes in their underlying conceptions of science. Thus, in the present section I shall concentrate on an admittedly roughly-outlined definition of the modern conception of science.

Characterizing the modern conception of sciences based on its trait of hypotheticity is, in two respects, part of contexts that reach beyond that characterization itself. In one respect it signifies an historical relationship: modern science, whose statements are anything but arbitrary, is certainly not hypothetical in an absolute sense but it *is* when compared to the former emphatic classical claim to truth. On the other hand, hypotheticity alone is not sufficient for defining the modern conception of science. Besides regarding other features concerning theory sophistication, modern science differs from the classical conception of science with respect to altered institutional and socio-cultural circumstances. The latter two illustrate historically why the transition from the classical to the modern conception of science is dated as occurring in the nineteenth century.[127] The result is a catalogue of criteria, which, although it does allow an evaluation of the scientific underpinnings supporting Helmholtz's mechanism, is still far from being a comprehensive conceptual characterization of modern science, which cannot be provided here (Section 4.1).

Indeed, the concept of modern philosophy of nature does not merge into the concept of science as well as it did for classical mechanism. While during early modern mechanism themes of science and philosophy of nature largely coincided, in the eighteenth century differentiation sets in when natural science, increasingly organized in special disciplines, raises ever fewer questions phrased as philosophy of nature. Historically, scientism in the sense meant here, which applies to both the classical and modern philosophy of nature, also appeared in ways other than as the legitimatory basis of those traditions.[128] This also holds for the nineteenth century, when mechanism in the form of so-called scientific worldviews gains significance that was not an issue in the specialized research on which it rested.

[127] Particularly **Diemer** (1964), (1968) and (1970), **Schnädelbach** (1983), **Redner** (1987) and **Pulte** (2005) have made suggestions for defining a modern concept of science of the sort that began to emerge in the nineteenth century. While A. Diemer outlines "tendencies of the modern conception of science" with a list of criteria (reflectivization [sic] positivization [sic], eliminating metaphysics, increasing autonomy, operationalizing, problematization, conditionalizing, hypothetization, propositionalizing and inter-subjectivity, cf. **Diemer** (1968), p. 36, and (1970), p. 15), H. Schnädelbach speaks of a "new master model" (**Schnädelbach** (1983), p. 107) or a "modern understanding of science" (loc. cit. p. 110) that "makes science dynamic under the empiricist banner" (loc. cit. p. 114), for which Helmholtz is an example of its "inductivist variant" (loc. cit. p. 111). On applying the modern concept of science to physics, see **Schiemann** (1996a).

[128] In my opinion, scientistic philosophy of nature must mean neither both denying that defining the main features of the entirety of nature is a self-contained philosophical task and viewing it as a totally scientific matter, nor acknowledging that questions in the philosophy of nature have their own self-contained character but are nonetheless reducible to empirical problems. Thus I disagree with Schnädelbach, who calls the first of these two strategies scientism and the second a regulative idea for scientism (**Schnädelbach** (1983), p. 123).

Today's contemporary understanding of the notion of a worldview often clouds the fact that this phenomenon (not limited to philosophy of nature) first emerged during the nineteenth century and was characterized idiosyncratically by that era's historico-cultural circumstances. Mechanism did not become a "worldview" until the nineteenth century. Therefore, I shall close this section with a few remarks on the idea of a worldview (Section 4.2).

4.1 Hypotheticity as a Mark of the Modern Conception of Science

According to both the classical and the modern conception of science, statements that are postulated for specific purposes without proof of their truth are called hypotheses. The modern conception of science does not differ from the classical notion chiefly by the fact that an altered concept of the hypothesis became widespread. The crucial difference is that a propositional form known since antiquity now became evaluated in an entirely new way. The classical conception knew – roughly – two forms of hypothetical propositions, each of which were interpreted as an expression of the human incapacity for knowledge: the lack of truth was either impossible to overcome, due to the world's divinely willed inscrutability or it was temporary and would have to be replaced (sometime in the future) by true statements. For the classical system of scientific knowledge there was no way that hypotheses could be considered one of its integral components. Modernity eliminates not only this antagonism between science proper and hypotheticity by adopting hypotheses as a firm element of scientific knowledge: compared to the classical notion it turns the evaluation of science entirely around by allowing the hypothetical character of a statement or an entire set of propositions to become the epitome of science proper.

In its form, the classical understanding of science follows binary propositional logic: in "categorically deductive" systems of knowledge, statements, claiming to be directly empirically testable, are derived by a deductive method from true premises or laws of nature.[129] I shall contrast the hypotheticity of modern science with the way it was classically understood in three ways: (a) in terms of the character of laws of nature, (b) as related to the status of logical-mathematical calculus, and (c) regarding experience as the touchstone of scientific statements.[130] I shall outline these three aspects from a contemporary standpoint and also mention, in anticipation of what is to follow, the link that can be made to Helmholtz's science and its environment.

(a) The modern understanding that laws of nature are hypotheses belongs to the empiricist and (less important for our purposes) conventionalist tradition.

---

[129] Cf. Chapter 2 for the classical concept of science and on the relationship between binary propositional logic and classical science see **Leinfellner** (1967), p. 36ff.

[130] This trichotomy is taken from **Stegmüller** (1972ff.).

David Hume established that no necessary connection exists between facts, so that the laws of nature, which normally, given their syntactical form of conditional universal propositions, claim necessity for a not-finite range of application, cannot be founded on induction.[131] The modern concept of science, as I presuppose it here, radicalizes Hume's position to say that the validity of physical law statements is neither definitely verifiable nor immune to invalidation through confrontation with conflicting experience.[132] Of course, their validity can be quantified using probability calculus (which, however, the classical notion of science considers unscientific) and be ascertained in examination methods to the extent that acknowledged expressions of physical laws are attested, within their own domains, a great degree of prognostic reliability. Inductive confirmability (Carnap), falsification confirming strategies (Popper) and being applicable for prognosis (Goodman) – just to mention the most important approaches – have all been suggested as methodological underpinnings for this sort of verification.[133]

In the nineteenth century, John Stuart Mill, whom Helmholtz mentions repeatedly[134] did the groundwork for fundamentally questioning the absolute validity of natural laws. While Mill still clearly does distinguish laws (the inductively discovered causal nexus holding between phenomena) from hypotheses (an empirically as yet unverified explanatory cause), he does not demand that lawful causation be of unlimited universal validity.[135] Finally, Helmholtz's own numerous comments on the concepts of force and law provide a wealth of material for studying his concept of science.

(b) The modern conception of science eliminates the a priori validity of logical statements that classical science took for granted by undermining the difference between analytic and synthetic sentences. Willard Van Orman Quine's thesis, widely accepted in contemporary philosophy of science, states that no conditions for analyticity can be explicated: the dichotomizing of language-analytic sentences, which are always logically true and the synthetic ascertainments of fact, whose truth is always a factual matter, finds no support.[136] Analytic sentences are subject to revision just as ascertainments of fact are. Quines thesis, which strictly speaking deals only with the status of logic and therefore must be isolated from the question of the

---

[131] **Hume** (1748).

[132] The latter can also be called the "fallibility" (i.e., the susceptibility to error) of universal hypotheses and should be distinguished from both "falsificationism" and "fallibilism", which are both defined much more narrowly and would, thematically, belong to hypotheticity as described under (c). While falsificationism claims that logically a single observation sentence (such as $\mathrm{V}\,x\,(Ax \wedge \neg Bx))$, whose verification is assumed, can "falsify" a universal hypothesis (such as $\Lambda\,x\,(Ax \Rightarrow Bx))$, fallibilism implies that observation sentences, too, are fallible. See also **Andersson** (1981).

[133] **Carnap** (1966), **Popper** (1935) and **Goodman** (1975). These authors call law-like universal propositions of nature hypotheses and have, when discussing their logical status, begun approvingly with the problem of induction raised by Hume.

[134] See Chapter 5, footnote 247.

[135] **Mill** (1869 ff.), Vol. 3.

[136] On this see particularly **Quine** (1960), p. 118 ff.

validity of mathematical truths, can be seen in a more general context that shatters a priori presupposed formal truths. Besides logic, in my opinion these include mathematics (including geometry). Negating the distinction between the analytic and the synthetic may then be taken for the most recent blow in the demise of a priori validity for logical-mathematic systems of rules, whose destruction began in the nineteenth century.

With respect to research on Helmholtz, what comes particularly to mind in this connection is how Bernhard Riemann raises and phrases the problem of the principles of geometry. Helmholtz discusses it in his empirical foundation of geometry. Other than Kant, he contests the apriority and other than Riemann, the hypotheticity of geometry.[137]

(c) While both aspects of hypotheticity just mentioned are expressions of empirical orientation for the modern concept of science, the third aspect erodes empiricism's classical assumption of truth: it rejects the assumption that empirical facts are the ultimate universally valid touchstone for knowledge. Among the reasons for this is the circumstance that under research conditions involving an increased use of equipment (see further below) it becomes problematic to determine what counts as an empirical basis. If what is observable remains the only admissible scientifically utilizable experience, as is commonly the case in contemporary theory of science, demarcating what is from what is not observable will always remain arbitrary.[138]

The crucial modern argument for the hypotheticity of empirical reference, however, lies not in the instrumental conditions of research practices but in the indubitable linguistic form of all observation that can be utilized scientifically.[139] At the linguistic level, so the argument goes, we find that empirical reference is subject to inevitable relativization. If we wanted, namely, to avoid all relativity of the empirical content of scientific statements, we would need to design a universally valid linguistic system where what is observed can be isolated in undefined descriptive terms (for predicates and under certain circumstances also for corresponding objects) unequivocally logically linked to all other terms and sentences of that ideal language. But since any attempt to design that sort of meta-language terminates in infinite regress[140], linguistic definitions are only valid relative to other equally legitimate languages. Consequently, we must assume that scientific empirical evidence is irreducibly replete with discrete, linguistic-theoretic elements and itself not guarded from vitiation.

Regarding Helmholtz's work the consequences are important that the linguistic conception of scientific experience has for the relationship that holds between theory and experience, two of which I intend to highlight and shall call, following

---

[137] For Helmholtz's work on non-Euclidian geometries see Section 6.2.

[138] On the concept of observability see **Stegmüller** (1973 ff.), Vol. II.1, p. 189 ff. and 296 ff.

[139] The linguistic form of modern science is an expression of its intersubjectivity (see **Hacking** (1984), p. 167 f.).

[140] **Quine** (1969), p. 65 ff.

Quine[141], the "underdetermination" of theories: the impossibility of gleaning a universal empirical standard examination is reflected

- First, in the hopelessness of choosing between different competing statements or theories on the grounds of "crucial experiments". An object of experimental investigation is always part of a complex theoretical network that influences every test, without itself being subject to examination.[142]
- Second, in the circumstance that we cannot rule out the possibility of theories being formulated in different linguistic ways, yet still agreeing in terms of their descriptive statements.[143]

These two theses can be applied retrospectively to nineteenth century mechanism: they match the notion that mechanistic reductions (like those of thermodynamics or electrodynamics) can neither claim to be absolutely true, nor must they be false. Instead, they can be in empirically undecidable relations to other comparable theories, making mechanism one hypothesis among others.[144] Scientific explanations and theories that, according to this approach, no longer claim to mirror essential structures of reality can be called models.

All three of the aspects mentioned – the hypothetical character of physical law statements, the loss of apriority for formal rules and the underdetermination of theories – together compose a view of science that, on a theoretical level, substitutes a plurality of possible theories for the classical monopoly on truth. Of course, claims to validity for physical laws and theories are subject to relentless empirical testing methods. But this neither excludes a plurality of theoretical approaches within a range of application, nor can contradicting observational statements alone refute theoretical statements. Indeed, without losing its mark of being characteristic of modern science, theoretical plurality can be made relative by indicating additional non-theoretical features.

At the theoretical level, I would like extend the notion of hypotheticity to include the more comprehensive feature of "conditionalizing".[145] Conditionalizing here does not mean making an issue (as in Galileo's physics) of what is naturally possible but not yet realized; it means the general tendency to stress the issue of conditions of validity for scientific statements and – compared to classical science – awarding the conditionality of the given greater significance.[146] Besides hypotheticity, which is to

---

[141] See **Quine** (1960), p. 146f., **Quine** (1970), **Quine** (1975), **Bechtel** (1980) and **Kirk** (1973).

[142] In the theory of science criticism of the "experimentum crucis" propounded especially by F. Bacon (**Bacon** (1620), II, 36, but also by **Descartes** (1984a), p. 60) can be traced back to Pierre Duhem (**Duhem** (1904f.), p. 351) and was revised in the twentieth century particularly by Quine (for instance in **Quine** (1953), p. 27ff.). See generally **Lakatos** (1974).

[143] Logically, the linguistic difference can even involve incompatibility. See **Stegmüller** (1978ff.), Vol. II, p. 258ff.

[144] In the history of science this idea is advocated for the final third of the nineteenth century by P. Feyerabend (**Feyerabend** (1978), p. 104.

[145] I take this term from **Diemer** (1968), p. 53ff.

[146] For Diemer the concept also includes what is naturally possible.

be understood as the most significant result of this general trend, there are other conditionalizing phenomena:

– The uprating of empirical initial and boundary conditions compared to physical law hypotheses in scientific explanations[147]
– The inclusion of pragmatic contexts in the theory of scientific explanations[148]

The relevance of such external conditions for validity grows as modern sciences become increasingly specialized. The scientific investigation of a particular phenomenon need by no means depend on whether its domain interplays theoretically with others. All that is required is that the object's domain can be communicated in scientific language, which consists foremost of demarcating itself from other domains. Thus the classical claim to comprehensive knowledge of nature has been abandoned and substituted by the qualitative diversity of scientifically processed objects of study. This means that compared to classical scientific knowledge, modern scientific knowledge is local and the theoretical diversity tied to hypotheticity is limited to the particular domains of specific phenomena.

Just as classical science, so can modern science also not be defined solely in terms of theoretical criteria. The environment of experimental research practice constitutive for scientistically founded classical science is essentially preserved in the modern concept of science.[149] But starting in the nineteenth century, the social organization of experimental research underwent fundamental change: it was withdrawn from the personal control of individual scientists and reorganized under the supervision of either the state or private enterprise. That transformation process, which Helmholtz significantly helped to get under way in Germany[150], can be described as making science both (1) autonomous, and (2) functional.

1. In Germany the trend towards making science autonomous, implied by the modern concept of science, becomes conspicuous, historically speaking, at the turn of the nineteenth century chiefly at the level of institutional academics. The traditional

---

[147] Historically, these conditions were not explicitly mentioned until they became an indispensable part of the explanans in twentieth century theory of science explanation (standard formulations were provided by **Popper** (1935), p. 31, and **Hempel and Oppenheim** (1948)). The distinction between scientific law conjectures and the set of initial and boundary conditions pertaining to them, a distinction which has since been predominant in the theory of science, has recently started to get blurred: the form of laws itself can depend on the choice of conditions. See, for instance **Küppers (Ed.)** (1987).

[148] Until the 1970s, the theory of scientific explanation was strictly isolated from pragmatic contexts (cf. **Stegmüller** (1973 ff.), Vol. I, p. 1 ff. and 940 ff.). The proof that explanations are ambiguous, however, made it indispensable to include such contexts, the knowledge of which is necessary for exactly understanding what scientific questions and answers mean at all. Cf. **Schurz (Ed.)** (1988).

[149] On viewing experiments as causal contexts of action see Chapter 2. The limitation concerns disciplines of modern science that do not rely on experiments but basically, for instance, on methods of observation.

[150] Cf. Chapter 5.

academic arrangement, upheld from the time universities were established in the Middle Ages and formative well into the eighteenth century, was the division into faculties[151], consisting of one propaedeutic faculty of philosophy and three higher faculties of medicine, law and theology. In the nineteenth century coexisting functional systems, marked by increasing heterogeneity, developed into individual autonomous disciplines and replaced the older hierarchy.[152] In the course of that development, what we call physics today was, on the one hand, split off from subjects like "applied mathematics" and "natural philosophy" and on the other, split up into a number of departments (such as theoretical and experimental physics) and special fields (such as mechanics, thermodynamics and so on).[153] This specialization was reinforced institutionally and corresponds at the theoretical level to the specificity of modern science mentioned above.

The newly created disciplines, departments and fields were autonomous in several respects: in terms of the theory of science, the mathematical, scientific, engineering, medical and legal disciplines emancipated themselves from the dominant faculty of theology, which in classical science was responsible – often in collaboration with the faculty of philosophy – not only for the dogmatic formulation of the claim to truth but also for enforcing the pertinent standards of proper science. Inasmuch as they were organized by the state, disciplines were guaranteed academic freedom.[154] Socially, the differentiation process went hand in hand with turning research professional, a phenomenon reflected in the middle-class status of scientists, for whom, as public servants, tenured career ladders were officially installed and professional associations and lobby groups sprang up.[155]

2. What I call functionalizing science was also not characteristic of the modern concept until the latter half of the nineteenth century. The term denotes the continuous and ongoing influence that the sciences – particularly scientific engineering – has on technological progress, the ensuing rapidity of industrial development, the penetration of all of society with science, as well as resultant repercussions for the

[151] Note on translation: Historically, the faculty was a division of learning at the university, such as in the phrase *the faculty of law* (in contrast to the contemporary meaning of the body of teachers). It can also be translated as *school.*

[152] Cf. **Stichweh** (1984), p. 31 ff.

[153] The critical phases of the shift towards research in physics as we now know it took place not, as **Krafft** (1982), p. 88 ff., writes, in the first half but as **Cahan** (1985) has shown, in the latter half of the nineteenth century. According to Cahan the transition began when after 1830 Physical Cabinets (private laboratories where scientists performed experiments, because universities provided no such rooms) were gradually replaced by university institutes – a process that was completed about 1860. Cahan notes three distinct features of this development: setting up (university) laboratories, creating professorships with assistant staff and institutionalizing research alongside the teaching program. Between 1870 and 1895 institutes typically began to purchase physics instruments, provide laboratory courses for students, set up a human resources office for institutional research and obtain publicly financed annual funding.

[154] **Schnädelbach** (1983), p. 35 ff.

[155] On professionalizing science see **Beer and Lewis** (1963), **McClelland** (1985) and **Turner** (1971).

organization and the self-image of the sciences themselves. Although early modern Baconian science was guided by an interest in technological feasibility, not until the nineteenth century did science become impellent for productivity and innovation. Among others, electricity supply is a most conspicuous example of a highly profitable economic sector based on principles derived from scientific work and whose products dramatically changed and continue to change, social and cultural habits.[156]

The repercussions on the sciences themselves were various. While the industrial applicability of scientific findings is usually already based on the technical form of experimental practice, this practice revolutionizes itself by using factory-manufactured equipment. Unifying methods of research and teaching and setting down norms for measurement systems and the rise in precision measurement in some areas, all constitute some of the most important aspects of that process.[157] To become more profitable and innovative, science itself becomes organized industrially. Scientific training becomes assimilated to training in any other trade, research planning becomes assimilated to commercial management, the desired application of results becomes a decisive argument for research projects and so on.[158] All of this happens either under the auspices of governmentally secured autonomy or in private training centers and research laboratories, which from the end of the nineteenth century onward, contribute increasingly to the production of scientific knowledge.[159] From the vantage point of cultural history one can say that science's self-conception and society's interest in that conception shifts from the *curiositas*, with which early modern science initially competed ideologically with the still dominant Middle Ages and moves towards aspects of practical relevance.[160] Science is no longer legitimated, as it was in the classical period, solely by the value of knowledge contents but by innovative practicability.

Making science autonomous and functional are two major historically contingent conditions, which, together with the empirical testing methods discussed earlier, define the scope within which scientific work is possible. In the theory of science, both of these tendencies constitute a procedural notion of science that once the classical claim to truth has been abandoned, abstains from any commitment related to content. This basically formal focus in the modern concept of science is irreconcilable with the classical concept: while the latter was guided by achievable goals

---

[156] On the complex relationship between science, industry and society in the nineteenth century see (for a socio-critical view) **Bernal** (1953) and on the importance of the electric industry (from a contextual standpoint) see **Hughes** (1983).

[157] For the evolution of nineteenth century experimental practice and measurement techniques cf. **Schimank** (1976), **Brachner** (1985), **Cahan** (1985) and **Steinle** (2005).

[158] **Plessner** (1924) coined the phrase "the industrialization of modern research".

[159] Behind this trend was the scientification of technology in the secondary and tertiary sectors of the business economy. On its commencement in the nineteenth century see **Manegold** (1976) and **Mauel** (1976).

[160] **Lübbe** (1986), p. 21 ff.

viewed as part of a finite process of progress, the modern concept of science sees itself as an endless evolution of generating, recognizing and solving problems and part of an unfettered continuous process of technical innovation.[161]

At the theoretical level, the procedural notion of modern science is no better satisfied than by its hypotheticity. Hypotheticity creates the elbow room for theoretical flexibility, which – seen from today's standpoint – alone does justice to the aspects of applicability and unconstrained future inherent to the modern concept of science. Nevertheless, more than this sort of correspondence for how the socio-institutional features of modern science connect to its theoretical characteristics is not known. Since in this case both of these kinds of features can be considered contingent marks of science, one cannot decide whether the insight on the hypotheticity (and conditionality) of science co-initiated the socio-institutional process of transition or whether conversely, that transition process initiated hypotheticity. Modernizing the general conditions for scientific work merely encouraged the modernization of scientific claims to validity. To this day classical claims to truth continue to be advocated, even within an autonomously constituted science functionally integrated within society. This holds all the more, as one may assume that the transition from classical to modern science is still an incomplete process for which, strictly speaking, one cannot know whether it will ever be terminated or is even only of an episodic nature. The fact that both of these aspects of modernization become conspicuous simultaneously in the nineteenth century indeed shows why it makes sense to locate the transition within that period.

In summary, four features of the modern concept of science distinguish it from the classical concept. These features reflect reactions to the loss of the claim to truth as well as the historical conditions of a complex socially transformative process: becoming conditional and hypothetical at the theoretical level corresponds to making science autonomous and functional at the institutional and social level. Obviously, only the theoretical features are crucial for legitimating a scientistic philosophy of nature confined to knowledge of nature alone.

4.2 Conceptions of Nature as Worldviews

The statements of modern scientistic philosophy of nature are at least as hypothetical as the scientific statements and theory on which they rely. Thus, the fate of modern mechanism is confronted with two extremes: either antagonistic findings or other circumstances lead to a loss of recognition as a hypothesis or it becomes acknowledged

[161] To this appraisal may be added the consideration that there are perhaps limits to the modern conception of science. One understanding of science that would extend beyond this one, could be, for instance, the notion that "progress becomes routine" (A. Gehlen), which would not evaluate science predominantly in terms of its capacity for advance. Also, external absolute limits, such as those posed by nature that is only to a certain extent exploitable, alterable and transformable, seem to be incompatible with the modern concept of science.

as one hypothesis among optional views of nature. Obviously, the first of these two options differs less than the other from the way classical science handles theories and scientific explanations. For while during the classical period theories were discarded when their weaknesses could no longer be tolerated, it was also not typical to consider different theories for one field as being equally true accounts.

The plurality characteristic of modern philosophy of nature is mirrored by the concept 'worldview' as it has come to be commonly understood today.[162] The word "view" implies having options among different representations and interpretations of something that is presented and these can be generated both from the perspective of the one creating the view and from the perspective of the person doing the viewing.[163] Today worldviews are ascribed not only to a populace or epoch but also, in an entirely individuated way, to persons, each of whom may have his or her own worldview or even simultaneously entertain different – compatible and incompatible – worldviews.[164]

Speaking of *worldviews* in the plural, however, does not really do justice to the historical context within which the notion arose after the early nineteenth century.[165] For in terms of cultural history, the rise of the concept was not accompanied by a desire for a variety of worldviews but instead was the result of endeavoring to create one single coherent interpretation of the world. The formative occasion was the collapse of absolute idealism. With it the last systematic total interpretation of reality was lost and this affected the intellectual atmosphere in Germany far beyond philosophy. The decline of Hegelian philosophy, which had, for a last time, attempted to overcome the loss of identity induced by the deterioration of religious traditions[166] led in the nineteenth century – so charged with scientific endeavor – to a will to compensate the loss of orientation by recourse to easily generalized scientific findings. The idea was to establish underpinnings similar to those provided by religious belief but based on scientific findings. Worldviews that arose with this intent were certainly more closely tied to the classical ideal of science than to the research practices developing before the very eyes of its proponents. In a review of the nineteenth century, Friedrich Paulsen emphatically pointed out that there

[162] German: *Weltbild.* **Grimm** (1854 ff.), Vol. 14-I.1 (1955), col. 1552 f.

[163] For Helmholtz's concept of pictures cf. Sections 6.3.1.6 and 7.1.3, excursus "Mechanic Theories as 'Pictures' of Reality".

[164] Examples (from literature) can be found in **Grimm** (1854 ff.), loc. cit. An exceptional use of the term in the singular is the title *Die Mechanisierung des Weltbildes* by **Maier** (1938) and **Dijksterhuis** (1956).

[165] **Grimm** (1854 ff.) loc. cit. col. 1552, states that from the onset of the nineteenth century the word 'Weltbild' [worldview] "is common in various usage and has become a technical term, yes, even a catchword particularly for modern psychology and philosophy". Based on a random sample inspection of philosophical encyclopedias by **J.G. Walch** (4th edition 1775), **W.T. Krug** (2nd edition 1832 ff.), **Fr. Kirchner** (2nd edition 1890), **R. Eisler** (2nd edition 1904 and 4th edition 1929), **W. Brugger** (3rd edition 1950) and **H. Schmidt** (21st edition 1982), in German-speaking regions the word *Weltbild* was not acknowledged as a technical term until towards the end of the nineteenth century.

[166] Cf. **Habermas** (1974).

existed this search for a binding view of the world. The expectation that nineteenth century scientific research faced was to be:

> An all-round and completely sound worldview and worldly wisdom based on conclusive thinking. Religion and theology had provided this to earlier generations. In the eighteenth century philosophy inherited the task [...] the last heir of pure reason was Hegel. But then a new generation, as skeptical about [pure] reason as their parents had been about belief, turned to science: accurate research will put us on firm ground and provide a true worldview.[167]

Within this scheme the concept of a worldview stands for a post-Hegelian and tendentiously anti-philosophical interpretation of the world emerging from the environment of "accurate research". It developed in two basic phases[168]: the first was the historicist reduction of pure reason to history, which, in place of reason, was to become a normative fundament.[169] Early historicism thus distracted the sciences from idealism and prepared the way for a general permeation of modern conceptions of science.[170] In the second phase the notion of the worldview began for the first time to become relevant to the natural sciences: fully aware of their power to shatter tradition and alter the world, scientists strove to glean a new interpretation of the world from their research findings, which, by their own standards, had to be more than simply a pictorial, perspective description of it. Thus mechanism, at the onset of its renaissance in the mid-nineteenth century, considered itself not one worldview among others but as knowledge of the world authenticated by scientific research findings and therefore exclusively valid. Helmholtz's early conception of nature will be an example of this.

That the worldviews could not hold up to their initial promise was an insight first gained in the later nineteenth century. Paulsen continues:

> But science cannot accomplish that; it becomes increasingly clearer, science does not lead to an all-inclusive worldview that satisfies the imagination and the mind; it finds only a thousand fragmentary bits of knowledge, some of which are only passably supported, particularly in the natural sciences aiming to establish a foundation at least for technology and others, such as those of the historical sciences, that are perpetually questionable, ceaselessly subject to re-evaluation. The result is a feeling of disappointment: science cannot satisfy the hunger for knowledge [. . .].[171]

The resignation of these words is still shaped by the spell of horror at losing binding truths.[172] But acclimatization to modernity also increases acceptance for the consciously understood perspectivity of worldviews. The second part of this study will clarify the extent to which that transition took place within Helmholtz's own idea of nature.

---

[167] **Paulsen** (1902), p. 81 f.

[168] Following **Schnädelbach** (1983), p. 49 ff. and 88 ff.

[169] **Schnädelbach** (1983), p. 56.

[170] **Schnädelbach** (1983), p. 22 and 88.

[171] **Paulsen** (1902), p. 82.

[172] That horror is still effective when Hermann Lübbe states: "The disappointment of the expectation that religious worldviews could be replaced in a functionally equivalent way by scientific worldviews adequate to reality, is complete" (**Lübbe** (1986), p. 32).

Part II
Helmholtz's Mechanism at the Dawn of Modernity

Chapter 5
Helmholtz, a *Bildungsbürger*, Scientist, and Research Strategist

> *We may venture to observe from safe distance the whole figure of the intellectual giant as he sits on some lofty cliff watching the waves*
>
> (J.C. Maxwell on H.v. Helmholtz).

Before I embark, in the second part, on exemplarily reconstructing how the modern conception of science evolved, it seems advisable to introduce some aspects of the context in which our case study is embedded. Outlining biographical, scientific and cultural contexts may compensate two things. For one, it may balance the focus on one single aspect of Hermann von Helmholtz's work by indicating the extraordinary complexity of his entire production. The development of his conception of nature and science, being the topic of this case study and remaining deliberately largely unelucidated so far in these preliminary remarks, is only one aspect of his complete works.

The topic I shall delineate is to be introduced as part of a whole, into which it – more or less harmoniously – fits. This arrangement intends to aid the reader. Yet, it remains open to question, to what extent Helmholtz's production can be characterized as a well-ordered whole at all. As a natural scientist, Helmholtz not only occupied himself with topics as diverse as questions in geometry, physics, chemistry, physiology, aesthetics, sensory perception, epistemology and education, he also did so in a variety of ways. Some motifs – such as his mechanistic conception of nature – do recur and he adhered lifelong to a mathematical experimental method. Another uniform or unifying element was his endeavor to produce syntheses by applying findings from one field to other domains as well. His intellectual interests, as D. Cahan has noted, were transdisciplinary[173] and he is certainly one of the few researchers who still commanded a view over a wide range of scientific activity.

At the same time, Helmholtz did acknowledge differences between the objects of his research and the scientific methods suited to them. In part, because of this and partly for other reasons, in some areas his views and work were subject to considerable modification. Furthermore, he was influenced by sundry, sometimes even contrary factors. So my depiction of Helmholtz can only be consistent to a limited

[173] **Cahan** (1994a), p. 1.

extent. This introductory sketch intends to heighten awareness of opposing tendencies in his life, thought and work. My purpose is to point out how diversely his work can be interpreted, which in the past has been received with great controversy. Against this backdrop, my own reconstruction of Helmholtz's conception of nature and science, which I shall undertake in the following sections, cannot (also for this reason) be more than a perspective view, whose one-sidedness only then does not present a disadvantage, if it is kept in mind.

Who was Hermann von Helmholtz? A brief answer to this question is by all means possible: he was a physicist and physiologist, with important accomplishments in both disciplines. He enriched them both with an impressive number of fundamental insights, decisively shaped the way they see themselves today, applied their procedures to new areas of research and finally, leading the way, contributed to turning them into institutional laboratory sciences. Physics owes to Helmholtz above all the first mathematical enunciation of the law of the conservation of energy and the practice of physiological examination owes to him the ophthalmoscope. But while it is appropriate to call Helmholtz a physicist and physiologist, it would be wrong, on equal grounds, to call him an epistemologist or even a philosopher. The significance of his directive and likewise famous contribution to epistemology – above all his theoretical work in sensory perception and geometry – derives partly from the fact that it was written by a natural scientist and not by a philosopher.

A remarkable contrast to the diversity of his activities in physics and physiology is given – if the comparison be allowed – by a lack of complexity in his life.[174] His family background and the intellectual interests he later developed make him a

[174] An early representative depiction of Helmholtz's life and work is given by **Koenigsberger** (1902 f.). His study has not been surpassed in terms of wealth of material, biographical details and professional evaluation. Unfortunately, it does not measure up to contemporary scientific standards. Koenigsberger neglects to list sources for the documents he cites, many of which can no longer be located. Scrutiny reveals that he did not indicate omitted passages (cf. footnote 239) and some biographical details are not well-researched (cf. footnote 181). For an initial overview, **Turner** (1970 ff.), **Rechenberg** (1994) and **Hoffmann** (2000) are very suitable. Some new and unusual views of Helmholtz's life and work are contained in the commemorative article written by **Eckart and Gradmann** (1994). Voluminous secondary literature on Helmholtz's work, which Rechenberg for unknown reasons leaves, for the most part, unconsidered, is documented most completely in the bibliography compiled by **Cahan (Ed.)** (1994). Together with **Krüger (Ed.)** (1994) this collection provides an informative survey of the contemporary state of the art in Helmholtz's research (cf. **Schiemann** (1999a)). Not only do analyses in the history and philosophy of science critically appraise Helmholtz's professional achievements in physiology, the theory of music, physics, geometry, philosophy of nature and epistemology but also the moves of his career strategy and his refined and merciless rhetoric aimed at the competition (cf. above all **Turner** (1994a) and **Kremer** (1994)). I shall refer to some of these publications further on in my account. A valuable look at the circumstances of life and work in the period from 1837 to 1859 is also provided in Helmholtz's correspondence with his parents and his first wife, in **Helmholtz** (1990) and (1993), supplemented by excellent editorial introductions. For details on Helmholtz's activities at the *Physikalisch-Technischen Reichsanstalt* in the German empire see **Cahan** (1992) and **Kern** (1996), on "Helmholtz's shaping of the American physics elite in the gilded age" see **Cahan** (2005).

representative *Bildungsbürger* of the Biedermeier times. He came from a fairly inconspicuous lower middle class family in Potsdam, born as the son of Caroline and August Ferdinand Julius Helmholtz on August 31, 1821. His father, a veteran of the war of independence, held a position as teacher of philology at a secondary school in Potsdam from 1820 onwards.[175] Among the few recollections of his home that Helmholtz later publicly relates are philosophical disputes that his father, a firm adherent of Johann Gottlieb Fichte's philosophy, is said to have often had with colleagues. We know from the in part intense correspondence between father and son, which continued until the year Helmholtz's father died (1859), that his father tried in vain to convince Helmholtz to accept the idealistic standpoint.[176]

Between 1830 and 1838 Helmholtz attended the school where his father taught and was given a comprehensive humanistic education. He then took training as a physician at the Friedrich-Wilhelm-Institut in Berlin, an army medical institution led in a paramilitary fashion, which benefited offspring from families lacking the means for university studies. During this training, which also included lectures at the university, Helmholtz begins to study under Johannes Müller, the famous anatomist and physiologist, who recognizes this student's skills in physics and mathematics and very deliberately promotes him – quite in the modern sense of "exploiting talent reserves".[177] Müller, who at the outset of his career had adhered closely to romantic *Naturphilosophie* and skeptically opposed experimental method, was, at the time Helmholtz encountered him, an active proponent of physical-chemical methods in physiology. He did, however, to the dismay of his thoroughly materialistic students, retain a vitalistic position, holding that immaterial life forces are at work in organic nature.[178]

Under Müller, Helmholtz takes a degree in 1842, having presented a thesis in physiology in which he reports his first and also important discovery: his microscopic studies show that nerve fibers originate in ganglia cells.[179] After receiving his degree, Helmholtz works as a squadron surgeon and military doctor in Potsdam until 1848. His presumably rather leisurely occupational routine[180] allows him to pursue his own research work in laboratories led by Müller, Eilhard Mitscherlich

[175] On his parents, see **Cahan** (1993).

[176] **Helmholtz** (1886a), p. 314, **Helmholtz** (1891), p. 17. The correspondence that has been found to date is documented in **Koenigsberger** (1902 f.) and **Helmholtz** (1993).

[177] **Cahan** (1993), p. 11 ff.

[178] **Müller** (1824), **Müller** (1826a) and **Müller** (1833 ff.), p. 254 ff. For Müllers philosophy of nature see **Hagner and Wahrig-Schmidt (Ed.)** (1992), on his followers see **Cranefield** (1957) and (1966), **Mendelsohn** (1965) and **Lenoir** (1988), on their basic materialistic attitude, probably not wholeheartedly shared by Helmholtz, see: Du Bois-Reymond's letter of 1842 to Eduard Hallmann in **Du Bois-Reymond (Ed.)** (1918), p. 108. On Helmholtz's relationship to Müller cf. **Lenoir** (1992) and **Holmes** (1994).

[179] **Helmholtz** (1842).

[180] That the service performed by Helmholtz was not all too strenuous in times of peace is attested by **Eckart and Gradmann** (1994), p. 102, and **Tuchman** (1994).

(chemistry) and Gustav Magnus (physics) and he writes his first papers in physiology. For the first time he now becomes acquainted with a more intimate circle of Müller's students, with whom he will remain friends for the rest of his life.[181] Among them are, just to name the most important, Du Bois-Reymond, Karl Ludwig and Ernst Brücke. Like Helmholtz, they will later be leaders in framing physiology as study guided exclusively by the examination of physical-chemical processes.[182] Besides the way Helmholtz innovatively plans, excellently performs and skillfully mathematically evaluates experiments[183] he soon also exhibits the capacity to easily survey contemporary theoretical approaches of a field of research. His representative conspectus papers[184] reveal his typical search for uniform theoretical explanation. These papers are also part of his career strategy: they announce a claim to leadership in the discipline of physiology.[185]

In terms of the philosophy of nature, Helmholtz's early work indicates that he was influenced by the vitalists.[186] On the other hand, he struggles to outline an area that would, in principle, exclude effects of vitalistic life forces. The result of that effort is Helmholtz's first and unquestionably most important work in physics, his paper *The Conservation of Force* published in 1847, in which he develops his enunciation of the theorem of the conservation of energy.[187] Here his search for theoretical unity is guided by an ultimate goal and he develops an experimental program of research to realize that end. His ultimate goal is the most comprehensive mechanistic explanation of nature possible and Helmholtz professes it here, also for the first time. By applying the principle of conservation he hopes to find experimental clues to atomistically structured matter. The towering importance of the energy theorem will not be acknowledged by the scientific community until later, towards the end of the 1850s.[188]

When evaluating Helmholtz as a person it is relevant to note that while his writing of 1847 is related to the experiments he performed at the time, in that writing he does not mention newer insights gained from experiments. On the whole it is more a theoretical review and interpretation of scientifically known research results,[189] noteworthy

---

[181] Cf. Du Bois-Reymonds letter from December 25 and 26, 1845 to Eduard Hallmann, in: **Du Bois-Reymond (Ed.)** (1918), p. 123. Deviating from that, see **Koenigsberger** (1902 f.), Vol. 1, p. 44. For more details **Holmes** (1994).

[182] **Cranefield** (1957) and **Mendelsohn** (1965).

[183] **Olesko and Holmes** (1994), p. 52 ff.; **Kremer** (1990a), p. 237 ff. and 292 ff., however, doubts the relevance of Helmholtz's experiments for physiology.

[184] **Helmholtz** (1846) and **Helmholtz** (1847b).

[185] **Kremer** (1990b), p. XXII f.

[186] **Helmholtz** (1843) and **Helmholtz** (1845), p. 743. Of major importance for evaluating Helmholtz's early position on vitalism is the carefully argued dispute he had with Justus von Liebig. See **Lipman** (1966), **Kremer** (1990a), p. 237 ff., and **Brain and Wise** (1994).

[187] Cf. Section 6.1.

[188] **Helmholtz** (1891), p. 10 f., and **Planck** (1887), p. 89.

[189] **Bevilacqua** (1994), p. 303 ff.

both for its physico-mathematical approach and convincing rhetoric. In particular, when deriving the conservation theorem, Helmholtz hardly makes the reference to experimental research practice obvious. In secondary literature this has often been read as indicating an attempt to articulate a foundation for science metaphysically.[190] If one follows this interpretation, the result endows Helmholtz with peculiar twin faces as a metaphysically minded empirical scientist.

Helmholtz does not remain a physician for long. After a year of teaching anatomy at the Berlin Academy of Arts, where for the first time he was able to combine his knowledge in physiology with his interest in aesthetics,[191] in 1849 he reaches his first partial academic goal: as successor to Brücke he takes on a professorship in physiology in Königsberg. This dates not only the end of his period of education and training but also essentially the completion of framing his own scientific ambitions. For the domain of sensory physiology he later further specifies his project in his academic inaugural lecture on the *On the Nature of Human Sensory Perception* [*Ueber die Natur der menschlichen Sinnesempfindungen*], given in 1852. The program, which refers to Müller and outlined in that speech, is one intended to explain sensory physiological performance and remains the directive for his subsequent research.

Taking up the position in Königsberg he now began a life whose parameters appear to be fixed. In fact, it is a well-ordered, middle-class life, whose external circumstances are shaped by the demands of an academic career: the places of residence and the social contacts organized there by his first and second wife are determined by the universities where he obtains professorships with increasing reputation. Not only his theoretical but also his experimental work remains largely independent of local circumstances. Helmholtz quickly learns to compensate a lack of local scientific contacts by traveling often within the country and internationally. Noteworthy in this context are particularly his numerous trips to England, the first of which he took as early as 1853. In England he met, among others, Michael Faraday, George Stokes, James Clerk Maxwell and William Thomson.[192]

In order to exploit the contrast between the rather uncomplicated circumstances of his life and the complexity of his work, I shall first provide a brief outline of Helmholtz's further biography and then turn to the scintillating diversity of his work.

Just before leaving Königsberg, Helmholtz and Olga von Velten marry.[193] In Koenigsberg he completes further significant work in physiology: proof of the finite velocity of nerve impulse transmission and the invention of the ophthalmoscope, an instrument that easily makes the back of the eyeball viewable and which was extremely

[190] E.g. **Cassirer** (1957), **Heimann** (1974), **Bevilacqua** (1994).

[191] Cf. Helmholtz's *Probevorlesung über die Gesichtspunkte bei dem Unterrichte in der Anatomie für Künstler* (**Koenigsberger** (1902f.), Vol. 1, p. VIII) loc. cit. p. 95ff.

[192] For the travels to England cf. **Koenigsberger** (1902f.), Vol. 1, p. 193ff., 255ff., 372ff, and Vol. 2, p. 49ff., 197ff., 278ff. and 313ff.

[193] On Olga von Velten see **Kremer** (1990b), p. XII f.

important for the development of his career.[194] Helmholtz is aware of this instrument's utility for medical and scientific purposes and actively encourages its dissemination.[195] Success in his endeavors enables him to leave fairly secluded Königsberg after living there for six years. He becomes professor of physiology in Prussian Bonn.[196] But just as he, his wife and two sons settle down in Bonn, he receives the enticing offer of a chair at the University of Heidelberg, where renowned researchers Gustav Kirchhoff (physics) and Robert Wilhelm Bunsen (chemistry) teach. He had been promised a new building for his institute in Bonn but when construction does not commence on schedule, in 1858 he accepts the chair in Heidelberg. His wife Olga dies in 1859. In 1861 he marries Anna von Mohl, the daughter of a liberal member of the Frankfurt National Assembly and former minister of justice for the empire, Robert von Mohl.[197] They have three children.

1871 brings a second crucial turning point in Helmholtz's biography. He is appointed professor of physics in Berlin, where he remains until his death on September 8, 1894. Switching from physiology to physics was rather rare, even for the nineteenth century.[198] Furthermore, by becoming successor to Gustav Magnus, Helmholtz accepted one of the most prestigious chairs for physics in Germany. In scientific experiments, although not in continued active public speaking, he almost completely abandons physiological questions and concentrates, for a time, entirely on topics in physics, particularly in electro- and thermodynamics. As a result, this research lacks the innovative luster characteristic of work completed before 1870[199] but they are usually of fundamental importance within the context of contemporary issues. After 1888 Helmholtz is also the president of the newly-founded *Physikalisch-Technische Reichsanstalt*, an institute for furthering fundamental research in physics and technology and for devising and controlling scientific standards and instruments. This institution, which Helmholtz decisively helped to create, combines natural science and technology via the work on measurement instruments in a manner very characteristic of early modern and modern research.[200] In a history of the nineteenth century, Thomas Nipperdey writes that the fact that "the foremost physicist of the times [...] headed" this institution was a "breakthrough" "for research cooperation between the economy and science" and a sign that "science had given up resistance to economically oriented research".[201] As much as this may apply to the historical situation of German natural science, it certainly does not equally apply to Helmholtz,

---

[194] **Helmholtz** (1850a), **Helmholtz** (1850b), **Helmholtz** (1850c) and **Helmholtz** (1851).

[195] **Kremer** (1990a), p. XXIII f., and **Cahan** (1994b), p. 574f.

[196] Cf. the letter of recommendation by Alexander von Humboldt, dated March 24, 1855 in **Koenigsberger** (1902f.), Vol. 1, p. 249f.

[197] On Anna von Helmholtz cf. **Koenigsberger** (1902f.), Vol. 1, p. 371 ff., and **Siemens-Helmholtz (Ed.)** (1929).

[198] **Du Bois-Reymond** (1895), p. 45.

[199] **Cahan** (1994a), p. 3f.

[200] On Helmholtz's work at the *Physikalisch-technischen Reichsanstalt* cf. footnote 174.

[201] **Nipperdey** (1992f.), Vol. 1, p. 588.

who even in the context of his work for the *Reichsanstalt* was skeptical of orienting research towards the needs of application and refused to do so.[202]

Both of his most significant achievements – the mathematical enunciation of the theorem of the conservation of energy and the invention of the ophthalmoscope – saliently demonstrate that Helmholtz had extraordinary talent for both experimental and theoretical work. Further work reveals how closely these two sides of his gift for research were dovetailed: his conceptual work initiated and determined the arrangements of his experiments (thus, for instance, his early experiments in physiology also served purposes of the theoretical analysis of the concept of vital force),[203] while other experiments influenced the production of theories (thus his theory of acoustics does not take shape until combined with experimental practice).[204] His "ability to integrate mathematical, theoretical and instrumental elements into one complex, yet uniform structure" has been noted and that his work is marked by a "reciprocal constitution of theory and practice".[205]

But Helmholtz's experimental skill and his capacity for theoretical synthesis are two sides of a talent that can also be seen as a problematic combination. Not rarely do his far-reaching theoretical conceptions depend on experimental results capable of only limited generalization (Helmholtz errs, for instance, in believing that his early experiments in physiology provide a basis for refuting ideas of life forces or that the unconstrained validity of the second law of thermodynamics allows conclusions about the future heat death of the universe).[206] Some polemic passages in his writing tend to conceal a lack of empirical confirmation with rhetoric.[207] Conflict-laden issues like these clearly reveal distinct experimental and theoretical elements that tend to become obscured in research practice. Even for these cases of conflict, it still holds that Helmholtz worked neither theoretically nor experimentally without incorporating the relevant state of the art experiments and theories of his time.

His important work in physics includes, beside formulating the theorem of the conservation of energy, three others: his important mathematical solution for hydrodynamic equations, still in use today,[208] his research in electrodynamics, the result of which contributed crucially to the acceptance of the now generally accepted Maxwell theory[209] and his treatises on thermodynamics, in which he introduced a new physical

[202] I shall return to Helmholtz's position on the technical application of scientific findings when I discuss his conception of science. Cf. Sections 6.3.1.2 and 7.1.1.

[203] **Lenoir** (1982) and **Olesko and Holmes** (1994).

[204] **Vogel** (1994) and **Lenoir** (1994). Cf. also **Buchwald** (1994a) and (1994b), who believes to find, in Helmholtz's electrodynamics, a physical conception that is so thoroughly tied to laboratory practice that it loses its own theoretical nature (**Buchwald** (1994a), p. 344).

[205] **Vogel** (1994), p. 260, and **Lenoir** (1994), p. 111.

[206] **Kremer** (1990a), p. 237 ff.

[207] **Kremer** (1994) and **Turner** (1994a).

[208] **Helmholtz** (1858), cf. **Truesdell** (1968), p. 341 ff.

[209] Particularly **Helmholtz** (1870 ff.) and **Helmholtz** (1881). From the enormous amount of secondary literature, I mention here **Woodruff** (1968), **Kaiser** (1981) and **Buchwald** (1985). Cf. on these also footnote 214 ff.

quantity, the ("free") energy convertible into work in isothermal reversible processes.[210] Finally, the analogies he developed in the 1880s between physical conformity to laws and mechanical equations for motion are relevant for evaluating how the claim to validity in his conception of nature evolved. These analogies served to generate a mathematical formalism for deriving the empirical laws of electro- and thermodynamics from the mechanical principle of least action. In place of a mechanistic ontology, which Helmholtz had still employed when phrasing the enunciation of his theorem of the conservation of energy, we thus see an alignment with one mechanical equation towards the end of his productive career.[211]

Helmholtz's place in the history of physics has been and continues to be judged by historians very differently. Soon after his death some claimed that Helmholtz represents the completion of so-called classical physics and that from the start of the twentieth century onwards he is, particularly because of his orientation by the paradigm of mechanics, only of limited interest.[212] But particularly due to the theorem of the conservation of energy and contributions to the foundations of geometry, which we shall discuss later, Helmholtz has also been considered a pioneer in non-classical concepts of physics, in their abstention from the research of causes and for non-Euclidian theories of space.[213]

Particularly his work in electrodynamics has been a topic of controversy. The subject of these diverging evaluations is not Helmholtz's effectual acknowledgement of Maxwell's theory but the question of how his own theory relates to it. Does his own theory essentially remain within the scope of Newtonian physics, which knows only instantaneously effective central forces[214] or is it more characterized by an endeavor to approximate itself to Maxwell's field approach, according to which force can only spread with a finite velocity[215] or is it even an entirely independent conception that fundamentally surpasses both Newton's and Maxwell's physics and implies something still very topical today: raising the relationship between theory and experiment to a new level?[216]

More clearly than in physics, which had seen the transition towards becoming a laboratory science as early as the 1830s,[217] in physiology Helmholtz was the pioneer of a new period. He has become a symbol for accomplishing the orientation to

---

[210] Particularly **Helmholtz** (1882 f.). **Kragh** (1994) lists the important titles of secondary literature on this.

[211] Particularly **Helmholtz** (1884b) and **Helmholtz** (1886); on this topic see **Klein** (1972), **Bierhalter** (1981) and **Bierhalter** (1987).

[212] For instance: **Böhm** (1904), **Wien** (1919), **Schlick and Hertz**, in **Helmholtz** (1921), **Warburg et al.** (1922), **Lenzen** (1944), **Laue** (1944). On the concept of classical physics see **Schiemann** (1996a).

[213] For instance: **Duhem** (1904 f.), **Süßmann** (1990), **Koslowski** (1986), **Buchwald** (1994a).

[214] **Buchwald** (1985), p. 233 ff., and **Kaiser** (1994).

[215] **Woodruff** (1968).

[216] **Buchwald** (1994a) and (1994b), cf. footnote 204.

[217] See Section 4.1, footnote 153.

physical and chemical questions that remains characteristic of physiology to this day. Important impulses came from creating strictly experimentally organized tenets of physiology in the nineteenth century, impulses for both the further development of scientific research, as well as for the image of man that rests upon it.[218] Helmholtz's own research work predominantly on the physiology of the senses is thoroughly tangent to questions in psychology. As for his works in physics, secondary literature also contains a wealth of information on the topics of Helmholtz's studies in physiology, particularly on the analysis and explanation of spatial vision and on the perception of colors and sounds.[219] His most important findings entered into both of his monumental books, *Treatise on Physiological Optics* [*Handbuch der Physiologischen Optik*], published in 1856 and 1867 and *On the Sensations of Tone as a Physiological Basis for the Theory of Music* [*Lehre von den Tonempfindungen als physiologische Grundlage für die Theorie der Musik*], published in 1863.[220] In these books Helmholtz not only compiles the knowledge of his times on sensation and the sensory perception of visual and acoustic phenomena, he also, with authority, engages himself in existing debates and develops, on his own, directive explanatory approaches for special phenomena. While *Sensations of Tone* extends his scientific research directly to a topic in aesthetics, Helmholtz also does groundwork for an analogous extension for the arts in his *Treatise*. The teachings on color that he explains there become later, in the 1870s, the basis for lectures he gives *On the Relation of Optics to Painting* [*Optisches über die Malerei*].[221] Although Helmholtz's standpoint on aesthetics is also predominantly one of physics and physiology, he nevertheless stresses the basic limits of his investigations that could neither explain artistic production nor the psychological effect of aesthetic perception completely.[222]

There are good reasons for doubting that the empiricist theory of perception developed by Helmholtz within the context of his work on the physiology of the senses could, from the start (as is often assumed in secondary literature[223]) hold up to its epistemological claim. In his academic inaugural lecture, Helmholtz already formulates the central idea of his theory: sensations and respectively, the human perceptions that result from them, are subjective because they depend on the human

---

[218] On Helmholtz's place in the history of physiology and psychology see **Bohring** (1942), **Lenoir** (1982), **Kremer** (1990a) and **Hörz** (1994). **Engelhardt** (1972) draws attention to the importance of physiology for the development of natural science; **Rabinbach** (1990) stresses Helmholtz's importance for the idea of the human being.

[219] Individual papers collected in **Cahan (Ed.)** (1994), Part 1, which discuss Helmholtz's most important work on the physiology of the senses in great detail, contain references to relevant secondary literature. Detailed work done by **Hatfield** (1979), **Hatfield** (1990), **Turner** (1994b) and **Lenoir** (2006) is available on Helmholtz's investigation of spatial perception.

[220] **Helmholtz** (1856 ff.) and **Helmholtz** (1863).

[221] **Helmholtz** (1871 ff.).

[222] Cf. **Hatfield** (1994).

[223] E.g. **Erdmann** (1921), **Cassirer** (1923 ff.), Vol. III, **Schlick** (1922), **Lenzen** (1944), **Heidelberger** (1994a).

organism and not on the objects sensed, which only elicit those sensations. Nevertheless, an epistemological claim could only be derived from this statement if the subjective structure of sensation processing was among the conditions for the possibility of knowledge at all. But in his early work Helmholtz holds scientific finding for objective because it surmounts mere subjective conditions for perception and advances to the invisible mechanical causes of phenomena.[224]

It is precisely this claim to objectivity that clearly separates Helmholtz's work in geometry from that on the theory of perception. While the theme of his theory of perception is the constantly changing processing of sensations in everyday life, his thoughts on geometry concern the foundations of scientific and thus strictly valid, knowledge. Here he discusses, as one of the first physicists to do so, questions that have remained fundamental for knowledge in physics right up to the present day: from the elementary properties of – in his opinion – real solid bodies he concludes the possibility of the existence of non-Euclidian spaces and discusses, using thought experiments, their intuitive structural features. Indeed, these writings include aspects that are virtually the opposite of ideas generally accepted in physics today. Helmholtz doubts neither the Euclidity of real space, nor does he seriously question the possibility of achieving any desired degree of precision in geometrical measurements.[225]

Yet in spite of the different claim to validity for statements in everyday life and geometry, they are related in ways that can lead to other assessments of the relationship between the theory of perception and geometry. Structural analogies can be found in each of their rationales[226] and last but not least, the transition in Helmholtz's notion of science is expressed by relativizing the claim to validity for statements in geometry.

In contrast to his work in physics, physiology of the senses and geometry, Helmholtz explicitly expresses his thoughts on the theory of science and on scientific policy, almost without exception only in public talks. His widely regarded speeches, published in several editions during his lifetime, are of a thoroughly popular scientific nature. There is hardly an aspect of his own scientific work that he does not also elucidate in generally understandable terms and hardly a speech in which he does not make reference to it. But his speeches also reach beyond the topics of his own research. In them he develops approaches for a scientific interpretation of the world, reflects on the history of science, seeks to clarify the relationship between natural science and the humanities and phrases tasks and goals for science as a whole. Passages on the theory of science and scientific policy tend to be unsystematic and of an unmistakably propagandistic character. His criticism of other conceptions of science and nature – particularly Romanticism – are less well-founded, than many more attempts to discredit them. He knew how to use the extraordinary public influence he possessed as a renowned scientist. He was aware

[224] Cf. Section 6.3.1.1.

[225] For literature, see Section 6.2.

[226] **DiSalle** (1994).

of representing an influential group of researchers in science and he addressed that part of the *bildungsbürgerlich* audience that was very interested in and confident about scientists' opinions.[227]

Some of the background on the sounding-board effect of a general public with great confidence in science has been mentioned in the previous section, "Contours of Modern Philosophy of Nature". Along with thoroughgoing industrialization, in the second half of the nineteenth century the social and cultural relevance of natural science reached a hitherto unknown level. The system of science itself undergoes a profound transition, sketched above as making science autonomous and functional.[228] In terms of educational policy, this transitional process particularly upgraded natural science and technology in contrast to the traditional, chiefly humanistic areas of education – a transformation that is still continuing today but which in the second half of the nineteenth century was accompanied by fierce cultural debates. Within that setting, Helmholtz's speeches were meant to both increase interest in the sciences and also recruit finances.[229] This can be done with more success the more it caters to a growing public circle's need for worldview orientation.

The contents of Helmholtz's conception of science, this much can be said in anticipation of what shall follow, remain entirely within the framework of the Baconian program: knowledge that has been discovered experimentally and tested is knowledge worth learning and further developing. It teaches individuals what the world consists of, peacefully brings together people who learn and research and provides humanity with control over nature and thus with better material living conditions.[230]

---

[227] The first published releases of his talks, which were usually printed in various publishing organs immediately following the speech, were, as a rule, difficult for the educated audience to obtain. In order to alleviate this (see **Helmholtz** (1865), p. V), Helmholtz began to publish them, with a slight delay, summarized in three booklets that followed the first issue (**Helmholtz** 1865 (Booklet 1), 1871 (Booklet 2), and 1876 (Booklet 3))). An indication of the public interest in them (today the publisher cannot provide information on print run numbers) may be that when the third booklet was issued, the first two were already being reprinted for the second time. The last issue of the booklets to appear during Helmholtz's lifetime (the third issue of the first two booklets and the second issue of the third booklet) were compiled to become the new edition of **Helmholtz** (1884a) (see also Section 6.3.2, footnotes 535 ff.). Only in **Helmholtz** (1862) and **Helmholtz** (1877b) does he make no reference to his own work but discusses exclusively issues of educational policy and the theory of science. Particularly important for the theory of science are **Helmholtz** (1862), **Helmholtz** (1869), **Helmholtz** (1871b), **Helmholtz** (1878c) and **Helmholtz** (1878a). The only work on the theory of science known to me that was neither first published as a lecture nor later reworked to become a popular version is **Helmholtz** (1887a). On the propagandistic elements of Helmholtz's public appearances see **Lenoir** (1988), **Cahan** (1994b), **Cahan** (1994c) and **Cahan** (2006). References for Helmholtz's conception of science can be found in the introduction to this book, footnote 24, and in Section 6.3.1, footnote 403, and Section 7.1, footnote 625.

[228] See Section 4.1.

[229] **Lenoir** (1988).

[230] This characterization coincides for the most part with that of **Cahan** (1994b), who sees Helmholtz's conception of science as supported by the enlightened belief in its "civilizing power". By being fundamental knowledge, Helmholtz's science civilizes in four respects: (1) in terms of knowledge of nature, (2) command of nature, (3) understanding aesthetics, and (4) cultural life within a nation.

Helmholtz, however, does not fail to recognize the importance of education in the humanities and arts and not only does he not deny their own right but explicitly and repeatedly defends it.[231] Beyond the bounds of scientific knowledge lie, for him, the indispensable achievements of intelligent creativity and reasonable action. In his public speeches Helmholtz does not talk like a scientific technocrat but as a *Bildungsbürger* and occasionally as a citizen.[232]

In this role the physiologist and physicist is less representative of the research community of his times than in his scientific preoccupations. The trend among scientists towards concentration and limiting (life's) activity to ever more narrowly understood special domains, a trend that began in the eighteenth century, continues into the nineteenth century such that ultimately among the foremost natural scientists there are conspicuously few who take stands on the philosophy of science or even public matters. Helmholtz was not alone in engaging in popular-scientific worldview discussion – there were also Emil Du Bois-Reymond, Rudolf Virchow and Jacob Moleschott. But his extensive philosophical interests can probably be considered exceptional.[233]

This applies particularly to his struggle with Kant's epistemology, mentioned repeatedly in his speeches. By turning to transcendental philosophical issues, related above all to causality and the form of intuition for space, Helmholtz sustainably influences the intellectual discourse of his time. By speaking of Kant's epistemology at all, he turns, armed with the authority of a renowned personality in research, against the philosophical materialism spreading not only in science but throughout the larger pubic as well and which contests the independence of the mind and is skeptical of epistemological reflection.[234] His judgment expressed in 1855, namely that Kant's critique of knowledge had staked out the tasks for all future philosophy,[235] turns out to have a considerable effect on the history of philosophy. It contributes to the fact that philosophy, which in the period of restoration had lost a lot of credit, once again became reputable and it facilitates the rise of the Neo-Kantian movement, whose influence increases within the context of academic philosophy.[236] Nevertheless, as early as the second half of the nineteenth century, Kantians and Neo-Kantians evaluated the relationship between Helmholtz and Kant

---

[231] Particularly in **Helmholtz** (1862). See also **Hatfield** (1994).

[232] Helmholtz publicly expresses, as an exception, a presumably rather national-liberal attitude (**Hörz and Wollgast** (1971), p. XLVI, **Brocke** (1996) and **Cahan** (2006)), e.g., in **Helmholtz** (1878a), p. 216, and in **Helmholtz** (1877b). **Eckart and Gradmann** (1994), p. 103, write of Helmholtz's "remarkably unpolitical character" and emphasize that "his progressiveness [...] was limited purely to matters of science".

[233] Cf. **Merz** (1907 ff.), Vol. III, Part I, and **Schnädelbach** (1983), p. 39 ff.

[234] On the materialism of the nineteenth century and how it related to the sciences see **Lange** (1866), p. 587 ff., **Gregory** (1977), **Köhnke** (1986), p. 168 ff., and **Bayertz et al. (Eds.)** (2007), Vol. 1.

[235] **Helmholtz** (1855), p. 88, cf. Section 6.3.1.3.

[236] Cf. **Schnädelbach** (1983), p. 131 ff., and **Lehmann** (1953), Vol. 2, p. 57 f.

very differently.[237] The extent to which Helmholtz's philosophical ideas are influenced by transcendental philosophy is a topic of controversy in the history of reception that continues to be of debate. The dominant views are subject to a change that reveals more about the interest in reception than about the actual matter. After well into the 1970s many authors believed to discover clear Kantian influence at least in Helmholtz's early works, in recent years more voices have stressed a fundamental difference between Helmholtz's and Kant's positions.[238]

As mentioned in the introduction to this book, evaluating the relationship between Helmholtz and Kant is extremely important for classifying the development of the claim to validity intrinsic to Helmholtz's conception of nature and science. At this point I would like simply to mention how un-Kantian-like Helmholtz spoke of philosophy. I mention this to counteract the possible misunderstanding that for all his interest in issues of the humanities and aesthetics, Helmholtz ever abandoned his genuinely scientific way of thinking. His not pervasively but indeed repeatedly expressed resentment about studying historical texts condenses with respect to the state of philosophy during his times to the opinion expressed only privately, namely that with philosophers one has to do "normally [...] only with impotent bookworms" who "have never generated new knowledge and thus have absolutely no idea how it is done".[239] The only "justified rights of philosophy", as Helmholtz says in his rectorial address of 1862 are the critique of knowledge and the vaguely formulated normative task of "determining the standard for intellectual work".[240] This idea of philosophical work, containing no mention of transcendence, aesthetics, philosophy of nature or practical content matches his – also expressed behind closed doors – thought that it would help philosophy to give "a natural scientist with a bent for philosophy a professorship in philosophy".[241]

Alongside Kant's epistemology, Fichte's subjective idealism has also been suggested as having influenced Helmholtz.[242] This assumption rests on Helmholtz's alleged close tie to his father and some passages of his writings where he makes

---

[237] While **Lange** (1866) tried to link Helmholtz's position to the transcendental approach, **Krause** (1878), **Goldschmidt** (1898) and **Schwertschlager** (1883) stress the difference between Helmholtz's empiricism and Kant's idealistic system.

[238] **Riehl** (1904), **Erdmann** (1921), **Cassirer** (1923 ff.), Vol. III, **Kahl** (1951), **Hörz and Wollgast** (1971), **Heimann** (1974), **Hatfield** (1979), **Gehlhaar** (1991), **Buchwald** (1994a), **Schiemann** (1994) and **DiSalle** (2006).

[239] Letter to Rudolf Lipschitz dated March 2, 1881, in the *Deutsche Allgemeine Zeitung* (Gross Berlin), Morgenausgabe, No. 453, dated September 27, 1932 (reprinted in **Lipschitz** (1986), p. 131 f.) and without this sentence in **Koenigsberger** (1902 f.), Vol. 2, p. 163 f. On Helmholtz's resentment about studying historical texts and book knowledge in general, see e.g., **Helmholtz** (1874b), p. 423 f., **Helmholtz** (1878a), p. 218, **Helmholtz** (1878c), p. 171 f., but also remarks to the contrary in **Helmholtz** (1892).

[240] **Helmholtz** (1862), p. 164.

[241] Letter to Fick in 1875, undated, **Koenigsberger** (1902 f.), Vol. 1, p. 243.

[242] **Turner** (1977), **Heidelberger** (1994a), very early also in **Conrat** (1904), p. 251 ff. and 267 f., and **Erdmann** (1921), p. 42.

positive reference to Fichte for individual issues.[243] But Helmholtz explicitly confronted idealism with his own position, which after the 1870s he called "realism".[244] However, one may evaluate Helmholtz's relationship to Fichte, the comparison is instructive inasmuch as it reveals pragmatic aspects that Helmholtz felt were very important for explaining life's everyday mechanisms of perception and achievements in scientific experimental action.[245]

The fact that Helmholtz lists scientific experience as the foundation for claims to validity suggests that what influenced him was something entirely different from idealism. Accordingly, secondary literature has pointed out the connection between his ideas and those of English empiricism.[246] While Helmholtz is critical of David Hume, he refers often with assent to John Locke and John Stuart Mill.[247] As a result of the impetus in natural science and technology in Germany from the middle of the century onwards, Mill's empiricism, characterized by its positive attitude towards accurate knowledge and which had been important in England for the philosophical involvement with science from the start, attracted increasing interest also in post-idealistic Germany.[248] In terms of the history of ideas, the entanglement of empiricism and certain types of materialism played a considerable part. Disregarding his initially implicit and later explicit critique of philosophical materialism, Helmholtz was close to a kind of materialistic thought that rejected the involvement of independent spiritual principles for explaining nature and which was most clearly advocated by his college friend Emil Du Bois-Reymond. Nonetheless, Helmholtz's mechanism, which (most of the time) limits itself to explaining material and energy aspects of natural phenomena, does itself involve one immaterial principle, namely that of mechanical force.[249]

Accordingly, Helmholtz never linked his critique of vital forces to any overall condemnation of vitalism. One may trace this reservation back to the influence of his teacher Johannes Peter Müller, an influence which recently has gained more appreciation.[250] Müller's vitalistic "law of specific nerve energies", according to

---

[243] **Helmholtz** (1855), p. 89, **Helmholtz** (1856ff.), p. 193 and 456, **Helmholtz** (1878a), p. 216f., 219, 227, 238 and 241, **Helmholtz** (1885 ff.), p. 248 f., and passages taken from **Helmholtz** (1878a), **Helmholtz** (1891), p. 17, **Helmholtz** (1892), p. 358, and **Helmholtz** (1897 ff.), Vol. I.1, p. 14.

[244] **Helmholtz** (1878a), p. 238 f.

[245] On the relationship between Helmholtz and Fichte cf. Sections 6.3.1.1 and 7.1.3.5.

[246] **Heyfelder** (1897), p. 9, **Hamm** (1937), p. 35, **Hörz and Wollgast** (1971), p. XXXVII, **Heidelberger** (1994a), p. 486 ff.

[247] On Hume see **Helmholtz** (1856 ff.), p. 455; on Locke see **Helmholtz** (1856 ff.), p. 385, 455 and 593, **Helmholtz** (1868a), p. 320, and **Helmholtz** (1878a), p. 219; on Mill see **Helmholtz** (1856 ff.), p. 447 and 453, **Helmholtz** (1862), p. 178, **Helmholtz** (1885 ff.), p. 581, and **Helmholtz** (1897 ff.), Vol. I.1, p. 7 f., and in this book Section 6.3.1.4 and Chapter 8.

[248] **Poggi and Röd** (1989), p. 84 ff. and 127 ff.

[249] See Sections 6.1.1, Part α, and 6.3.2.1.

[250] **Elkana** (1974), p. 104 ff., **Lenoir** (1992), **Cahan** (1994a), p. 18, **Olesko and Holmes** (1994), p. 23 ff., **Vogel** (1994), p. 282 f., **Heidelberger** (1994a), p. 484, **Kremer** (1994), p. 245 ff. In contrast, Jacob and Thure von Uexküll wholly contest that Müller had a positive influence on Helmholtz (see, e.g., **Uexküll** (1947)). They contrast Müller's conception of nature one-sidedly with Helmholtz's alleged materialistic position.

which the quality of sensory perception is not related to the object perceived but to the quality of the sensory nerves, sets the (subsequently modified) stage for the development of Helmholtz's theory of perception.[251] But Müller's influence need not have been limited to specific contents. It was in his laboratories that Helmholtz learned his mastery of experiments. Müller's monumental *Elements of Physiology* [*Handbuch der Physiologie*][252] was a directive for great synthetic achievements in theory. In addition, his educational scope was exemplary beyond the narrow domain of research topics. Finally, Müller's own scientific development offered an impressive example of the inevitable incompleteness and creative candor of thought in times of change, which was also to become characteristic of Helmholtz.

When one looks at Helmholtz's vast and diverse scientific work, his lively academic and public speaking activities, the complex historical background of his times, some highly diverse influential factors and how the history of science and philosophy received it all partially with so much controversy, it does look, in fact, as if a final judgment, one that would do justice to the various aspects of his productivity, is hardly possible at all. Assuming neither that such a judgment can be reached, nor striving for one, in the following study I would like to do no more than look at one element of Helmholtz's work from a certain perspective. My evaluation of Helmholtz's conception of nature and science could only be mistaken for a total evaluation of his work if one loses sight of the presuppositions.

The object of study, namely, a limited selection of texts, has been fixed by both thematic and pragmatic viewpoints. Since almost all of Helmholtz's work deals with issues in natural science or refers to them, an overriding interest in the transition of his conception of science and nature is insufficient for gleaning criteria on which texts to use. Presumably, there is hardly a piece of writing or other activity of Helmholtz's that could not be used to elucidate this aspect. For pragmatic reasons, however, it is hardly advisable to draw upon Helmholtz's entire works for this purpose. His writings and books that appeared during his lifetime alone comprise a total of seven volumes.[253] Add to this the posthumously published lectures and volumes of correspondence[254] and literary remains that include 700 still unedited documents. If one wanted to use Helmholtz's scientific practice to delineate his conception of science and nature, one would have to also examine other sources about his activity, his relationships to other scientists, the furnishings of his laboratories and institutes and much more. In light of all this material, the relevance of which can only be controlled in individual cases, any selection at all is marked by subjectivity, unless its use can be justified by how it relates to Helmholtz's total production.

The present work will now concentrate chiefly on his public speeches. More than in any other part of his work, in speeches Helmholtz explicitly stated his position on

---

[251] Cf. Sections 6.3.1.1 and 6.3.2.2.

[252] **Müller** (1833 ff.).

[253] **Helmholtz** (1856 ff.), resp. (1885 ff.), **Helmholtz** (1863), **Helmholtz** (1882 ff.) and **Helmholtz** (1884a).

[254] **Helmholtz** (1897 ff.), **Helmholtz** (1990) and **Helmholtz** (1993).

issues in science and the theory of nature. By publishing them separately as *Popular Lectures on Scientific Subjects* [*Populäre wissenschaftliche Vorträge*] – after 1884 as *Lectures and Speeches* [*Vorträge und Reden*] – he created a method of public explanation that offers a wealth of material for the analysis of the contents and formal rationale of his mechanism.[255] To the extent that they are thematically directly linked to the task at hand and are available in a completed state, one can say that Helmholtz's speeches constitute a particularly appropriate object of study.

On the other hand, editorially severing his public statements from all of his other scientific work could raise the question of the extent to which fundamental differences existed between these two activities, such that if we are to concentrate on studying the speeches, this might suggest an inadequate characterization of Helmholtz's ideas. I mentioned above that his lecture activity incorporated the most diverse institutional, political and cultural circumstances of his time. His intentions were frequently propagandistic and the picture he sketched of natural science also served the purpose of enhancing its reputation, legitimating its growing social influence and recruiting financial support. Nonetheless, I do not believe that the ideological factors and thoughts related to research strategy in his public explanations lead to a distorted image of his mechanism when compared with that of his professional writing.

First, because Helmholtz's public reports on empirical findings from natural research and the procedures employed to make them are neither distortional for external purposes nor unacceptably simplified. On the contrary, in the case of the theorem of the conservation of energy we find that he publicly mentioned its universal validity only with reservation.[256] Second, it would be erroneous to believe that professional writing relevant to his rationale for mechanism might be free of ideological and strategic interests. For instance, his first two treatises on the foundation of geometry were obviously guided by a professionally poorly justified interest in proving the absolute validity of Euclidian geometry.[257] Third, in terms of his method of presentation, for the present thematic range of interest no clear demarcation is possible between professional writings and those of popular science. The *Treatise on Physiological Optics,* for example, meant for interdisciplinary use, also contains some sections of a popular scientific nature.[258] Fourth and finally, conceptions of science and nature are cultural phenomena, which might even be more adequately expressed in lectures and speeches than in professional publications, where cultural influence is often only implicitly at work.

Not all of his important works in the theory of science and nature written in a generally understandable way were also published as speeches during Helmholtz's

[255] Cf. footnote 227.

[256] Cf. Section 6.1.2, Part β.i.

[257] Cf. Section 6.2.

[258] In the 2nd edition of **Helmholtz** (1856ff.) one of them, namely §26, was accordingly substituted by a speech (**Helmholtz** (1878a)) delivered previously (cf. **Helmholtz** (1885ff.), p. 576ff.).

lifetime. Thus, I shall make reference to his programmatic academic inaugural lecture, which Helmholtz for unknown reasons did not include in his collection of speeches and also make reference to parts of the *Treatise* and *Conservation of Force* that were written in a popular science style. I shall only take recourse to his professional writings when the material offered by the popular science essays is not sufficient for a detailed reconstruction of the transition in his conception of science and nature. This relates particularly to the first and last phase of the development of his mechanism. In connection with some specific issues – for instance the validity of mathematical and physical axioms – I occasionally also make use of Helmholtz's personal correspondence.

Yet, the material used in addition to the speeches basically only completes the view of the development that has already been reconstructed by using them. I believe that focusing on the conceptions that Helmholtz himself expressed publicly provides a fairly complete and comprehensible depiction of the transition in his ideas, which could be linked to or contrasted with other parts of his work in further studies in the history and philosophy of science.

I shall begin with Helmholtz's first and for the remainder of his life clearest explanation of his mechanism, taken from the introduction of *Conservation of Force*. From the introduction and how it relates to the main part of the treatise I gain first clues for positioning his conception of nature. According to this it would belong, in terms of content, to the dual line of mechanism, formulating its claim to validity as exclusive and establishing that claim scientistically by legitimating reference to past and future empirical findings in natural research (Section 6.1: The Mechanistic Program of 1847).

The second related example confirms this assessment with respect to form. Helmholtz's empirical foundation for geometry, which I believe to be his most important accomplishment for mechanism besides the foundation of the conservation of energy, does not lead to relativizing formerly absolute validity but to programmatically affirming it. From the properties of mechanical solid bodies Helmholtz derives not only geometric systems of axioms but he also believes to be able to conclude that real space is Euclidean. Nonetheless, his work on geometry does contain some elements of a modern conception of science (Section 6.2: Mechanics – The Underlying Principle of Geometry).

Inasmuch as Helmholtz takes his mechanism to be the confirmed result of tested scientific practice, its claim to validity depends on how much scientific procedure can be separated from other ways of gaining knowledge. This issue goes beyond the narrower mechanistic theme and constitutes the leading theme for the following section. Helmholtz's speeches show that his conception of science underwent remarkable development in the 1850s and 1860s but still aimed for a consistent rationale for the classical conception of science (Section 6.3.1: Helmholtz's Conception of Science up to the Late 1860s).

I complete the portrayal of the period thus characterized as classical with a summary of statements on the contents of Helmholtz's mechanism (Section 6.3.2: Helmholtz's Classical-Mechanistic Conception of Nature).

My thesis for the transition in Helmholtz's mechanism within the context of classical and modern conceptions of science is that for Helmholtz the early 1870s indicate a clear cut. His speeches provide evidence of an unmistakable, almost sudden erosion of classical mechanistic positions. These are less related to statements about the theory of matter than to the formal foundations of its rationale and thus to his conception of science. An analysis of the further development shows how much Helmholtz can still relate to his former position and how much he is approaching a modern conception of science. It also turns out that this change of direction does not necessarily lead to abandoning his orientation to mechanics but rather implies changed conditions of validity for each of the philosophy of nature's orientations to science (Section 7.1: Helmholtz's Conception of Science from the Early 1870s on).

Changing his understanding of science excellently prepared Helmholtz to continue his program of explaining nature mechanistically. This had considerable influence on further development, which I shall proceed to demonstrate using an example from his work on the so-called "principle of least action" (Section 7.2: Helmholtz's Model-Theoretic Mechanism). After reconstructing the transition process I shall, in closing, as set forth in the introduction, discuss the spectrum of possible conditions and causes behind the transitional process. At that point I can and must abandon the hitherto merely occasionally slackened restriction to a certain selection of his works (Chapter 8: Conditions and Causes for the Change in Helmholtz's Conception of Science and Nature).

Chapter 6
Helmholtz's Classical Mechanism

6.1 Mechanistic Program of 1847

> *More than anything else, the law of the conservation of energy contributed to the diffusion of the 'mechanical conception of nature', being the opinion that all phenomena in observable nature are produced by purely mechanical effects of the smallest bits of matter [...]; indeed, a theory in physics was not yet considered a real explanation, until the phenomena it dealt with had been reduced to mechanics. This makes the general theorem of conservation of energy a special indispensable law of nature*
> (Carl Friedrich von Weizsäcker, *The Impact of the Law of the Conservation of Energy on Physics).*

For many reasons Helmholtz's treatise *Conservation of Force* stands out among the texts that should be examined for the purpose of evaluating his mechanistic conception of nature. In the Introduction to this treatise Helmholtz presents, for the first time, the conception of nature that he had previously expressed only reservedly within the context of some studies in physiology. In terms of a public statement, this clearly marks the inception of further development in his conception of nature. Here Helmholtz articulates the aims of his mechanism with clarity and resolve rarely found afterwards. He also provides arguments based on the theory of matter, allowing us to understand his early conception of nature as the continuation of an early modern line and foundational strategy of mechanism and to demarcate it from other positions. The latter leads to a new appreciation of the controversial relationship between Helmholtz and Kant. But the Introduction first reveals its relevance for the history of science when seen as a philosophical underpinning for the rest of the text, where Helmholtz explains his formulation of the conservation of energy and illustrates it with ample applications.[259] The seldom disputed high status still

[259] On the history of discovering the law of the conservation of energy, in which Helmholtz's text played a key part, see **Mach** (1872), **Planck** (1887), **Haas** (1909), **Kuhn** (1959), **Hiebert** (1962), **Scott** (1970), **Elkana** (1974), **Breger** (1982), **Winters** (1985) and **Schirra** (1991).

attributed in current science and technology to the theorem of conservation makes it indubitably the most significant document for grasping Helmholtz's mechanistic conception of nature.

Although his enunciation of the law of the conservation of energy still clearly exhibits traces of its mechanistic origins, the connection between Helmholtz's conception of nature and his part in the epochal discovery of the law was of lesser importance for how his treatise was received.[260] This reflects one way of understanding Helmholtz's theorem of conservation, regardless of its philosophy of nature parameters. The currently widespread phenomenological approach to the theorem of conservation is set up such that its unparalleled influence, extending far beyond the realm of natural research, need by no means be considered proof of an equally widespread mechanistic conception of nature. Since in mid-century Germany a variety of strong non-mechanical – be they romantic, be they positivistic – trends existed both within and outside of research, it is more accurate to say that acknowledging the theorem of conservation while ignoring how it was connected to mechanism was one of the very conditions that contributed to its dissemination.

One of the most extraordinary things about the development of natural science and the philosophy of nature in Germany during the third quarter of the nineteenth century is that to a great extent it was the growing recognition of the theorem of conservation that helped mechanism experience a mighty renaissance that was entirely unanticipated. It is against this backdrop that the severity of the crisis setting in towards the end of that century began to show. Analyzing Helmholtz's treatise may help to understand the conditions that contributed to the popularity of mechanism and its deep impact on science, if it can explain why the relative independence of the theorem of conservation did not weaken mechanistic explanatory power but instead, was crucial for its success.

The role the treatise had for Helmholtz's mechanism coincides with its significance to the history of science, which I shall not pursue further at this point. It is also crucial for reconstructing Helmholtz's conception of nature whether the treatise can be seen as a whole. In the following I would like to reconstruct the close relation between mechanism and energy conservation as a program that starts with classical-scientistic mechanism and that through the course of studies on energy acquires not only practical relevance for research but also develops its own reductionism-oriented heuristics. In order to do even approximate justice to the complicated web of assumptions involved in Helmholtz's philosophy of nature, how they are justified and how they relate to experimental research, which altogether underlie this brief description of the subject, I shall (a) first analyze only the conception of science and nature laid out in the introduction to the treatise and then (b) show how it is related to the enunciation of the law of conservation.

[260] This is particularly true for the debate that followed **Kuhn** (1959) on whether the law of the conservation of energy was discovered simultaneously by different scientists. **Elkana** (1974) and **Winters** (1985) discuss controversial inner-scientific aspects of the prehistory, for the most part without contemplating the philosophy of nature. **Bevilacqua** (1994) provides an interpretation of the entire treatise, which with respect to the introduction takes orientation from **Heimann** (1974). The impact the treatise had in the nineteenth century coincides with the diffusion of the idea of conservation.

6.1.1 Dual Mechanism

If one understands generally the metaphysical components of a rationale as the methodological and conceptual prerequisites of experience, then at first glance it might seem as if in the Introduction to *Conservation of Force* Helmholtz tries to offer a metaphysical justification for the mechanistic conception of nature explicitly advocated there. In a remark added to the Introduction in 1881 he suggests that this rationale might be valid a priori:

> *The philosophical explanations in the introduction* are more strongly influenced by Kant's epistemological views than I now prefer to acknowledge as correct.[261]

Throughout the history of the treatise's reception this remark has been taken as indicating the influence that Kant's metaphysics had on the contents of the introduction and – even more – on the direction of experimental research that Helmholtz advocated. The opinion carries considerable weight, because soon after the treatise was published, Helmholtz became a leading figure in the scientific community and also because the theorem of conservation was groundbreaking.[262] Nevertheless, without questioning Helmholtz's own assessment, in the following I would like to show that it does justice neither to the Introduction nor to its intentions to count them as among the metaphysical tradition of arguments for classical science, such as those paradigmatically advocated by Descartes, Leibniz and Kant. Closer examination reveals that Helmholtz explicitly corroborates the contents of the definitions of his mechanism with knowledge from mechanics and chemistry. Not until he has established a

[261] **Helmholtz** (1847a), p. 53 (Germ.)/49 (Engl.). Overseeing the regret it expresses, **Kahl** (1971) mistakenly translated this statement from the Appendix as: "The philosophical discussion in the introduction is strongly influenced by Kant's epistemoligical insights. I still consider these correct". (Wide letter spacing has been italicized. Helmholtz's emphasis for surnames will not be adopted.)

[262] The idea that Helmholtz reduced all effects in nature to central forces, because doing so alone satisfies the law of causality, can be traced back to Ernst Cassirer (**Cassirer** (1957), p. 92 f.). This tendency of scientists, apparent in Helmholtz "most clearly and decisively" (loc. cit. p. 92), is paralleled in philosophical attempts to legitimate underpinnings for science, as undertaken for example by Wilhelm Wundt (loc. cit. p. 94). Mechanical explanation gets deduced from a certain sort of concept of causality by doing "a kind of 'transcendental deduction'" (loc. cit. p. 95). Cassirer does not mention that Wundt defines causality – similar to **Mach** (1872) and in contrast to Helmholtz – independent of chronological order (**Wundt** (1866), p. 94) and under strict denial of final causes (e.g., loc. cit. pp. 32, 103 and 110). Helmholtz's central part of the rationale, defining the structure of those forces, lies beyond Wundt's metaphysics: "We here limit ourselves to [. . .] stating the contemporary empiricist fundamental opinion" (loc. cit. p. 41).

Heimann (1974) interprets Helmholtz's mechanistic conception of nature as the result of a metaphysical deduction that only makes sense within the context of Helmholtz's struggling with Kant's theory (for which there is no historical evidence). Despite critical objections, **Fullinwider** (1990), **Heidelberger** (1994a), **Heidelberger** (1994b), **Darrigol** (1994) and **Krüger** (1994) share this view. Exemplary for far-reaching consequences in the theory of science that can be drawn from Helmholtz's alleged Kantianism is the evaluation of his relationship with his student Heinrich Hertz, as discussed in **Janik and Toulmin** (1973) and **D'Agostino** (1975).

foundation for the concept of the element, borrowed from chemistry, does he feel justified in a central passage to think of the "universe" as split up into "elements".[263] He then applies principles of physical mechanics, traceable to Newton's formulation, to this approach to nature.

So if there had been any influence by transcendental philosophy, it would have asserted itself only in a very diffracted way. Within the suggested rough schema of two traditions of mechanistic rationales, namely a metaphysical and a scientistic justification, Helmholtz, with the reasons he gives for mechanism in 1847, places himself – as I would like to show – with sufficient clarity on the side of scientism. Furthermore, regarding content, I believe there are good reasons for seeing his mechanism, to which both concepts of matter and force are central, as not belonging among the dynamic variants established metaphysically by Leibniz and Kant but as being of a dual kind.

A closer look at the introduction also shows that the security of the validity that Helmholtz desires for his conception of nature by referring to scientific experience has its limits. His comments indicate that he does not see that limitation as a gain for theory-making in physics, as would be typical for a modern conception of science. His mechanism is not an envisioned model that might be on equal standing with other models; it claims to be "objective truth".[264] His claim to the truth of this argumentation is admittedly no less emphatic than any in transcendental philosophy. It is precisely the abandonment of this pursuit of absolute validity that will begin to emerge in Helmholtz's conception of science from the early 1870s. That approximation to a modern conception of science will remove him disproportionately further from Kant than would have been possible in the 1840s and would explain the remark quoted above, namely that the Introduction to the treatise was overly influenced by Kant's epistemology.[265]

In order to systematically determine when empiricism enters Helmholtz's first foundation- laying for mechanism, I would like to divide into two groups the presuppositions about physical sciences that he deals with there. The first relates to methodical and conceptual assumptions concerning research, which initially hold independently of experience (α), the second constrains that validity by subordinating the range of the method and purpose of research to knowledge that can be gained solely through experience (β). These presuppositions constitute the framework for defining Helmholtz's mechanism in the narrower sense (γ).

(α) The methodical presuppositions are based on a rigorous division of the physical sciences into two parts, one experimental and the other theoretic:

> The task of the physical sciences is to discover laws so that individual processes can be traced back to and deduced from, general rules. [...] The search for such rules is the task of the experimental part of our sciences. The theoretical part, on the other hand, seeks to ascertain from their visible effects the unknown causes of natural processes; it seeks to comprehend them according to the law of causality. We are compelled to and justified in this undertaking by the fundamental principle that every change in nature *must* have a sufficient cause.[266]

[263] **Helmholtz** (1847a), p. 5.

[264] **Helmholtz** (1847a), p. 7.

[265] **Krüger** (1994) interprets this statement similarly.

[266] **Helmholtz** (1847a), p. 3 f. (Germ.)/3 f. (Engl.) (Italics in the German original).

The division of scientific labor amounts to separating the concept of law from the principle of causality. Finding a cause behind visible effects is not the same as stating a general rule or a law. Laws that remain immediately tied to visible effects and thus actually have phenomenal character, are not seen as causes. For instance, the law of gravity is not the cause of free fall but merely the description of it. Whether other general rules or substantial entities may be causes is a question Helmholtz does not answer at this point. He limits himself to expressing his conviction that there must be "ultimate causes" "which operate according to invariable law".[267] Since he herewith goes beyond the range of the first group of assumptions, I would like to discuss that definition somewhat later.

It must be kept in mind that the (metaphysically) presupposed law of causality does not itself provide anything from which a mechanistic conception of nature (in either the broader or narrower sense) is derivable.[268] As a presupposition of theoretical physics, the law of causality implies only that the discipline cannot proceed other than to search for non-teleological causes for experimental laws – regardless of some plausible acausal constitution of nature, which, while it might not be comprehensible using physics, may be so using other disciplines. For physics this excludes both a teleological approach to nature and explaining contingent events.

In comparison to the merely weak definitional assumption of causality in the philosophy of nature, the conceptual presuppositions that Helmholtz subsequently discusses, calling them "abstractions", seem to implicate dual mechanism (in the broader sense). Helmholtz reduces all scientific terminology to the two basic concepts of matter in motion and force as related to matter. The fact that he limits the variability of matter to spatial movement particularly indicates mechanistic ontology:

> Science treats the objects in the external world according to two different abstractions. In the first place, considering them apart from their effects on other objects and on our sense organs, it regards them merely as existing objects. As such, they are called *matter* or *material*. To us, matter in itself is inert and without effect; we only distinguish spatial distribution and quantity (mass), which is assumed always to be constant. We cannot ascribe qualitative differences to matter itself, for when we speak of different kinds of matter we always consider these to be differences in effect, that is, differences in the forces of matter. Thus matter itself can undergo no change other than a spatial one, namely movement.[269]

This ontology is dual inasmuch as each of these concepts of matter and force includes definitions not reducible to the other. For example, Helmholtz does not define matter solely as that which moves in space. He also attributes a second property to matter, namely the conservation of mass, which he in contrast to Leibniz's foundation for dynamic mechanism does not reduce to the effects of forces. Finally, the property of motionlessness also indicates that matter has the quality of being mechanically inert.

---

[267] **Helmholtz** (1847a), p. 4.

[268] On the conceptual distinction between mechanism in the narrower and broader sense see Section 1.1.

[269] **Helmholtz** (1847a), p. 4 f. (Germ.)/4(Engl.).

Compared to matter, force is the epitome of qualitative variability and is not derived from the movements of matter but is assumed to be a basic characteristic of nature. Helmholtz continues:

> Objects in nature are not, however, effectiveless. Indeed, we are acquainted with them only through their effects upon our sense organs, since it is from these effects that we infer objects which produce them. Hence, if we wish to make actual application of the concept of matter, we can do so only by ascribing force to it, that is, by adding a second abstraction from which we wanted to abstract previously, *the capacity to produce effects*.[270]

While here he does introduce the conception of force as an independent concept, calling it a "capacity to produce effects" is very vague. One is left to simply assume that effects of force are supposed to evoke "qualitative differences" and not simply be effects of pressure and impact, as stated in the materialistic foundations for mechanism. This means that from the spectrum of the previous traditional lines in mechanism, dual mechanism remains as the next closest conception. But in light of the wide meaning given here for the concept of force, this can only be mechanism in the broader sense, because these forces need not be the forces of "pure" mechanics, whose effects are confined to the mere interference of inert motion. Instead, all effects are permissible, as long as they are related to a substrate that corresponds to the specifications for matter.

It is remarkable that Helmholtz here distinguishes two different effects of "objects in the external world": visible effects on "our sense organs" and invisible effects "on other objects". We achieve "acquaintance" [Kenntnis] with "objects in nature" only by inferring (from the effects they produce on our sense organs) the "objects which produce" those effects.[271] This weighty assertion made by Helmholtz in founding the introduction to his concept of force deserves two comments. Note first that the inferred agent (or object) affects not only on the sense organs but also on other objects. Applied to natural research, this would mean separating scientific objects into two groups: in the one would be research dedicated to investigating the immediate affectation of human senses (in terms of physiology and perception); in the other, research of all other phenomena involving inferences from the visible effects to the invisible (effects of the objects among themselves). This could also be characterized as distinguishing – as mentioned above – between law and cause: laws of nature refer to visible effects and causes refer to invisible ones. Note also that this raises the question of whether Helmholtz's claim that we achieve "acquaintance" with nature only via the sense organs can be applied to the study of nature at all. Does this "acquaintance" constitute scientific knowledge? Regarding the "theory of signs" (which I shall elaborate later on), at this point I have my doubts.[272] In the theory of signs Helmholtz will later say that the sense organs provide perceptive

---

[270] **Helmholtz** (1847a), p. 5 (Germ.)/(Engl.) p. 4. (Italics in the English translation).

[271] **Helmholtz** (1847a), p. 4 f.

[272] Cf. Section 6.3.1.1.

consciousness only with signs or symbols, which have no more similarity to the external world than personal names have to the persons whom they designate. Depreciated in this way, "acquaintance" is reduced to being mere subjective information just barely sufficient for everyday orientation, which is, in principle, to be distinguished from scientific knowledge about laws of nature and theoretically established causes.[273]

One can say that by way of pre-scientific experience, Helmholtz illustrates nature's general trait of achieving qualitatively dissimilar effects. This broadly understood meaning of the concept of force weakens the contour of the mechanistic conception of nature provided previously by "matter in motion" and allows thinking of conceptual presuppositions independent of any specific ontology. Since it is about ideas that are at the root of all science, it should also cover non-mechanical branches. 'Matter' becomes the term for every kind of existence and 'force' the term for every kind of (even non-spatial!) variation on that existence. Helmholtz appears to have understood the presuppositions in this way, when he continues:

> It is evident that the concepts of matter and force cannot be applied separately to nature. Pure matter would be indifferent to the rest of nature [...]; a pure force would be something which exists and yet does not exist, for that which exists we call matter.[274] It is equally erroneous to interpret matter as something real and force as an empty concept to which nothing existing corresponds; both, on the contrary, are abstractions from reality, formed in precisely the same way.[275]

This passage reveals not only the general meaning of both concepts but also their realistic and by all means non-transcendental philosophical character.[276] In contrast to Kant, who saw the concepts of matter and force as conditions for experience related to possible objects of external experience, Helmholtz deduces these two concepts as "abstractions" from an absolutely given reality, i.e., a reality that exists independently of being known. Kant would say that Helmholtz's concepts do not abstract from the given, as they should, were their validity to be based on pure thinking but that they are abstracted from the given.[277] True, for Helmholtz whatever can be known is unquestionably determined by the external world. But bereft of any foundational derivation, the concepts constitutive of knowledge appear to have

---

[273] In one of his final publications, Helmholtz explicitly distinguishes *"Kenntnis"* (acquaintance) from *"Wissen"* (knowledge): **Helmholtz** (1894a), p. 540, and **Helmholtz** (1885 ff.), p. 598; cf. Section 7.1, footnote 694.

[274] This remark is presumably aimed at romantic *Naturphilosophie* influences that were still lively when this memoir was written. Cf. **Wise** (1981).

[275] **Helmholtz** (1847a), p. 5 (Germ.)/4 (Engl.).

[276] As **Heyfelder** (1897) commented: "*Helmholtz's* worldview is conform to what *Kant* called transcendental realism and to which he consciously contrasts his own transcendental idealism" (loc. cit. p. 55 – italics in original).

[277] **Kant** (1900 ff.), Vol. II (*De mundi sensibilis atque intelligibilis forma et principiis*), p. 394 (§6).

been selected arbitrarily. Why does science study the objects of the external world using the "abstractions" listed by Helmholtz and not other or entirely different categories? The next group of presuppositions that define certain sorts of experience as absolute shows that it would be erroneous to assume that Helmholtz left it to the liberty of the inquiring mind to decide what should be abstracted and what not.

Based on what has been said, one is now in the position to evaluate the main thesis of a study presented by P. M. Heimann, which in past decades has been of considerable importance for discussing the relation between Helmholtz and Kant.[278] Starting from an analysis of the Introduction, Heimann believes to find evidence of four "structural analogies" existing between Helmholtz's conceptual presuppositions and Kant's *Metaphysical Foundations of Natural Science* – analogies that he elaborates corresponding to the four main parts of Kant's treatise.[279] It can be shown that such analogies either do not exist or that those which do exist are insignificant. First, Heimann is wrong in claiming that Helmholtz thinks of matter, as Kant does, as only what is movable in space.[280] Heimann neglects to note[281] that Helmholtz also attributes to matter, as mentioned above, the properties of inertia and the conservation of mass. Second, Kant does not "supplement"[282] the concept of matter with a concept of force. In line with his dynamic mechanism and in contrast to Helmholtz's dualistic conception, Kant derives the properties of matter from basic forces.[283] Third, in haste one may think it an analogy that in the third main part of the treatise, namely on mechanics, Kant also acknowledges the principles of Newton's mechanics, which Helmholtz simply takes for granted from doing research (one should note that for reasons related to the possibility of materialistic mechanism as not advocated by Helmholtz, Kant does not explicitly mention Newton's Second Law).[284] But for the special relationship between the two thinkers, their common acknowledgment of Newtonian principles is in no way significant: with few exceptions those principles were shared by practically everyone interested in natural research subsequent to the second half of the eighteenth century. Since fourth, it is questionable whether Helmholtz's brief remark that we can only know about nature by studying the effects of its forces is related to scientific knowledge at all, the perhaps remaining remote existent similarities to Kant's fourth main part, namely phenomenology, are hardly sufficient evidence of Kant's influence on Helmholtz.

(β) The presuppositions subsumed in the second group deal with the range of physical method and defining the purpose of natural research. These can be developed using the

---

[278] **Heimann** (1974) (see footnote 262), on the critical debate cf. **Fullinwider** (1990) and **Krüger** (1994).

[279] **Heimann** (1974), p. 228.

[280] **Heimann** (1974), p. 228.

[281] **Heimann** (1974), p. 218.

[282] **Heimann** (1974), p. 228.

[283] Cf. Section 3.3.2.

[284] Ibid.

question raised by Helmholtz in the Introduction, namely whether insight in nature corresponds to the human capacity for knowledge[285]:

> Whether all nature must be completely comprehensible or whether on the contrary there are changes which lie outside the law of necessary causality and thus fall within the region of spontaneity or freedom [...]. In any case it is clear that science, the goal of which is the comprehension of nature, must begin with the presupposition of its comprehensibility and proceed in accordance with this assumption until, perhaps, it is forced by irrefutable facts to recognize limits beyond which it may not go.[286]

In contrast to his treatment of the first group of assumptions, here Helmholtz considers the possibility of a factual impediment that could be insurmountable in empirical research and for the causal explanation of nature. But Helmholtz hardly wanted to address freedom as a specifically human capacity. He was more likely thinking of real effectual principles that in general distinguish animate nature from the inanimate and which comply not with a causal but with a teleological context.[287] Even without any natural spontaneity of the animate externally limiting the range of what can be researched in terms of causes, Helmholtz is certain that within that range the principle of causality cannot be indefinitely applied to experience. By separating causes into those that are variable and invariable, i.e., by using a time-related distinction, he arrives at the concept of "ultimate cause":

> The proximate causes, to which we refer natural phenomena, are themselves either invariable or variable; in the latter case, the same fundamental principle [of causality – G.S.] compels us to seek still further for the causes of the variation and so on, until we arrive finally at ultimate causes which operate according to invariable law and which consequently produce under the same external conditions the same effect every time. Thus the final goal of the theoretical natural sciences is to discover the ultimate invariable causes of natural phenomena.[288]

Thus the concept of causality is enhanced to include a second meaning that is characteristic of Helmholtz's conception of science and nature: while in the first case he claimed that "every variation in nature" has "a sufficient cause"[289] he now introduces "causes" (that in turn differ from – yet more fundamental? – laws), which themselves are not based on causes, because they are attributed the capacity to effect changes, without themselves being subject to change. Because of their invariability, "ultimate causes" could also simply be called conditions. They are those real conditions to which natural phenomena, inasmuch as they are comprehensible, are, with ultimate certainty, to be reduced. They indicate a program of reduction that can

---

[285] Faced with the possibility that the diversity of external nature may exceed the capacity of the human mind, Kant buttresses the transcendental principles of the unity of knowledge, cf. **Kant** (1781), B 681 ff., **Kant** (1790), p. 410 ff. (§78), and also **Schiemann** (1992).

[286] **Helmholtz** (1847a), p. 4 (Germ.)/4 (Engl.).

[287] This interpretation is plausible because in another passage Helmholtz explicitly applies the concept of freedom to the animal kingdom (**Helmholtz** (1856 ff.), p. 454, cf. Section 6.3.1.4, footnote 491, and quotation mentioned there).

[288] **Helmholtz** (1847a), p. 4 (Germ.)/4 (Engl.).

[289] **Helmholtz** (1847a), p. 4 (as footnote 266).

approximate a solution, not a priori but by means of scientific experience. Ultimate causes cannot be postulated; they are "a part of nature" and must be "discovered" by science.[290] By designating their identification "the final goal", Helmholtz expresses his conviction that the task is accomplishable. With respect to this final goal and considering the possibility that nature may have acausal traits, the character of methodological presuppositions changes profoundly. These presuppositions are now subject to an anti-metaphysical directive, for which the range of causal method is definitively terminable by experience.

Not only do both dual ontology and a realistic conception of reality distinguish Helmholtz's conception of nature from Kant's dynamic mechanism, their different ideas on causality reveal even more clearly the difference between their approaches. For Kant neither can causality be restrained by experience, nor are final causes conceivable without contradiction.[291] Compared to Helmholtz, Kant is indubitably more consistently mechanist. He sees nature as constituted causally without exception and no phenomenon exists that could not be reduced to mechanical causes.

(γ) The presuppositions in the second group, which in themselves also do not favor a mechanistic conception of nature, are tied to the conceptual specifications of the first group only in one respect: since ultimate causes "produce effects", they must be forces. So the statement Helmholtz uses to proceed to the definition of his kind of mechanism in the narrower sense adds nothing new to his former explanations – and is thus not a turn towards a dynamic notion of nature:

> This requirement [that natural phenomena should be traced back to inalterable ultimate causes – G.S.] may now be restated: as ultimate causes, forces which do not vary in time should be found.[292]

In contrast, Helmholtz's following statement does include a new thought, something that hitherto has only been conveyed implicitly – and clearly marks a break within the Introduction:

> In science we call material objects with unchangeable forces (indestructible qualities) chemical elements. But if we think of the universe as consisting of elements with inalterable qualities, the only possible changes in such a system are spatial ones, that is, movements. [...] the forces acting can only be motive forces [...].
>
> Put more precisely: natural phenomena should be traced back to the movements of material objects which possess inalterable motive forces that are dependent only on spatial relations.[293]

Here Helmholtz recurs for the first time to vocabulary from research practice. From the concept of the chemical element, an established term in experimental science of his time, he extracts preliminary definitions for his mechanistic programmatic, which previously had been outlined only sketchily. Against the backdrop of the

---

[290] As in footnote 288.

[291] **Kant** (1781), B 232 ff. (2nd Analogy of Experience) and B 472 (3rd Antinomy).

[292] **Helmholtz** (1847a), p. 5 (Germ.)/5 f. (Engl.).

[293] Ibid. (as Section 1.1, footnote 29).

scientifically well-tested concept of the element, Helmholtz's subsequent deliberation ("but if we think of") is less a thought experiment, than much more the claim of having consistently followed through, in all its consequences, an elementaristic conception of nature already prompted by research. Up to this point it had not been necessary to conclude that "forces" must be "motive forces". No longer is every kind of force permissible as basic reciprocal action but only the spatial change-inducing force of mechanics.

But narrowing the concept of force also subtly shifts the meaning of the term "element". For the chemistry of the times, the term commonly designated the fact that there is a limit to the diversity of qualities, insofar as those qualities correspond to chemical substances. The terms 'substance', respectively 'element', were used to denote chemical substances assigned to a single kind of hypothetically presupposed atoms. Neither the hypothesis of atoms nor the term 'element' necessarily involved ontological assumptions about the structure of matter.[294] Helmholtz's concept extends considerably beyond this confined meaning: elements are defined not only by their chemical qualities and combinational relations, they are physical objects and ultimate parts of divisible matter. Although Helmholtz does not use the term 'atom' in this passage, it is clear that matter can only be decomposed discretely into movable elements. Since all change is a question of the forces, the corporeal particles that remain left over as elements also cannot differ from one another in terms of shape. Elements have thus become atoms, have become what is indivisible in matter. As the text proceeds, when he discusses the motion that causes heat, Helmholtz suggests that the concept of the atom must not be identical with the concept of the chemical element:

> If it be permitted to try to make the concept of this [heat – G.S.] motion still clearer, a hypothesis derived from Ampére's view seems in general best suited to the present state of science. If we think of bodies as being made up of atoms, which are themselves composed of different particles (chemical elements, electricities, etc.), then three kinds of motion may be distinguished in such atoms: (1) displacement of the center of gravity, (2) rotation around the center of gravity and (3) displacement of the particles of the atom relative to one another.[295]

Just as Helmholtz thinks that the divisibility of matter is finite, he also thinks that atoms can be split into spatially separate components of varying quality. His hypothesis concerns not the inner structure but the general existence of atoms. For this reason it is plausible that for Helmholtz the concept of 'element', related to that of the atom, was also – in the sense of being a physical object – merely hypothetical. Yet what 'hypotheses' are remains widely undefined in this context. Not until the 1870s does Helmholtz explain what he means by 'hypotheses'. As it stands, at this point he thinks the ambiguity of hypotheses, i.e., whether or not they are true, is not something that cannot be repaired, because he evaluates them according to

[294] Cf. **Ströker** (1967), p. 156ff., **Nye** (1984), p. XV, and **Whitt** (1990).

[295] **Helmholtz** (1847a), p. 24 (Germ.)/22 (Engl.).

standards of progress in scientific knowledge ("best suited to the present state of science"). Obviously he holds the hypothesis of chemical elements to be so secure that he can take them as the basis of his reductionist objectives.

The notion of the discrete decomposition of matter, which followed the pattern of the concept of the element, is only one of two inputs for the more accurate definition of forces that Helmholtz pursues in the Introduction. The other is that he tacitly accepts physical mechanics' formalism, further limiting the concept of force to merely central force. How Helmholtz arrived at this definition for the structure of "motion forces" can be reconstructed as follows.

The starting point is to think of elements as extended particles. Similar to his hypothesis concerning atoms, Helmholtz attributes mechanical properties to particles, properties that do not differ from those predicated of visible solid bodies. Thus their motion is "only possible compared to delimited quantities of space" and can "occur only as a change in the spatial relation of at least two material bodies relative to one another".[296] Consequently, these are objects whose mass remains constant within their surface boundaries, as is the case for all bodies, which is why he calls them "complete masses".[297] Helmholtz now derives the properties of the forces by applying the formalism of point mechanics to the extended, though invisible, elements:

> The force, moreover, which two complete masses exert upon one another must be resolved into forces of all of their parts upon one another. Thus mechanics considers the forces of material points, that is, the forces of points of space containing matter, to be basic.[298]

"Material points" is the expression used in an eighteenth century method in mechanics for mathematically describing the distribution of mass in a given macroscopic body.[299] The volume a body takes up can be geometrically reproduced by spatial points, each of which is appointed a certain factor for mass. For this to work, the body must be of a sufficient size, such that the number of points can be mathematically thought of as being infinite and the finite total mass of the body results from integration.

For Helmholtz – in contrast to Kant's theory of matter – the relation of extended material elements to material points is not a matter of metaphysical deduction but a job for mechanics. While his idea of extended elements is taken from the concept of elements in chemistry, he resolves them into material points by a mechanical procedure that apparently requires no further explanation. But in comparison to the discipline's scientific meaning of the term, here, too, a shift in meaning occurs: Helmholtz interprets the geometrical constructions of mechanics ontologically; matter's discrete structure iterates itself one last time at the core of its elements.

That the forces starting from a spatial point might depend on a particular direction in space is something that Helmholtz does not consider, neither for extended

[296] **Helmholtz** (1847a), p. 5f.

[297] **Helmholtz** (1847a), p. 6.

[298] **Helmholtz** (1847a), p. 6 (Germ.)/5 (Engl.).

[299] Cf. **Truesdell** (1989), p. 51f.

elements, nor for material points. But he does exclude it for material points by assuming that the forces related to them do not depend on direction. He also limits possible interaction to changes in pairs. The force operating between any two points thus occurs independent of the presence of further points in space. After all, non-extended points can experience no rotation and forces operate only as repulsive or attractive mechanical motive forces. Overall, these requirements amount to the idealized view of Newton's formulation of mechanics and so it is plausible that Helmholtz adopted them as an ascertained component of successful theory production in mechanics. From them the features of the law of force between two points (respectively, of what operates on a pair of points) follow effortlessly: it is a function that depends on distance and effects a change of velocity along the line connecting two points. He calls forces having these qualities central forces.[300]

Material points and the motive forces assigned to them constitute dual mechanism (in the narrower sense) as paradigmatically advocated by Newton and Boscovich. Newton, too, assumed that the submicroscopic particles he postulated were subject to attractive and repulsive forces. Roger Boscovich rounded off this approach by letting his universal law of force, which defined attraction and repulsion between two points as a function of their distances, operate homogenously in all directions in space.[301]

Nevertheless, in contrast to Boscovich, Helmholtz neglects to mention details about the strength ("intensity") of such forces. Thus it remains open whether a universal law of force operates between the material points or – corresponding to the various chemical elements – various elementary forces exist. The only clue that Helmholtz provides is the fact that he without exception uses the plural when speaking of force. At any rate, the task for science is limited to determining the strengths of those forces:

> The task of the physical sciences is therefore ultimately to trace natural phenomena back to inalterable forces of attraction and repulsion, whose intensity depends on distance.
>
> [...] once the reduction of phenomena to simple forces has been completed and if simultaneously it can be proven that this reduction is the only one that the phenomena allow, then this will be established as the conceptual form necessary for understanding nature and we shall be able to ascribe objective truth to it.[302]

It would be a misinterpretation of the entire Introduction to conclude from this last phrase, which mentions only forces and not matter, that Helmholtz's programmatic established a kind of dynamic mechanism. Defining matter cannot and need no longer be included as a task for natural research. The conservation of mass, its inertia and the restriction to spatial change can no longer be topics of research, because they are preconditions of conceptual premises. The discrete distribution of matter in space is itself part of the structure of forces and needs no explicit mention.

---

300 **Helmholtz** (1847a), p. 6 and 9. Cf. **Buchwald** (1994a), p. 366.

301 Cf. Section 3.2 and **Schiemann** (1997), p. 107ff..

302 **Helmholtz** (1847a), p. 6f. (Germ.)/6 (Engl.).

On the other hand, the strength of operative forces cannot be inferred from the qualities of matter. The fact that forces can be of varying strength is fixed by the concept of force and requires no further grounding.

In summary, the Introduction can be analyzed as follows.

First, Helmholtz's mechanism, inasmuch as it can be reconstructed from the few pages that comprise the introduction, can be seen as belonging to the dual line of mechanism in the narrower sense. In terms of content, this at first removes him from Kant's metaphysics of matter, which is characterized precisely by the fact that it reduces the concept of matter to "moving force" [*bewegende Kraft*], utilizing a concept of force that goes beyond the specifications of mechanics. Helmholtz's conception of nature excludes every even merely implied possibility of materialistic mechanism, which both Kant and Leibniz had permitted as an explanatory approach for natural research.[303]

Second, the claim to truth that Helmholtz asserts for his conception of nature is clearly intrinsic to this initial programmatic declaration of his objectives. Within the prescribed framework of a causal explanation of nature he says it should be the "final goal" of theoretical physics to "discover" the central forces of particular strength as "ultimate causes"[304] and to "complete" the reduction of natural phenomena to these forces.[305] This reduction should also be "the only one that the phenomena allow", meaning that it constitutes not one explanation among many but should have absolute validity. Inasmuch as the discrete structure of matter is only premised hypothetically, progress in knowledge means precisely to transform hypotheses into true statements. This closed conception of progress, aimed at a true system of knowledge about nature, is typical for a classical conception of science. By adhering to it Helmholtz – seen from today's viewpoint – formally falls back behind Kant. Kant, who thought that the complete identification of the structure of basic forces is possible a priori, took the objective of systematic comprehensive knowledge of nature to be a "regulative idea", to which research could only nearly approximate.[306]

Third, analyzing the Introduction offers the first evidence that Helmholtz strove to establish the validity of his conception of nature not metaphysically but scientistically, through reference to scientific practice. Besides adopting the concepts of contemporary research (elements and mechanical forces), one important sign is that his conception of nature and the mechanistic program of reduction concur.[307] For Helmholtz, as shown by the passage just quoted, research does not obtain truth about nature's ultimate causes until it proves that these causes are the only ones "that the phenomena allow". This not only binds solving the theoretical problem of

---

[303] Cf. **Schiemann** (1997), p. 115 ff. and p. 124 ff.

[304] **Helmholtz** (1847a), p. 4.

[305] **Helmholtz** (1847a), p. 7.

[306] **Kant** (1781), B 697 ff.

[307] For the distinction between mechanism as philosophy of nature, which refers to nature as a whole and mechanism as a scientific program of reduction applicable to isolated cases, see Section 1.1.

reducing phenomena to central forces to experience, as also laid out in the general definition of theoretical physics ("seeks to ascertain from their visible effects the unknown causes of natural processes"); the task itself is now subject to scrutiny through experience. Helmholtz does not yet have the "necessary conceptual form" (displaced matter and central forces) required, whose application to experience ("objective truth") would simply need confirmation. Instead, the necessity of that conceptual form must first be "established" empirically. This holds also for the law of causality, which is why its validity is not simply certain but may be constrained by further advance in the knowledge of nature.

Thus, although Helmholtz cannot yet be certain of the truth of his conception of nature, he does not doubt that absolute validity is attainable within finite time. His unbroken confidence cannot be explained by the rudimentarily developed a priori premises suggested by the Introduction, because no specific objectives can be derived from them. It seems, rather, that Helmholtz views his ideas as confirmed by the state of the art in research. In fact, he was able to support his ideas with inner-scientific developments that fomented mechanistic approaches to nature and which were all the more significant, as towards mid-century the cultural repute of experimental research had risen considerably compared to the first half of the century.[308] These developments included a decline in the lingering influence of romantic and thoroughly anti-mechanistic philosophy of nature,[309] the discrediting of the material theory of heat (important for Helmholtz as a physiologist), the nascent triumph of the undulation theory of light and the successful utilization of the hypothesis of atoms. While all these factors promoted confidence in the obtainability of truth, the Introduction's euphoric tenor suggests that Helmholtz felt he had contributed decisively to discovering a complete explanation of nature. It is supposed to consist of the relation between the Introduction and the theorem of the conservation of energy, which he then proceeds to describe. As I shall show below, Helmholtz believes to have derived the theorem of conservation (whose enormous relevance he presumably sensed while writing the treatise) from mechanistic ontology. He hopes that proving its applicability to various phenomena in physics will retroactively confirm mechanism.

It must be kept in mind, however, that no reference to any specific research practice can substitute systematic arguments for a conception of nature's claim to validity and that in the Introduction Helmholtz provides no such reasoning. Particularly natural research, if it is to guarantee the truth of mechanism, is in no way distinguished from other forms of knowledge. Accomplishing that would require identifying the relationship between perceptible and non-perceptible causes, respectively forces and also specifying what makes scientific procedure distinct from other methods of knowledge acquisition. In public lectures and speeches

---

[308] Cf. **Lange** (1866), p. 512 ff., **Merz** (1907 ff.), Vol. I, p. 158 ff., **Schnädelbach** (1983), p. 89 ff., **Knight** (1986), **Poggi and Röd** (1989), p. 90 ff.

[309] Cf. **Engelhardt** (1979), p. 161 ff.

Helmholtz soon started to respond explicitly to such issues. His answers provide material for analyzing his classical concept of science (cf. Section 6.3.1: Helmholtz's Conception of Science up to the Late 1860s).

6.1.2 The Energetics Heuristics in Mechanism

Now that the Introduction to *Conservation of Force* has suggested that Helmholtz's argumentation for his dual mechanism's claim to truth is scientistic, the further details on the theorem of the conservation of energy are suitable for more accurately defining the relation between the mechanistic conception of nature and experimental research, as Helmholtz saw it in this early publication on physics. First it needs mentioning that regarding energy conservation Helmholtz responded differently to a transcendental metaphysical argumentation underlying science than he had done for the mechanistic conception of nature developed in the Introduction. He never explicitly related his late confession of having been too strongly influenced by Kant in his early years to the parts of the treatise that follow the Introduction. On the contrary, in 1871 he comments on his popular lecture *On the Conservation of Force*[310]:

> It is also obvious that these [basic concepts of physics for work and its immutability – G.S.] are abstract concepts of a very special kind. Even such a great mind as I. Kant had difficulty understanding them, as his polemics on Leibniz reveal.[311]

While Leibniz established the mechanical conservation of energy metaphysically,[312] Helmholtz's much more encompassing theorem of conservation is closely related to empirical scientific research results. The physical dimension of energy that Helmholtz continues to call "force" in both the treatise and its title[313] can in contrast to the concepts of elementary matter and central force, be used directly for measuring things (particularly mechanical work and heat) and their conserving properties can be examined experimentally in physical processes.

As the nineteenth century proceeds, precisely this phenomenal character of energetic transformations will become the paradigm of positivistic critique of mechanism. So-called "energetics" rejects the assumption of invisible atoms and their forces, wanting to see propositions about nature restricted to theoretically processing measurable dimensions.[314] Did Helmholtz, by formalizing the theorem of the

---

[310] **Helmholtz** (1862/3).

[311] **Helmholtz** (1871), p. VI f. Here Helmholtz is presumably referring to Kant's first treatise *Gedanken von der wahren Schätzung der lebendigen Kräfte und Beurtheilung der Beweise, deren sich Herr von Leibniz und andere Mechaniker in dieser Streitsache bedient haben* … (1747).

[312] See, for instance **Leibniz** (1904 ff.), Vol. 1, p. 277 (Letter to l'Hôpital dated 15 January, 1696), and **Schiemann** (1997), p.113 ff.

[313] My comments will take for granted that Helmholtz was able to formulate the relevant relations of energy conservation so clearly that for the most part the expressions he uses (vis viva [*Lebenskraft*], tensional force [*Spannkraft*], etc.) can, wherever they occur, be substituted by energy terms.

[314] Cf. Section 7.2, footnote 889.

conservation of energy, himself prepare the demise of his conception of nature later induced by advocates of energetics? Is there any connection at all between the elementary mechanics hidden behind phenomena and the experiments in physics to which the idea of the conservation of energy refers? Can the treatise of 1847 be seen in its entirety as evidence of his mechanistic conception of nature or do we know about that conception only from the few pages of the Introduction?

In the following I would like to propound the thesis that for formulating the theorem of conservation generally, Helmholtz in part develops a concept of energy independent of mechanics and that therefore his treatise itself suggests mechanism's limited explanatory power. However, within the spectrum of mechanistic views of nature it is precisely and conversely the relative independence of the concept of energy, as I would like to show, that helps to accomplish the program of reduction in which physical phenomena are to be traced back to mechanics. Due to the lack of empirical scientific confirmation, the basis of reduction (being dual-mechanistic ontology) cannot yet be considered proven. The observation of the conservation of energy, which does not depend on the rigid conception of mechanics, may, according to Helmholtz, provide the first reliable clue to the features of elementary forces he is seeking. From the erroneous assumption that the conservation of energy holds only for phenomena completely reducible to mechanical central forces, Helmholtz draws, from the conservation of energy, a conclusion about the kind of reciprocal effects operating among submicroscopic elements:

> It follows further and conversely that in all actions of natural bodies upon one another, wherever the principle given above [i.e., the theorem of the conservation of energy – G.S.] is capable of general application to the smallest particles of those bodies, such central forces must be regarded as the simplest and most fundamental ones.[315]

The proof that the conservation of energy (assuming it is valid for only a limited submicroscopic domain) does apply to certain phenomena provides an independent standard for recognizing just what can be reduced to mechanics and allows – as it were – a first look behind the scenes of what is measurable. In this sense the conservation of energy may be understood as a heuristics of mechanism and together they constitute a research program that will completely explain nature.

At the outset of the Introduction, Helmholtz says that his "derivation" of the theorem of the conservation of energy could "begin from two vantage points":

> Either from the principle that it is impossible to obtain an unlimited amount of force capable of doing work as the result of any combination whatsoever of natural objects, or from the assumption that all actions in nature are reducible to forces of attraction and repulsion, the intensity of the forces depending solely upon the distances between the points involved.[316]

As for the theory of matter, the Introduction lays the foundation for the second option, which I would like to describe as explaining the theorem of the conservation

[315] **Helmholtz** (1847a), p. 9 (Germ.)/8 (Engl.).

[316] **Helmholtz** (1847a), p. 3 (Germ.)/3 (Engl.).

of energy in terms of Newtonian mechanics and mechanism (α). The trouble with this explanation is that it is tied to a specific mechanical concept of energy too narrow for the observations in energetics that Helmholtz undertakes in the treatise. In contrast, the other option that Helmholtz mentions for deriving the law, namely applying the impossibility of a *perpetuum mobile* to the physical quantity of work (force), implies a wider concept of energy. This concept has only a formal similarity to mechanics, namely to the formulation of mechanics proposed by Lagrange.[317] But if the scientistic rationale for the validity of this concept is not to rest on analogy in mathematical structure alone, one must look beyond formal similarity and examine the experimental practice of research. I believe this is why Helmholtz justifies acknowledging the validity of the concept as both a generalization of all previous experience with energetic transformations and as the appropriate standard for measuring such transformations (β). The fact that he overlooks the difference between the concepts of energy is also clear right from the outset of the Introduction, where he stresses that the two derivations are "identical".[318] His comments in the treatise and in later speeches on the theorem of the conservation of energy given throughout the 1850–1860s show that he starts with a one-sided mechanistic understanding of energy and turns it into heuristics for his own conception of mechanism (γ).

(α) Helmholtz utilizes the Newtonian theorem of the conservation of energy "for implementing our law [of the conservation of energy – G.S.] very generally".[319] He starts with "any number of material points" in a three-dimensional Cartesian coordinate system, among which the central forces defined in the Introduction operate. To every single point in this system he applies both of Newton's first two laws, summarized into one equation: every component of force is equal to the product of the mass coefficient and the change of velocity in the direction in which the force operates. Newton's third law (equality of action and reaction) is accounted for by the structure of the central forces. From these assumptions then necessarily follows the theorem of the conservation of the total energy for the point system,[320] where total energy is split between the purely velocity- and mass-dependent (kinetic) energy of the material points and the (potential) energy resulting (motion independently) from the reciprocal attraction, respectively repulsion, of the points. In terms of mechanics, the conservation of energy for a point system having central forces means that the change in the kinetic part is inversely equal to the change in the potential part.[321]

It is a case of a rigorous logical deduction so typical of the classical concept of science, where the theorem of conservation follows from mechanical laws, i.e.,

[317] Y. Elkana was the first to show that in the treatise Helmholtz used the equivalent formulations of mechanics made by both Newton and Lagrange. See **Elkana** (1974), p. 17 ff., 50 and 129 f.

[318] **Helmholtz** (1847a), p. 3.

[319] **Helmholtz** (1847a), p. 13.

[320] Cf. **Schiemann** (1997), p. 38 ff., where conversely the structure of forces resulted from presupposing certain properties of conservation.

[321] **Helmholtz** (1847a), p. 14, Equation 4.

Newton's three axioms and the constraints characteristic of the mechanical system. A mechanical system's energy is by no means an independent quantity: it must be inferred. Since this sort of approach begins with the inner structure of a system defined by the spatial-temporal coordinates of its parts and corresponding forces, it falls in line with Newton's concept of mechanics that relies on the knowledge of force functions. Helmholtz limits this explanation to mechanical systems that fit his theory of matter. In terms of the theoretical assumptions about nature made in the Introduction, this suffices for a universal foundation of the conservation of energy. Wherever nature can invariably be described by mechanically mobile and interactive elements, the theorem of the conservation of energy is also valid without exception.

As a consequence of the duality of force and matter, the specifically mechanical conception of energy consists of severely dichotomizing the total energy and distinguishing the kinetic part from the potential one.[322] The basic state of inert matter, its uniform motion, is attributed with kinetic energy independent of position and only by adding forces to that state does one get potential (position-dependent) energy. Both features are of major importance for Helmholtz's account: his formal treatment of the conservation of energy begins with the mechanical theorem of the conservation of the "vis viva" [*lebendige Kraft*], which refers one-sidedly to matter in motion:

> If we inquire into the mathematical expression of this principle [the impossibility of a mechanical *perpetuum mobile* – G.S.], we shall find it in the well-known law of the conservation of "vis viva". [...] if any number whatsoever of material points are set in motion solely by the forces which these points exert upon one another, or by forces which are directed towards fixed centers, then the sum total of the vis vivae is the same at all times.[323]

The first mathematical elucidation of the principle of conservation is also dedicated to the "vis vivae".[324] As he continues, Helmholtz divides the most important force of energy into those that are kinetic and those that are potential. This results in free and latent heat.[325] Analogously, he describes electric and magnetic static processes each using terms of motion and position dependence respectively.[326]

The mechanical concept of energy does not take it as a general physical state that can depend on different variables but as a special expression of matter in motion or at rest. One consequence of such an interpretation is that the occurrence of quantities of energy is seen as direct evidence of the existence of substance. This notion, however, is inadequate particularly for electromagnetic processes imaginable in a vacuum. In his treatise of 1847 Helmholtz circumvents these difficulties by only marginally mentioning optical phenomena and by limiting electrodynamic processes to those of

---

[322] Cf. also **Planck** (1887), p. 136, and **Frank** (1932), p. 328 f.

[323] **Helmholtz** (1847a), p. 8 f. (Germ.)/7 f. (Engl.).

[324] **Helmholtz** (1847a), p. 10, Equation 1.

[325] **Helmholtz** (1847a), p. 24.

[326] **Helmholtz** (1847a), p. 29 and 45.

closed currents[327] but later he views the conservation of energy demonstrated for the vacuum as an unmistakable indication of universal ether substance.[328] Without equating energy with (material) substance, all of his life Helmholtz believed that energy cannot occur without matter in motion.

(β) The mechanistic explanation of the theorem of the conservation of energy elucidates two things: why it is valid for mechanical systems and why it must be valid in this form throughout nature. The latter, however, contradicts the notion that non-mechanical energy is essentially distinguishable from mechanical energy: the one is neither tied to the movement of matter nor need it be split into two kinds. Helmholtz's accomplishment was to provide a formal representation of the theorem of conservation that implies a self-contained concept of energy, guided by mechanics only in terms of mathematics. His approach rests on Lagrange's formulation of mechanics, inasmuch as he replaces the concept of force with that of potential.[329] Lagrange's formulation makes the concept of potential, together with kinetic energy (which seems no longer independent on the position of the objects in question), the point of departure for all calculation. Potentials acquire a key function for generalizing methods in mechanics because they can also be determined for non-mechanical phenomena, for which force functions are (still) unknown.

Helmholtz alludes to this function: in the first pages of the treatise he carefully explains what is meant by potential energy and introduces a special term to express it: tensional force.[330] By furnishing the quantity of tensional force with a negative sign, he is able to set up the theorem of conservation as a principle of summation analogous to the conservation of mass.[331] While "vis viva", respectively, kinetic energy, is used for all forms of matter in motion, "tensional force" includes all other kinds of energy. When applying the dichotomy of energies taken from the Newtonian tradition to non-mechanical forms of energy, tensional force gains major significance. Hence, the first formulation of the theorem of conservation no longer mentions material points:

> Hence, *the sum of the vis viva and tensional forces present is always constant.* In this most general form, we can call our law *the principle of the conservation of force.*[332]

Helmholtz discusses both electrostatic and magnetic static events by proceeding from tensional force to expressions of potential and then, for the first time in the treatise, using the concept of potential.[333] In both cases he substitutes the definitional

---

[327] **Helmholtz** (1847a), p. 19f. and 47.

[328] For the ether issue, see Section 6.3.2, footnote 591.

[329] In the treatise Helmholtz does not mention this formalism but he does indicate having occupied himself with analytic mechanics in the period between 1838 and 1842 (**Helmholtz** (1891), p. 10).

[330] **Helmholtz** (1847a), p. 11 f., **Kuznecov** (1970) and **Laue** (1947) consider Helmholtz's introduction of the term 'tensional force' *("Spannkräfte")* the treatise's real merit.

[331] **Planck** (1887), p. 36f.

[332] **Helmholtz** (1847a), p. 14 (Germ.)/14 (Engl.) (italics in original).

[333] **Helmholtz** (1847a), p. 29f. and 45f.

contents of mechanical expressions with non-mechanical quantities. Thus he substitutes "electric" or "magnetic" "mass elements" for mechanical masses,[334] now using the term "mass" in a merely transferred meaning. For electrostatic energy Helmholtz can discuss the conservation of energy without saying anything about the nature of electricity.[335]

What happens is that the universally valid mathematical formulation of the conservation of energy results from transferring one and the same formal structure from originally mechanical equations to non-mechanical phenomena that exhibit certain regularity or conformity to natural laws [*Gesetzmäßigkeiten*]. Once the mechanical theorem of conservation has been deduced from the specifications of a system of material points and explained in those terms, the general principle is gained via analogy, which can no longer be called an explanation. A different kind of explanation thus follows: it no longer states a law but concerns instead individual cases of energy conservation. The law now becomes an unquestioned major premiss for a purely logical deduction, with which under given boundary conditions individual events can be deduced, respectively predicted.[336] Since experimental research must rely precisely on such prognoses, this form of explanation is better suited to research practice than the theoretical deduction of the mechanical law of energy.

Helmholtz provides varying reasons for the general concept of energy not restricted to mechanical work: on the one hand, he understands it as the inductive product of scientific experience (i) and on the other, as a pure definition of a relation of natural forces that may (metaphysically) precede scientific experience (ii). Both interpretations can be found in the treatise of 1847 but emerge more distinctly in speeches published later.

(i) The inductive foundation for the law of the conservation of energy is related to the postulate of the impossibility of a *perpetuum mobile*, which Helmholtz rightly claims to be equivalent to the theorem of conservation. He mentions this equivalence in the second sentence of the treatise but later mitigates it in two ways. First, straightaway he stresses that it is identical to his own assumptions about the theory of matter[337] and second, he demonstrates it using only an example from mechanics, where he must presuppose the reversibility of the changes in states, i.e., he can only claim limited generality for it.[338] Not until later speeches does he insist on the universal significance of this postulate based on scientific experience. He reports of eighteenth but also early nineteenth century attempts to design a *perpetuum mobile*, for which scientists devoted themselves extensively to non-mechanical forms of energy and how those could be transformed into mechanical forms. None of the known forms of energy in inanimate nature were neglected in the search for perpetual motion.[339] Helmholtz thus

---

[334] **Helmholtz** (1847a), p. 29 and 45.

[335] **Helmholtz** (1847a), p. 29 ff.

[336] **Helmholtz** (1847a), p. 36 f., cf. **Elkana** (1974), p. 127, and **Planck** (1887), p. 42.

[337] Cf. quotation from footnote 318.

[338] **Helmholtz** (1847a), p. 7f., and **Planck** (1887), p. 140 f., draws attention to the assumed reversibility.

[339] **Helmholtz** (1854a), p. 59.

understands the postulate of the impossibility of perpetual motion as generalizing past experience of not finding any facts that contradict it. In other words, the theorem of impossibility is not something that can be proven by experience; it is a necessary experiential axiom that applies generally to all imaginable forms of energy:

> But, warned by the futility of former experiments, the public had become wiser. On the whole, people did not seek much after combinations which promised to furnish a perpetual motion but the question was inverted. [...] If a perpetual motion be impossible, what are the relations which must subsist between natural forces?[340]

In this rare phrasing Helmholtz expresses reservations about the validity of deriving the conservation of energy from mechanical central forces. The mechanical explanation for energy is true only under the condition that no perpetual motion has yet been discovered. This radical stating of conditions, which admits a possible refutation of the theorem of the conservation of energy, clearly expresses the empirical character of the general concept of energy. Nevertheless, its surprising proximity to the modern conception of science[341] is not yet effectual for Helmholtz (at this time), because he dispels his reservations with a mechanical explanation that offers him the concluding proof for the postulate:

> The possibility of a perpetual motion was first finally negated by the law of the conservation of force.[342]

Even in the 1860s, however, the "objective truth" of the idea of central force was still as feebly proven as it was in 1847. For Helmholtz the further identification of central forces depends just as much on the development of empirical knowledge, as on the confirmation of the theorem of the impossibility of perpetual motion. In spite of the existing and – in his opinion – completed mechanical explanation of energy and disregarding the euphoric tenor of his writing, Helmholtz is aware that the validity of his statements is conditional. If one compares the various wordings he used to describe the theorem of the conservation of energy after 1847, one finds that he speaks of its validity as being subject to previous experience and he consistently points out that additional investigations are necessary. Only gradually does he wager to extend the validity of the theorem from the inorganic to the organic.[343]

This context reveals another feature of the general concept of energy gleaned from experience: continuous scrutiny of the theorem takes for granted that processes that do not conserve energy must always be recognizable. While that possibility loses relevance as Helmholtz's research projects move forward, today the assumption of constraints on and partial violations of the theorem of the conservation of energy are paradigmatic for the fundamental theory of physics.

---

[340] **Helmholtz** (1854a), p. 62 (Germ.)/26 (Engl.).

[341] On conditionalizing as a feature of the modern concept of science see Section 4.1.

[342] **Helmholtz** (1862/3), p. 228 (Germ.)/126 (Engl.).

[343] **Helmholtz** (1847a), p. 51, **Helmholtz** (1854a), p. 75 and 65, **Helmholtz** (1869), p. 386.

(ii) One approach to understanding a concept of energy in terms of relations is the notion of mechanical work that Helmholtz focused on as the general unit for measuring energy conversion.[344] In 1847 Helmholtz uses the immediately measurable quantity of work that is so important for technical applications to phrase the wording of his theorem of the conservation of energy, as mentioned above: "that it is impossible to obtain an unlimited amount of force"[345] and more clearly in a later speech:

> *It is a universal character of all known natural forces that their capacity for work is exhausted in the degree to which they actually perform work.*[346]

Work is performed by or on systems and the increase or decrease in the quantity of work can be measured without knowing anything about the inner structure of those systems. It is decisive, rather, to identify the system's adiabatic boundaries, where energy transformation takes place and define a uniform physical dimension, which may be chosen arbitrarily, as long as it does not depend on the course of changes of state. Nevertheless, this phenomenological definition sufficient for the purposes of thermodynamics already includes the transformation of system-immanent energy into outside work and thus itself presupposes the theorem of conservation.

(γ) Of course, one finds no indication that Helmholtz selected the dimension of work as a unit of measurement arbitrarily. In the pamphlet from 1847 he defines work by equating it directly with kinetic energy.[347] He understands the equivalence of work and other forms of energy not merely as a mathematical relation of physical quantities but as an expression of an essence common to all forms of energy, of which mechanical work is distinguished from all the others by the fact that it alone makes visible what is covertly active in them: matter moved by inertia and force.[348] The reciprocal conversion relation of heat and mechanical work is paradigmatic for such ontologically interpreted equalizations, where apparently the idea of energy taken from Newtonian mechanics shines through once more. Helmholtz believes he can virtually derive the mechanical nature of heat from this relation.[349] He illustrates this by explaining frictional heat: if the energy of a system with friction, for instance a wheel positioned such that it rotates, remains constant and consists solely of motion, then the decrease in the visible (rotating) movement must be equivalent to the increase in invisible (thermal) movement caused by friction.[350]

---

[344] On Helmholtz's concept of work see also Section 6.3.2.3(iv).

[345] Cf. Footnote 316.

[346] **Helmholtz** (1862/3), p. 226 (Germ.)/124 (Engl.) (wide letter spacing has been italicized).

[347] **Helmholtz** (1847a), p. 8 f.

[348] **Helmholtz** (1869), p. 379.

[349] **Helmholtz** (1862/3), p. 217 f., similarly in **Helmholtz** (1854a), p. 64, **Helmholtz** (1861), p. 569, and **Helmholtz** (1897 ff.), Vol. VI, p. 175.

[350] For instance **Helmholtz** (1862/3), p. 214 ff.

In the mechanistic program of reduction the equivalence of work and other forms of energy is accordingly of central importance. Twenty years after his discovery, which meanwhile had been brilliantly confirmed, Helmholtz summed it up:

> It has actually been established, then, as a result of these investigations, that all the forces of nature are measurable by the same mechanical standard and that all pure motive forces are, as regards performance of work, equivalent. And thus one great step towards the solution of the comprehensive theoretical task of referring all natural phenomena to motion has been accomplished.[351]

In contrast to the mechanical explanation of the theorem of the conservation of energy, this appraisal contains the opposite outlook. Instead of coming to a conclusion about the universal validity of the conservation of energy by starting with a mechanical theory of matter, here the theorem of conservation becomes the basis for the reduction of natural phenomena to mechanics. The erroneous notion that central forces (which until late into the 1870s Helmholtz held to be the only forces that conserve energy) are absolute becomes an energetics heuristic for mechanism. As early as 1847 he mentions in the pamphlet, the "heuristic importance" of the law of the conservation of energy[352] claiming that it could "provide a guideline for experiments".[353]

6.2 Mechanics – the Underlying Principle of Geometry

> Thus *geometry* is founded in the practice of mechanics and is nothing other than that part of the *entirety* of mechanics that purports and proves the art of exact measurement
> (Isaac Newton, *Mathematical Principles of Natural Philosophy*).

Besides establishing and enunciating the theorem of the conservation of energy, Helmholtz's studies in geometry are another example of the integrating strength of the classical mechanism he advocated. This work rates among his most important epistemological accomplishments and has remained relevant to the current debate on the fundamental principles of geometry.[354] In it Helmholtz infers fundamental geometric axioms from mechanical properties of "solid bodies". By "solid bodies" (*feste Körper*) he means objects of physical mechanics, whose immutable shape

---

351 **Helmholtz** (1869), p. 383 (Germ.)/214 (Engl.).

352 **Helmholtz** (1847a), p. 53.

353 **Helmholtz** (1847a), p. 7. Examples of heuristics rules based on energetics transformations can be found in **Planck** (1887), p. 110ff. and 123f.

354 **Süßmann** (1990), p. 214ff., provides a comprehensive and updated list of mathematical literature dealing with the "Riemann-Helmholtz Controversy on Space". For the current discussion of Helmholtz's basic principles of geometry in the history and philosophy of science see **Wahsner** (1992), **DiSalle** (1994), **Carrier** (1994), **Schüller** (1994), **Volkert** (1996) and **DiSalle** (2006). See also **Nagel** (1939), **Boyer** (1956), **Torretti** (1979) and **Mainzer** (1980), p. 134ff., on how the conception of geometry changed throughout the nineteenth century.

and free mobility constitute the "factual foundation of geometry".[355] Thus this work could also be considered a part of his conception of nature and science guided by the principles of physical mechanics.

Nevertheless, I shall not draw upon his work in geometry chiefly to demonstrate how my concept of mechanism applies. More important is its significance for evaluating the claims to validity and the foundational strategies that Helmholtz associates with his mechanism. At first he primarily views his own work as an alternative to Bernhard Riemann's suggestion that fundamental principles of geometry are hypothetical. Contrary to Riemann he defends the claim that systems of geometric axioms are true, thus positioning himself firmly within the classical conception of science. However, his arguments for it differ from traditional ones. He substitutes the traditional Kantian claim that the truth of geometric axioms is proven by compelling evidence, whose validity can be known subjectively and intuitively, with precisely that relentless reference to experience that he appears to count as given "facts" based on the mechanical properties of real solid bodies. Although from the start he leaves no doubt that he objects to Kant's conception of geometry, not until later writing, when explicitly using geometry as a paradigm for a critique of metaphysics, does he stress the polarity existing between his own empiricist grounds and the geometric implications of transcendental aesthetics of space.

Nevertheless, it would not be uncontroversial to say that his understanding of science is scientistic and classical. With certain reservations regarding content, one can only say that his work reflects that notion of science up until the early 1870s. Up until that point he had written only three pieces devoted exclusively to issues in geometry. In them Helmholtz develops for the first time empirical foundations, the final version of which will later become part of the speech titled *On the Origin and Significance of Geometrical Axioms* [*Ueber den Ursprung und die Bedeutung der geometrischen Axiome*]. These items, written and in part published between 1868 and 1870 will be the focus of interest somewhat later.[356]

---

[355] **Helmholtz** used this title – *Die thatsächlichen Grundlagen der Geometrie* – in (1868b), p. 616. See also **Helmholtz** (1868c), p. 621, and **Helmholtz** (1870), p. 19.

[356] **Helmholtz** (1868b), (1868c), (1869a), (1870) and (1870b). The first printing of **Helmholtz** (1868b) contains the apparently erroneous statement that Helmholtz delivered the speech in 1866 (i.e., prior to the publication of Riemann's work) and submitted it immediately for publication in the *Proceedings of the Society for Natural History and Medicine in Heidelberg* [*Verhandlungen des naturhistorisch-medicinischen Vereins zu Heidelberg*]. However, the chronological order of the speeches published by the society and **Koenigsberger** (1902 f.), Vol. 2, p. 125, too, indicate that Helmholtz probably gave the lecture on May 22, 1868. Additional evidence found by **Volkert** (1993) and compiled for the purpose of settling the question of the date also suggests this. Another controversial date is **Helmholtz** (1870). Published in **Helmholtz** (1876), this is actually a heavily reworked and enhanced version of a lecture held for the "Heidelberger Society of University Lecturers" [*"Docentverein zu Heidelberg"*] in 1870 (cf. **Helmholtz** (1884a), Vol. 2, p. 1, and **Koenigsberger** (1902 f.), Vol. 2, p. 126, and further **Helmholtz** (1884a), Vol. 2, p. V). The extent of subsequent changes can only be estimated approximately by comparing the text to an English rendition of the first section that was published in 1870, namely **Helmholtz** (1870b).

In the comments Helmholtz makes after 1870 in various public speeches and accompanying appendices containing geometrical explanations, he no longer modifies the relation he initially postulated between geometry and experience but limits himself to elucidating and interpreting it.[357] These comments show that meanwhile only minor steps separate him from abandoning the unconditional claim to truth and accepting a modern notion of geometry. According to my definition, a conception of geometry can be considered modern if its statements can be attributed the same hypothetical character that is attributable to all scientific statements in the modern meaning of science and if that character is interpreted as resulting from the undeniable reference that statements of geometry make to experience.[358] Since I particularly wish to focus on how Helmholtz's idea of mechanism evolved under the general tension existing between the classical and the modern understanding of science, I shall discuss his later remarks further on when I comment on the transition in his concept of science, which began in the 1870s. This will also allow me to do justice to Helmholtz's outstanding achievements in geometry in a section of their own.

In his own words, it was while "investigating the spatial aspect of the field of vision" that Helmholtz happened upon "the question of the origin and nature of our general intuitions about space".[359] While initially Riemann's publication prompted him to publish his own ideas as an alternative approach, his thoughts about a physiological "theory of vision" also contain a first link to empirical foundations for geometry – a link that is overlooked in secondary literature on his geometry. In the final contribution to the *Treatise* published in 1867 he explains why "real spatial relations" are not represented correctly by everyday perception:

> As to the representation of space-relations [...] still only in a limited way; for the eye gives only perspective surface-images and the hand reproduces the objective area on the surface of a body by shaping itself to it as congruently as possible. A direct image of a portion of space of three dimensions is not afforded either by the eye or by the hand.[360]

Now, in order to deny ordinary perception the capacity to represent correctly, one must first have an idea of what extended reality would be like, if an objective representation of it were to be had. Thus, Helmholtz assumes that nature manifests a particular spatial structure, probably Euclidean, which is also at the bottom of the dual ontology intrinsic to his notion of mechanism. One of his rather

---

[357] **Helmholtz** (1878a), (1878b) and (1878c), p. 186.

[358] Cf. Section 4.1.

[359] **Helmholtz** (1868c), p. 618, cf. also **Helmholtz** (1868b), p. 610, and **Koenigsberger** (1902 f.), Vol. 2, p. 138 f. **Richards** (1977) and **DiSalle** (1994) discuss the conditions of the origin of spatial intuitions in terms of the theory of perception and **Cassirer** (1944) discusses the consequences they have for scientific theory. See Section 6.3.1.2 for Helmholtz's conception of geometry prior to 1867.

[360] **Helmholtz** (1856 ff.), p. 445 f. (Germ.)/23 (Engl.).

incidental remarks in this context reveals his attempt to make the assumption of an objective structure of space plausible by postulating that it is a natural basis required for promoting the development of the human capacity for spatial representation.

> We must be on our guard against saying that all our ideas of things are *false*, because they are not equal to the things themselves and that hence we are not able to know anything as to the *true nature* of things. [...] If there were not a number of similar natural objects in the world, our faculty of forming generic notions would indeed not be of any use to us. Were there no solid bodies, our geometrical faculties would necessarily remain undeveloped and unused, just as the physical eye would not be of any service to us in a world where there was no light.[361]

Perhaps Helmholtz did not have the discipline of geometry in mind when referring to "our geometrical faculties".[362] It seems plausible however, that it too is not founded solely on subjective and constitutive requisites of perception but instead depends on material conditions (the existence of solid bodies) in an independent world that we assume exists. Here Helmholtz begins to part questions of geometry from those of perceptual processes and establishes an objectively analyzable point of reference (solid bodies) from which it may still be possible to discover the "real nature of things". He gradually proceeds from questions related to physiology and quotidian perception to those of physics and it begins to become clear that this separation of issues will become one prerequisite for securing geometry's claim to truth.[363]

The traditional absolute validity of Euclidean axioms was thoroughly shaken by Riemann's approach.[364] In an *habilitation* speech (also published in 1867) delivered on a topic suggested by the, in his own day, legendary Friedrich Gauss[365] and titled *On the Hypotheses which lie at the Foundation of Geometry* [*Über die Hypothesen, welche der Geometrie zugrunde liegen*][366] Riemann shows how both non-Euclidean and non-three-dimensional spatial structures can be deduced

[361] **Helmholtz** (1856 ff.), p. 446 f. (Germ.)/24 (Engl.).

[362] The next sentence, namely, says that "our geometrical faculties" need be "neither complete nor accurate", which corresponds poorly to the conception of science advocated by Helmholtz at that time.

[363] **DiSalle** (1994), who sees a close connection between Helmholtz's geometry and theory of sensory perception, obscures the distinction between Helmholtz's initially assumed absolute validity of the statements of geometry and a merely relative validity of evidence provided by the senses. In contrast, **Carrier** (1994), p. 284, rightly remarks that in terms of justification for claims to validity, geometry and perception are two entirely different matters.

[364] Helmholtz undermined that validity to a greater extent than did Riemann, who died young; it was he who made Riemann's ideas popular and instigated a public debate involving scholars sympathetic to Kant's epistemology. That debate did not occur during the period discussed here. The most important document is **Erdmann** (1877); see also **Krause** (1878) and **Schwertschlager** (1883).

[365] **Jungnickel and McCormmach** (1986), Vol. 1, p. 170 ff. and 174 ff.; **Klein** (1926 f.), Vol. 2, p. 165.

[366] **Riemann** (1867). The edition prepared by Hermann Weyl, Berlin 1919 (2nd edition: 1921) includes a commentary on this work. For Riemann's geometry cf. **Torretti** (1979), p. 82 ff.

mathematically, if we think of space as being one special case of a manifold determined by definition.[367] There are no limits to the number of dimensions possible in space thus defined, each of which permits a variety of geometries; Euclidean geometry being then merely one among much other equally valid geometries. Riemann, like Helmholtz, is convinced that the question of which geometries and systems of dimensions are actually realized in nature is a question answerable only empirically, as long as mathematical definitions for spatial structures can be applied to nature's physical objects:

> Space is only a particular case of a triply extended magnitude. But hence flows as a necessary consequence that the propositions of geometry cannot be derived from general notions of magnitude but that the properties which distinguish space from other conceivable triply extended magnitudes are only to be deduced from experience. [...] there may be several systems of matters of fact which suffice to determine the metric relations of space – the most important system for our present purpose being that which Euclid has laid down as a foundation. These matters of fact are – like all matters of fact – not necessary but only of empirical certainty; they are hypotheses.[368]

By demonstrating that Euclidean geometry is merely one of other possible systems for "determining the metric relations of space", which, due to its empirical content, must itself be considered a hypothesis, Riemann creates – in comparison to the classical understanding of the formal truths of mechanism – an entirely new situation. Up to that point there had been no reason to seriously question the Euclidean structure of real space that underpins the formalism of Newtonian mechanics. Its evident validity prior to all experience had been assumed and confirmed by unprecedented success in scientific and technological practice.[369] Taking it for granted had been undermined by Riemann's thesis; and the truth of the mechanistic conception of nature was now dubitable.

Besides this fundamental challenge, other concrete objections arose to the mechanism that Riemann, too, endorsed.[370] While both Riemann and Helmholtz assumed that physical

---

[367] Riemann presupposes the invariability of line segments (i.e., the (positive, definite) square differential form $(ds)^2 = \Sigma\, g_{ij}\, dx^i\, dx^j$ (sum of i,j = 1 to n) with coordinates $x^1, \ldots, x^n$ and the coefficients g_{nm} as a continuous position function), which serves to define measuring relations (i.e., metrics) in a space defined as a continous manifold. A line segment is a generalized form of the Pythagorean theorem: $a^2 = b^2 + c^2$ (for $g_{nm} = 1$ and $n = m = 2$).

[368] **Riemann** (1867), p. 8 (Germ.)/653 (Engl.).

[369] I. Kant and J.S. Mill are among those who entertained undeterred esteem for Euclidean geometry. On the one hand, Helmholtz refers to Kant's transcendental aesthetics for the standard contemporary conception of space. On the other hand, as we know from the works of J.S. Mill (cf. **Torretti** (1979), p. 256ff.), British empiricism, then gaining ground in Germany (Cf. Chapter 5, footnote 248) held fast to the unshakeable truth of Euclidean geometry. But we know from F. Gauss's studies and experiments that the possibility of non-Euclidean geometry was by all means imaginable (cf. **Stäckel** (1863ff.)).

[370] An underlying mechanistic attitude is alluded to in **Riemann** (1867), p. 22; see also **Jungnickel and McCormmach** (1986), Bd. 1, p. 176.

space is three-dimensional,[371] Riemann – unlike Helmholtz – explicitly points out the problem of an uninterrupted transition from one geometry to another, a problem particularly important for areas of the smallest extension relations. Mathematically, these various geometries can be assigned so-called spatial curvatures with a value that is fixed for each one of them. A distinction is made between flat, i.e., Euclidean geometry (curvature value equals zero), spherical geometry (positive curvature) and hyperbolic or pseudo-spherical geometry (negative curvature).[372] A curvature value that is not constant in real space would have disastrous consequences for the theory of matter used in mechanical atomism:

> The questions about the infinitely great are for the interpretation of nature useless questions. But this is not the case with the questions about the infinitely small [...]. In the natural sciences [...] we seek to discover the causal relations by following the phenomena into great minuteness, so far as the microscope permits [...].
>
> If we suppose that bodies exist independent of position, the curvature is everywhere constant [...] But if this independence of bodies from position does not exist, we cannot draw conclusions from metric relations of the great, to those of the infinitely small.[373]

These lucid statements from the closing paragraph of this concise treatise indicate Riemann's serious doubt that classical mechanics is applicable to the tiniest particles of matter. For if curvature was not constant down to the level of the most minute orders of magnitude, more than just Newtonian laws would lose their validity. The existence and mesh of more than one kind of geometry would be a phenomenon demanding an explanation no longer deliverable by the formalism of mechanics, for mechanics knows of no dependence of physical magnitudes of mass on that of space.

Although Helmholtz never addressed this difficulty[374] his papers on the theory of space can nonetheless be read as a counter-draft in defense of the claim to truth not

---

[371] "That space is an unbounded threefold manifold, is an assumption which is developed by every conception of the outer world" (**Riemann** (1867), p. 21 f. (Germ.)/660 (Engl.). "We, as inhabitants of three-dimensional space" (**Helmholtz** (1870), p. 15).

[372] The spatial curvature K that Riemann introduced is a generalization of a Gaussian surface curvature and can be defined by the equation $K = \Omega/F$ mit $\Omega = \sigma - \pi$, where σ is the sum of the angles of an infinitely small triangle and F is its area. In general, the curvature function of a manifold depends at one point only on the coefficients g_{nm} (see footnote 367). See, for instance, **Klein** (1926 f.), Vol. 2, p. 170 ff.

[373] **Riemann** (1867), p. 22 f. (Germ.)/660 f. (Engl). If the size and shape of bodies were not independent of their positions, spatial determination of bodies would depend on the fluctuating curvature of space. In these paragraphs Riemann (unlike Helmholtz) also considers that the presence of matter in space might determine the curvature (loc. cit. p. 23) and thus anticipates the crucial basic idea of the theory of general relativity (cf. **Farwell and Knee** (1990), p. 116). Accordingly, he is already thinking of "reworking" mechanics: "The answer to these questions can only be got by starting from the conception of phenomena which has hitherto been justified by experience and which Newton assumed as a foundation and by making in this conception the successive changes required by facts which it cannot explain." (**Riemann** (1867), p. 24 (Germ.)/661 (Engl.)).

[374] Of course, Helmholtz may have been so thoroughly convinced that his theoretical ideas on matter were correct that at that time he could not seriously doubt that mechanical laws of motion also validly apply to sub-microscopic phenomena. The insouciant transition from one magnitude of order to another in his theory of space's mathematical calculations seems to suggest it was so.

only of geometry as a theory of magnitudes[375] but also of mechanism as a conception of nature and a program of reduction. To do so, Helmholtz had to reverse Riemann's procedure. Whereas Riemann started with mathematical definitions unrelated to physical objects and first made the application of various mathematically feasible geometries depend on experience, Helmholtz proceeds from the opposite direction:

> Riemann starts by assuming the [...] algebraical expression which represents in the most general form the distance between two infinitely close points [...]. I, on the other hand, starting from the observed fact that the movement of solid figures is possible in our space, with the degree of freedom that we know, deduce the necessity of the algebraic expression taken by Riemann as an axiom.[376]

Helmholtz establishes the mechanical precondition for the free mobility of solid bodies on what he considers to be characteristic of geometry, namely that statements in geometry are not part of any "pure theory of space" but have to do with "magnitudes" that are, indeed, measurable.[377] All geometric measurement rests, according to Helmholtz, "on the assumption that the measuring instruments (solid bodies) actually are invariable in form". Only under that condition can we use them to perform elementary measurements, by physically moving the standards around.[378]

This also expresses a crucial prerequisite for guaranteeing geometry's claim to truth: the solid bodies moved around mechanically for purposes of measuring are not supposed to "really" change shape. How important is this requirement? Does it reflect a natural feature of the objects themselves? Or is it rather, or even, essential for existing bodies that they deviate from this expectation that the subjects place on them whilst performing measurements? Helmholtz prefers, as he did in the *Treatise,* the first option by stating that:

> Natural bodies, in their purest and undisturbed state, do not even in fact correspond exactly to the abstract notion we have obtained of them by induction. Taking the notion of solidity thus as a mere ideal, a strict *Kantian* might certainly look upon the geometrical axioms as propositions given, *a priori* [...] But in that case [...] the axioms of geometry would only express what follows analytically from the concept of the solid geometric figure needed for performing the measurement.[379]

Even when Helmholtz mentions the disparity between scientific concepts and the natural phenomena themselves and perhaps even finds it worthy of explanation, that

---

[375] **Herneck** (1973), p. 349, correctly notes that Helmholtz's choice of title *On the Facts which lie at the Foundation of Geometry* [*Ueber die Thatsachen, die der Geometrie zu Grunde liegen*] (**Helmholtz** (1868c)) seems like a "deliberate antithesis" to Riemann's *On the Hypotheses which lie at the Foundation of Geometry* [*Ueber die Hypothesen, welche der Geometrie zu Grunde liegen*] (**Riemann** (1867)).

[376] **Helmholtz** (1870), p. 19 (Germ.)/236 f. (Engl.).

[377] **Helmholtz** (1870), p. 29.

[378] **Helmholtz** (1870), p. 23.

[379] **Helmholtz** (1870), p. 30 (Germ.)/244 (Engl.) (Helmholtz's emphasis).

does not let him question the certainty of his own concepts vouched for by induction. Since the concept of the shape of an unchanging and freely mobile solid body is abstracted from reality, for Helmholtz it seems nonsensical to want to surpass it with any "ideal" of the same.[380] He thus rejects that "the notion of solid geometrical figure might [...] be conceived as transcendental"[381] (synthetic a priori), thereby replacing metaphysical underpinnings with scientistic induction.

Nevertheless, within the context of his writing on geometry, Helmholtz does not explicate the theory of induction so central to his foundation for science. He does so in public speeches and lectures to which I shall return later. At this point it may be said that Helmholtz only unconditionally endorsed absolute universal validity for inductively derived axioms until the last years of the 1860s. Inasmuch as he considers geometry an empirical science, truth for the axioms in geometry, too, begins eroding as the process of hypothesizing and the subsequent new understanding of science sets in. This process vaguely emerges in his writing on geometry. For if not only abstract figures but also varying states belong to the bodies (objects), then variability, instead of abstractness, might be the distinguishing feature of geometry and geometry must forfeit its claim to accuracy. After all, Helmholtz says that even bodies in their "purest and undisturbed" state are not conform to the concepts from which he derives the axioms of geometry. At this point the empiricist program based on "facts of observation" could itself undermine the truth claim asserted before.

If one more carefully examines Helmholtz's reasons for the reality of immutability of form and free mobility, it turns out that he only escapes relativizing the claim to validity via scientistic reference to the scientific practice of physical mechanics. Mechanics is the discipline that can determine to which "degree of freedom" (as he says in the passage quoted above) solid bodies may move. That moving bodies do not change their shape, i.e., are independent of position, stems from the idea of absolute space intrinsic to Newtonian mechanics. Helmholtz is also certain that solid bodies are independent of position, because experience provides proof:

> The proposition, that the mechanical and physical properties of bodies are, other circumstances remaining the same, independent of position, such a system of propositions has real import which can be confirmed or refuted by experience but just for the same reason can also be gained by experience.[382]

The option of either refuting or producing propositions empirically is fully within the framework of the classical notion of how to obtain ultimate falsification or

---

[380] **Wahsner** (1992), p. 35, notes that Helmholtz "actually equates measurement demands with facts" but does not deal with the philosophy of science background behind Helmholtz's theory of induction that is crucial for understanding this passage. On induction see Chapter 2, Sections 6.3.1.2, 7.1.3.2 and 7.1.3.4.

[381] **Helmholtz** (1870), p. 30.

[382] **Helmholtz** (1870), p. 30 (Germ.)/244 (Engl.). This wording was probably added after 1870 while revising the second part (cf. footnote 356) and matches an expression used on page 22 of the same publication. For its further fate, see Section 7.1.3.4, footnote 730.

verification for empirical statements. That a mechanical statement might at some future time be overturned is only an abstract possibility compared to the plenary validity it has as a statement derived by induction. By calling the properties of solid objects, without which we can perform no measurements, "observed facts", Helmholtz wants to express precisely that: the scientific practice of mechanics guarantees that those properties are real.[383]

Based on this, Helmholtz determines, as a criterion for the solidity independent of motion, that the shapes a body acquires when it is in motion must be congruent or that the transition from one shape to the next be shape-preserving.[384] He then derives an equation from the mathematical expression for the congruence relation, which is comparable to Riemann's axiom.[385]

Nevertheless, he does not reduce Riemann's axioms entirely to mechanics and this incompleteness itself contains his defense for the mechanistic program against objections that Riemann raised. Helmholtz allows for no spaces having varying curvature, in which Newtonian mechanics would no longer be valid.[386] The congruence he presupposes is only possible in space having constant curvature.[387]

---

[383] **Mehrtens** (1990), p. 54 ff., sees Helmholtz's empiricist approach as an example of an emergent "counter-modernity" within the context of the controversy over the foundations of mathematics. He contrasts Helmholtz's approach with the fact that Riemann isolates physics from mathematics, something he sees as constituting nascent "modernity". In this context, "modernity" includes "divorcing [mathematical – G.S.] signs from what they signify and making them externally conceivable objects". This raises the "question of the other in mathematics", an issue that "particularly" motivates "counter-modernity" (loc. cit. p. 106). I think both parties in this debate, namely the "modern" and the "counter-modern" as defined by Mehrtens, should be considered parts of a process that can be called modern whenever it waives claims to absolute truth.

[384] For the concept of congruence see **Volkert** (1996). Hugo **Dingler's** (1934) critique of this attempt to define the concept of the solid body within the framework of empirical physical geometry has been of considerable consequence for the history of the topic. Dingler argues that a priori ideas are indispensable, even if we assume, as Helmholtz does, that rigid bodies (objects) are needed as instruments for measuring lengths. Dingler's reference to Helmholtz secured the latter's eminent importance for the constructivist theory of geometry that is effective to this day. On the relationship between Helmholtz and Dingler see also **Carrier** (1994), p. 285 f.

[385] His formal procedure is to assume the existence of a system of coordinates and then deduce the definition of an infinitely small three-dimensional solid body from the observation of congruencies within a freely mobile system of points: there must be an equation representing the coordinates of congruent pairs of points, such that the equation remains invariable despite their movement in space. From this he derives the analytic expression for a line segment (**Helmholtz** (1868b), p. 614 f., **Helmholtz** (1868c), p. 621 ff., and **Helmholtz** (1870), p. 19 ff.). Helmholtz's computations contain mathematical errors related to the transition from infinitesimal to finitely extended dimensions, which cannot be discussed here. Soon corrected by Sophus Lie, group theory provided a mathematical formalism that today is an integrated component of quantum mechanics (**Lie** (1888 ff.)).

[386] So-called "Helmholtz spaces" are merely a subset of so-called "Riemannian spaces". They correspond to Riemann's globally isotropic spaces, where rays emanating form one point are geometrically equivalent (**Süßmann** (1990), p. 214 ff.).

[387] **Kanitscheider** (1971), p. 172 f. In order to avoid this constraint, Hans Reichenbach, following Helmholtz, introduced the concept of a "rigid body" (**Reichenbach** (1928), p. 40).

Helmholtz was not forced to see this constraint as a theory-salvaging trick or a flaw. From his standpoint it was simply the necessary consequence of the real and universal mechanical properties of solid bodies:

> Our axioms are, indeed, the scientific expression of a most general fact of experience, the fact, namely, that in our space bodies can move freely without altering their form. From this fact of experience it follows, that our space is a space of constant curvature but the value of this curvature can be found only by actual measurements.[388]

If the constant curvature of space itself follows from the immutability of solid bodies, then curvature present in space, if it is empirically observable at all, will indubitably be measurable to any desirable degree of precision and with unrestrained validity. The result of such measurement, however, would be irrelevant for the claim to truth asserted by the mechanistic view of nature. With reference to work done by Rudolf Lipschitz,[389] Helmholtz may safely assume that any curvature of space is irrelevant for the validity of mechanics, as long as there is only one of them and that one universally governs the entirety of physical space:

> The laws of motion, as dependent on moving forces, could also be consistently transferred to spherical or pseudospherical space.[390]

Now we could summarize the reconstruction of Helmholtz's foundation for geometry as follows: in a mathematically not entirely unobjectionable way, Helmholtz deduces various metrics for physical space from mechanical properties of solid bodies. He assimilates geometry to a kind of scientific research based on mechanics and it is this science that guarantees the absolute validity of all the statements in geometry, just as it qua reduction is supposed to guarantee the absolute validity of all scientific statements. The apodictic bond between mechanics and Euclidean geometry is loosened such that mechanics can now be applied to non-Euclidean geometries. However, inasmuch as different constant space curvatures can be considered equivalent, one can continue to calculate according to Euclidean principles, as before.

As if that were not already a brilliant accomplishment for his classical mechanicism, Helmholtz is occasionally tempted to go one step further: he suggests that one can perhaps conclude the reality of Euclidean geometry from the empirically reliable form of the laws of mechanics. His writing is marked by a struggle to simultaneously acknowledge non-Euclidean geometries while attempting to prove that they do not exist. That endeavor, which has frequently been subject to criticism, can be interpreted as a reaction to another aspect of his conception of geometry – one that

---

388 **Helmholtz** (1870b), p. 130.

389 **Lipschitz** (1870).

390 **Helmholtz** (1870), p. 21 f. (Germ.)/238 (Engl.). Helmholtz's transferring classical mechanics (stated in terms of Euclidean geometry) to non-Euclidean space does not work. For example, in spherical space translations cannot be distinguished from rotations or momentum distinguished from angular momentum (cf. **Blaschke** (1942)). Of course, this constraint does not alter the fact that the validity of Newtonian mechanics does not imply that (physical) space is Euclidean; cf. footnote 399.

already alludes to a later relativization of claims to validity and that might have been induced sooner, had Helmholtz not resisted it with his one-sided orientation to mechanics.

In his first publication on the problem of space, Helmholtz thinks, based on an unquestioned assumption that space is infinite, that he can directly prove that space is Euclidean. In error, he assumes that there are only two types of curvature, namely flat and spherical and he attributes – which would be correct based on his position that only these two types exist – infinitely extended space to flat curvature and finite physical space to spherical curvature.[391] It is not his failure not to have seen that the cosmological assumption of infinity cannot be proven that later forces him to correct his views but the fact alone that he at first overlooks the possibility of a third kind, namely, hyperbolic curvature. In contrast to spherical curvature, hyperbolic curvature is also compatible with an infinite extendedness of space.[392]

The analytical enunciation of three kinds of curvature reveals the loss of a uniquely valid geometrical representation for physical space. In publications that follow, Helmholtz accepts that loss as much as possible and for the first time partially abandons science's classical claim to discovering – as he puts it – the "true circumstances".[393] In a famous analogy he illustrates the possibility of equally well representing space using differing systems of geometry by describing a thought experiment in which the world – as it were – is viewed from the outside in mirrors that are curved differently. The "image of the world" provided by a convex mirror would correspond to the image one would get if one viewed it from a positively curved ("spherical") world. Seen from a negatively curved ("pseudo-spherical") world, our image of the world would look distorted as it would on the surface of a silvered globe.[394] The equivalence of these perspectives anticipates the plurality of theories that will become so typical of the modern conception of science.[395] There no longer exists one true depiction of the world but various depictions depending on standpoint, none of which is closer than any other to the truth about the actual structure of space. In terms of geometric representations, the question of the one truth thus becomes unanswerable and meaningless.

However, this surrender of truth, or, respectively, gain of plurality does not infringe on the unrestricted correctness of the principles of mechanics, which Helmholtz believes still apply for the various geometries. Furthermore, in his closing description of geometry published in 1870 he continues, contrary to his own

---

[391] **Helmholtz** (1868b), p. 613.

[392] **Helmholtz** (1869a), p. 617. **Dingler** (1934), p. 353, rightly calls this amendment "a crucial event". It does not however, as Dingler suggests, mark the end of hope of discovering an a priori foundation for geometry (cf. Section 6.3.1.2), which Helmholtz must have abandoned at least by the time he wrote his first treatise on geometry in 1868. It does, rather, mark the surrender of an assumption from which Euclidicity would have followed by necessity.

[393] **Helmholtz** (1870), p. 25, cf. text for footnote 361.

[394] **Helmholtz** (1870), p. 24ff.

[395] Cf. Section 4.1.

beliefs, to pursue his initial attempt to prove that real space is Euclidean. His few scattered remarks can be interpreted as an attempt to reverse the conceded abandonment of the claim to truth for geometry, in order to comply with the classical claim that one exclusively valid system can explain all of nature. No longer is it the infinity of space but now it is the hitherto uncontested applicability of physical mechanics from which the "true circumstances", which will be expressed in a Euclidean way, will follow. Inquiry into the structure of space only makes sense if the answer is provided by mechanics, or as Helmholtz writes:

> Indeed, I cannot see that such a question would have any meaning all, so long as mechanical considerations are not mixed up with it.[396]

This remark from a lecture remains somewhat in the dark. One clue to understanding it is perhaps provided by implicit assumptions made for his above-mentioned mirror thought experiment. They emerge in his statement that "we should have to change our system of mechanical principles entirely" if we were to view the "space in which we live" as (either positively or negatively) curved.[397] In such a system "even the proposition that every point in motion, if acted upon by no force, continues to move with unchanged velocity in a straight line" would no longer hold.[398] He tacitly assumes, however, that this change would only occur in the unreal case where the measuring instruments would not be subject to the same curvature as the objects being measured. Instead of realizing that the validity of the mechanical law of inertia does not imply the Euclidicity of real space,[399] he comes to precisely the erroneous conclusion that:

> The axioms of Euclid [...] have been verified by experience to that degree of precision, which practical geometry and astronomy have reached hitherto and, therefore, there is no doubt that the radius of curvature of our space, if it should be spherical or pseudo-spherical, is infinitely great, when compared with the dimensions of our planetary system.[400]

In his work on geometry Helmholtz parallelizes the experience of space with the experience of time. In his epistemology (as I shall next explain), the experience of time accurately represents how time really is.[401] As long as reasons support the assumption that the mechanical law of inertia is true, this analogy permits using an intuition shaped by Euclidean principles as a representation of the real structure of space. "Representation" here does not mean a pictorial, perspective view of the world. It means an unambiguous and indisputable account of what according to the laws of nature must be the case.

---

[396] **Helmholtz** (1870), p. 25 (Germ.)/241 (Engl.) in analogy cf. p. 29.

[397] **Helmholtz** (1870), p. 29.

[398] Ibid.

[399] Cf. e.g., **Mittelstaedt** (1976), p. 66f.: "The question of the geometric structure of empty space can thus not be answered empirically. The geometry that we find in experiments depends on the definition of the standard being used and on the changes that standard undergoes as a result of the fields of force that exist in that space".

[400] **Helmholtz** (1870b), p. 131, and also **Helmholtz** (1870), p. 23.

[401] Cf. Section 6.3.1.5.

It is the only undistorted account of the world that satisfies the claim to being true knowledge of nature and is alone accessible for experimentation.

Overall, Helmholtz's foundation for geometry is one chapter in the success story of classical mechanism. Yet it is a chapter that apparently already contains elements of something along the lines of a modern conception of science. First of all, Helmholtz mentions the fact that real objects, whose features are supposed to constitute the basis of geometrical axioms, must not necessarily match the concepts from which he derives the traditional claim that geometry is true. Secondly, he acknowledges almost without reservation that there may be equivalent ways to represent physical space.

It is also one of the final chapters. The oncoming crisis that will evoke the doom of classical mechanism is already casting shadows. It has become less evident that mechanics applies to dimensions of what is smallest and it is no longer certain that all mechanical calculations can be based on absolute (Euclidean) space. Significantly, in his studies on space Helmholtz avoids the question of whether space is void or full of matter. Material bodies – of any dimension whatsoever – exist in space whose geometrical qualities are derived from the mechanical qualities of those bodies. But the idea that material things, as a consequence of their very presence, modify space and that physical states can be attributed to space itself, were already part of the theory of electrodynamics developed by Michael Faraday in Helmholtz's day.

6.3 Helmholtz's Classical Conception of Science and Nature

I have mentioned a number of preliminary reasons for considering Helmholtz's early conception of nature a variation of classical scientistic mechanism. In his treatise *Conservation of Force* Helmholtz develops a mechanistic program that claims to provide the one and only possible explanation of nature. He takes truth substantively: the final aim of theoretical science is the discovery of the "ultimate causes" to which all causally stated phenomena should be reducible and which Helmholtz defines as mechanical central forces. He concludes this claim to truth not from metaphysical assumptions but from experimentally obtained scientific knowledge, particularly from physics and chemistry. His program is impressively confirmed both by his explanation and investigation of energy-conserving processes and by the empirical foundations he defines for geometry. From the outset he is confident that science itself will empirically limit the scope of the causal method it presupposes.

Nevertheless, none of the texts examined thus far deal explicitly with the criteria that true knowledge must fulfill nor with the feature of scientific experience that Helmholtz uses to establish his kind of mechanism. However, in order to safely relate his mechanism to the classical conception of science it is necessary to investigate two things: do the truth criteria he demands satisfy the conditions of universality and completeness? Do his ideas of the differentiae and the establishing function of scientific experience satisfy the criteria for scientistic legitimation laid out by the early modern line of mechanism? I mentioned that for scientism the differentia of the empiricist basis of legitimation are its experimental context, the quantifiable form of empirical data and the concept of progress guided by final goals.[402] The question now is whether Helmholtz, in a comparable way, distinguishes scientific experience from other (particularly everyday) kinds of experience, to which extent he makes scientific experience the standard for the validity of scientific and particularly mechanical statements, what kind of validity he attributes to those statements and finally, whether he formulates a goal for the advance of knowledge that is both achievable and satisfies the classical demands.

These conceptual definitions for Helmholtz's early mechanism describe the starting point for the further development of his mechanism, which is what I intend to characterize. Slightly ahead of what is to come, for now it can be said that for the most part the process of transition beginning in the 1870s is related to distinctions that Helmholtz made in the early phase of developing his mechanism. That is true of both the contents of definitions in his dual mechanism and the formal aspects of his scientistic arguments. A peculiarity of the entire development, however, is that corrections of content are few compared to corrections of form. Helmholtz does not abandon his mechanism; he changes its kind of validity. Within the setting of growing critique of mechanism in scientific circles of the second half of the nineteenth century, changing the concept of science also salvaged the theory: it allowed Helmholtz to retain mechanism in a trade-off for the absolute claim to truth, or – in keeping with the metaphor – for a reduction of its truth value.

6.3.1 Helmholtz's Conception of Science up to the Late 1860s

When reading Helmholtz's lectures and speeches given until the late 1860s one notices that while his thought remains within the realm of the classical conception of science, it is already characterized by continuous change.[403] Without his conception

---

[402] See Chapter 2.

[403] Most developmental aspects of Helmholtz's life and work discussed in various places in secondary literature do not concern the changes in his conception of science within this period. **Hatfield** (1994), p. 547 ff., alone points out that the modification of Helmholtz's definition of the relation between natural science and the humanities begins in the 1850s with his work on the *Treatise on Physiological Optics* (**Helmholtz** (1856 ff.)). **Vogel** (1994) also sees changes in Helmholtz's theory of perception as related to his work on the *Treatise,* but these would only have had consequences for his theory of science if Helmholtz had, at that time, already understood his theory of sensory perception as scientific epistemology. **Darrigol** (1994) mentions merely that the transition he noticed in Helmholtz's conceptions of the "comprehensibility of nature" began before the late 1860s. **Krüger** (1994) – also in terms of the "comprehensibility of nature" – and **Schiemann** (1994) outline only the beginning of how Helmholtz's mechanism developed and its end. In other words: a detailed reconstruction of that development is still wanting. **Heidelberger** (1994a), p. 474 ff., and (1994b), p. 179 ff., assumes that Helmholtz's conception of science first begins to change as of 1869. For further secondary literature on Helmholtz's conception of science see footnote 24 in the introduction to this book.

of science – presupposed from the onset – itself becoming questionable, traces of a struggle with its argumentation are noticeable, a struggle that does not end until the late 1860s. It is in connection with a finding that Helmholtz proudly presented in an opening speech (mentioned above) for a convention of natural researchers and doctors in Innsbruck in 1869 and that harbingered the subsequent jolt in the classical conception not in the least. On the contrary, finally mechanism's claim to truth seemed legitimate because it was based on scientific knowledge.

This pursuance of the "classical" line of thought can be roughly split into two phases: the first phase begins with the *habilitation* lecture held in Königsberg in 1852, when he distinguishes the objectivity of scientific knowledge as being the opposite of the subjective witness of human sensory perceptions. The latter are merely "symbols" for a reality with which they have absolutely no similarity and that can only be known through science. Somewhat later Helmholtz corroborates that opinion in his critique of Goethe's studies on nature (Section 6.3.1.1). At the onset of the 1860s the contrast between objective and subjective knowledge had become the basis of his first more encompassing statement on the philosophy of science: in a speech titled *On the Relation of Natural Science to Science in General* [*Ueber das Verhältnis der Naturwissenschaften zur Gesammtheit der Wissenschaften*] he compares the methods of the humanities to the subjective processing of sensory perceptions, in order to then contrast them with the logical method of the natural sciences. Within this context he introduces, as basic concepts of his conception of natural science, the concepts of law, "logical induction" and experiment. He deems complete knowledge of nature the intent of all research. (Section 6.3.1.2).

The beginning of the *second* phase is less easy to date. Overall it is marked by a partial retraction of the opposition between scientific knowledge and subjective witness of sensory perception. Helmholtz's elaboration of causality in the lecture *On Human Vision* [*Über das Sehen des Menschen*] given in 1855 can be read as heralding the improvement (Section 6.3.1.3). In the last text delivered for his *Treatise* and which appeared in 1867, he depreciates the contribution that logical induction makes to knowledge, if by it one means merely making generalizations about what is discovered through observation. Compared to mere observation he now stresses the importance of experimental procedure by defining it as the sole adequate use of causal natural research. This cancels the distinction he had made in 1847 between effective causes and experimentally identifiable laws.[404] Now not only mechanically moved matter and central forces but empirical laws can also be seen as causes of phenomena. Nevertheless, the range of causality that has now come to the fore as the guaranty for the claim to truth and that is limited by empiricism remains the same (Section 6.3.1.4). For in Helmholtz's realism causality is not the sole guarantor of truth. In the *Treatise* he perfects the appreciation of sensory perception by, for the first time, attributing to it not only the capacity to symbolically reproduce the world as it is but also to represent it in a clear and undistorted manner. Admittedly, there is some indication that the representation relation cannot be

[404] Cf. Section 6.1.1, Part α.

recognized without employing experimental method. Without science the witness of sensory perception only satisfies a pragmatic concept of truth (Section 6.3.1.5). Later Helmholtz rephrased the opposition between sensory perception and scientific knowledge by contrasting signs with images and then linking the formal criteria of his concept of science with the content definitions of his conception of nature (Section 6.3.1.6). The latter will be the topic of the next section.

6.3.1.1 Scientific Objectivity Versus Subjective Perception

Helmholtz's *habilitation* lecture delivered in 1852[405] *On the Nature of Human Sensory Perception* [*Ueber die Natur der menschlichen Sinnesempfindungen*] is devoted to the problem of truth, just as he thereafter repeatedly made truth the focus of his reflections on the theory of the physiology of perception. He inquires as to "how at all our sensory perceptions can correspond to the objects we perceive".[406] The answer is an initial elaboration of Helmholtz's so-called theory of signs – a theory of the interpretation of sensory perception's physiological and psychological processes. Helmholtz basically maintained this theory all of his life without changing it; it is of fundamental importance for his position in the philosophy of science.[407]

Using vision as an example, he explains his initial assumptions: human perception attains no objective image of real objects. The objective image is the one identified by science:

> In answer to the question 'what is light?' recent physics asserts that it is the quivering, undulating propagation of a hypothetical medium called light ether [...]. The assumption that light consists of vibrating motion is [...] indeed not an arbitrary hypothesis of physics [...] What we have is the fact that light moves not simply as direct rays [currents] but as alternating undulation.[408]

Not only does human light perception know nothing of these physical properties, it also depends on special physiological factors resulting in the circumstance that

[405] This lecture (first published in *Königsberger naturwissenschaftlichen Unterhaltungen* (1852), p. 1 ff.) was not included in the collection of *Popular Lectures on Scientific Subjects* [*Populäre wissenschaftliche Vorträge*] (**Helmholtz** (1865), (1871) and (1876), partly translated in **Helmholtz** (1895)) and was not reprinted until 1883 in Vol. 2 of the *Scientific Treatises* [*Wissenschaftliche Abhandlungen*] (**Helmholtz** (1882 ff.)).

[406] **Helmholtz** (1852), p. 593.

[407] Helmholtz explained his theory of signs [*Zeichentheorie*] both in *Treatise on Physiological Optics* (**Helmholtz** (1856 ff.), p. 193 ff., 442 ff. and 797 ff., and **Helmholtz** (1885 ff.), p. 576 ff.) and in popular speeches (especially **Helmholtz** (1868a)).

The theory of signs is part of Helmholtz's theory of perception that covers the entire physiology of it and which has therefore been discussed particularly in connection with work on the theory of perception. **Conrat** (1904), **Erdmann** (1921) and **Hatfield** (1990) are fundamental. **Lenoir** (1994), **Turner** (1994a), **Turner** (1994b) and **Vogel** (1994) provide insight into the current state of research in the history of science for this field.

[408] **Helmholtz** (1852), p. 593 ff.

"light sensation in general corresponds neither in scope nor quality to the external object, namely light".[409] This is confirmed by comparing the various human senses, each of which reacts in its own way to external stimulation. Thus the senses of touch and vision experience heat rays each in its own specific way, the one as warmth, the other as light.[410]

In his theory of signs, which in this lecture is confined to discussing sensations of light and color, he extends the difference between subjective sensation and objective identification of the physical object to cover all five senses: he postulates a modality of sensation for each sense that excludes any transition among the senses and is entirely independent on external stimulation. Thus the eye, even when struck or stimulated electrically, always only sees "light" and not whatever exists objectively in reality.[411]

The reasons Helmholtz gives in this lecture for understanding the subjectivity of visual perception in this way are remarkably – and I shall return to this point later[412] – in line with the vitalist tradition and attribute the nature of each of the five kinds of sensation to the specific features of the sensory nerves affected. It is the "law of specific nerve energies" that Helmholtz adopts from his teacher Johannes Müller:

> Light sensations exist that are not produced by light stimulation and light exists that cannot produce the sensation of light [...]. Thus we can conclude irrefutably, as Müller did, that the idiosyncrasies of light sensation [...] depend on the special activity of the optic nerves.[413]

The central statement of his semiotic theory is that perceptions are subjective due to the constitution of the relative nerves. Since they are essentially different from the perceived objects, which are described objectively by science, Helmholtz calls sensations and the perceptions related to them "symbols" and compares them to the signs used in language.[414]

Now, it is crucial for the development of his conception of science that he defined the relation between perceptions and perceived objects in very different ways. His *habilitation* lecture reduces that relation to a minimal connection. Not only are the two not similar, the one also does not represent the other. The only relation that Helmholtz (very vaguely) admits is one of equality or inequality such that in a certain way the manifold and presumably changes in reality, too, are linked to the world of symbols:

> [Sensations] are only symbols for relations in reality; they have with the latter as little or as much similarity or relationship as the name of a person or the way that name is written has do with that person himself. Whether their appearance is the same or different tells us whether we are perceiving the same object and properties or different objects and properties,

---

409 **Helmholtz** (1852), p. 606. In this lecture Helmholtz does not clearly distinguish perception from sensation. For this distinction see Section 6.3.1.3.

410 **Helmholtz** (1852), p. 605 f.

411 **Helmholtz** (1852), p. 602 ff., and later e.g., in **Helmholtz** (1869), p. 442.

412 Cf. Section 6.3.2.2.

413 **Helmholtz** (1852), p. 605. Cf. e.g. **Helmholtz** (1856 ff.), p. 233, and in several other places.

414 **Helmholtz** (1852), p. 608 f.

> just as when in stories about unfamiliar people and cities we know from the same or different names, whether the narrator is talking about the same or different ones. But that is all they do. From sensations we learn just as little about the real nature of the external circumstances designated by them as we learn about unfamiliar people and cities by hearing their names.[415]

At any rate, the diversity of reality is not reduced in sensation. While one name can be used to designate different people, different objects or events should be represented by different corresponding sensation signs. Helmholtz wants to exclude another possibility, namely that one and the same object or event might be given different names by different individuals who perceive it; he does this by clearly limiting the analogy between sensation signs and language. He explicitly stresses that sensation signs are not determined by convention, as holds for the signs of language but by the natural physical constitution shared by all humans. That common natural basis explains why people – in spite of the symbolic nature of their sensations – navigate their way so well in the world.[416] But the vague relation between symbol and object is by no means sufficient for knowledge of reality. Reality remains alien and inaccessible to everyday perception. What people, who, for the practical purposes of everyday life rely successfully entirely on perception that has proven trustworthy for those purposes, believe to be the real world, is merely a world of symbols that is not necessarily connected either causally or in terms of correspondence in time to the reality of which it is not a part.

That this is the case is something known by science, which has access to reality not dependent on the witness of sensory perception. Helmholtz continues:

> And the physicist, who teaches these relations of reality through a different indirect method takes on the part of the one who seeks to describe to us and teach us how those people and cities look and are [...]. Imagine [...] further, how our ideas of the perceptual world would be without the symbolism of our senses, if we were capable of directly perceiving what the physicist approximates in a long chain of conclusions, namely everywhere nothing but always the same uniform action of attracting and repulsing molecular forces, no diversity other than the meager variation of numerical proportions, no light, no color, no sound, no warmth.[417]

The world of the physicist can also be understood as a world of signs: mathematical signs and equations, names for invisible molecular substances, geometrical movements of mass points in the dark and cold void–this is the bleak world of mechanism allegedly at the bottom of all natural phenomena. However, the symbols of the physicist do more than those of the senses. They represent knowledge of the causes and permit, according to Helmholtz, a "description" of that reality unattainably remote from everyday knowledge. The physicist is like a narrator who not only knows the correct, yet unfamiliar names of foreign cities but has also received so many reports that he has a realistic picture of what they are like.

---

[415] **Helmholtz** (1852), p. 608 (cf. analogously **Helmholtz** (1853), p. 41 f.).

[416] **Helmholtz** (1852), p. 608 f. This naturalistic explanation will later be replaced by a theory of learnable "indirect inferences", cf. Section 7.1.3.2.

[417] Ibid.

This still very rudimentary characterization of science already exhibits the main features of Helmholtz's "classical" conception of science, which will remain definitive throughout the second phase: scientific knowledge is the sole objective knowledge and legitimizes its claim to truth by being strictly distinct from sensory perception witness. At the end of the 1860s Helmholtz continues to insist that an "idea [i.e., a perception–G.S.] and the thing conceived evidently belong to two entirely different worlds".[418] On the one hand, characteristic of the first phase is the lack of a representational relation between sensation and the external world. On the other hand, Helmholtz neglects the difference between objects and their properties. He assumes as a matter of course that as individual items objects are accessible via their properties. In the second phase this one-sided reference to the object will become less important than the description of relations that hold between the properties of objects.

I would not like to conclude this discussion of Helmholtz's first public speech without mentioning the correspondence between him and his father that in the history of Helmholtz studies has often given occasion to wonder how much Helmholtz's ideas were influenced by Johann Gottlieb Fichte's philosophy.[419] His biographer Leo Koenigsberger says (source unconfirmed) that Helmholtz wrote to his father, an ardent follower of Fichte's philosophy, of his intention "to empirically illustrate Fichte's fundamental view of sensory perception".[420]

Just as Helmholtz mentions no such task in the lecture in question, neither do his subsequent publications contain any indication of solving it.[421] In addition, for the period in question – that is, up until the late 1860s – he mentions the philosopher only on the rarest of occasions. Only once does he claim that his physiological investigations coincide with Fichte's philosophy, without stating a reason for it.[422] In the late 1860s, with barely concealed hostility, he mentions Fichte in connection with Hegel and Schelling: their "idealistic systems" had had "very little import on the theory of sensory perception".[423] An explicit, albeit not elaborate, struggle with Fichte's theory does not take place until the 1870s, when Helmholtz becomes aware of the problems of his own realistic position and contrasts it as an equally justified and opposing alternative to an idealistic position. Because that contention is important for the development of the conception of science, I shall return to it later.[424]

Soon after giving his *habilitation* lecture Helmholtz sharpens (one could say: expounds the problems of) the contradistinction between scientific knowledge and sensory perception by no longer merely claiming that science has superior knowledge of

[418] **Helmholtz** (1856 ff.), p. 443.

[419] **Conrat** (1904), p. 251 ff., **Erdmann** (1921), p. 42, **Hörz and Wollgast** (1971), p. XXXVII, **Meyering** (1989), p. 117, **Turner** (1977), p. 57, **Heidelberger** (1994a), p. 463. Helmholtz's explicit references to Fichte have been documented in Chapter 5, footnote 243.

[420] **Koenigsberger** (1902 f.), Vol. 1, p. 169.

[421] Thus **Erdmann** (1921), p. 42, rightly notes that the intention expressed to his father "probably remained just a plan".

[422] **Helmholtz** (1855), p. 89.

[423] **Helmholtz** (1856 ff.), p. 456.

[424] Cf. Section 7.1.3.5.

reality. He now explicitly includes the critical and wholehearted investigation into sensory perception among the tasks of science. He himself delivers that critique at a meta-level of reflection on the theory of science. In a speech given in 1853 *On Goethe's Scientific Researches* [*Ueber Goethe's naturwissenschaftliche Arbeiten*] he rightly describes Goethe as one of the investigators into nature who perform science "without leaving the sphere of sensory perception".[425] Such a position is unacceptable for physical science:

> This demand looks most attractive but is essentially wrong in principle. For a natural phenomenon is not considered in physical science to be fully explained until you have traced it back to the ultimate forces which are concerned in its production and its maintenance. Now, as we can never become cognizant of forces *qua* forces but only of their effects, we are compelled in every explanation of natural phenomena to leave the sphere of sense and to pass to things which are not objects of sense and are defined only by abstract conceptions [...].
>
> [But to the natural philosopher] the impressions of sense are not an irrefragable authority; he examines what claim they have to be trusted; he asks whether things which they pronounce alike are really alike and whether things which they pronounce different are really different; and often finds that he must answer, no![426]

In *Conservation of Force* Helmholtz detailed which "ultimate forces [...] concerned in its [...] maintenance" are those that this definition for the concept of explanation, so typical of mechanism, addresses: mechanical forces that operate along the connecting line between any two "material points", the intensity of which depends exclusively on the distance between those points. The search for these central forces, which alone determine what is comprehensible in nature, does not require entering "the sphere of the senses" but instead leaving it. The physicist does not do so without first convincing himself – in thorough examination – that taking that step is right.

Here Helmholtz not only rejects the exclusive employment of sensory perception for investigating nature. He also assumes that it is possible, at least to a limited extent, to conduct research in physics without recourse to sense data. Nonetheless it would be absurd to think that by abandoning what is given in sensation he advocates a metaphysical foundation for science, i.e., one that starts with pure thinking. For according to Helmholtz, the effects of the forces to which phenomena are to be reduced are also "perceived". The context of the speech readily indicates what he means by scientific research independent of sense data. While the theme of the speech is the contrast between Goethe's theory of color and physical optics based on Newton, the methodological issues of the controversy concern the status of experiments. Helmholtz resolutely pleads the case for experiments, which Goethe dismisses. Only the experimental procedure of physics allows the observation of elementary force effects, making it indispensable for science's claim to objectivity.[427] Significantly, he neglects mentioning that no experiment is accomplished without

[425] **Helmholtz** (1853), p. 40.

[426] **Helmholtz** (1853), p. 40f. (Germ.)/13 (Engl.).

[427] **Helmholtz** (1853), p. 34ff.

sensory perception, that in every measuring of physical dimensions clocks and standards ultimately need to be read. Thus in the end, Helmholtz's critique denies that measuring procedures essentially rely on sensory perceptions.

6.3.1.2 Logical Versus Aesthetic Induction

Helmholtz's effort to distinguish scientific experience serves to scientistically justify the claim to truth expressed programmatically in his definition of scientific explanation as given above. Each individual phenomenon should be reducible, with certainty, to a finite number of mechanical causes. Nevertheless, not only does the classical conception demand absolute validity for individual explanations but in addition, the entirety of phenomena must fit into a system of natural explanation that since early modern times is expected to be, in principle, attainable by human understanding.

Based on the distinction between scientific knowledge and insight established only through immediate sense perception, in an academic speech *On the Relation of Natural Science to Science in General*, delivered in 1862 on the occasion of taking up the pro-rectorate in Heidelberg, Helmholtz publicly[428] professes his pursuit of classical science's all-encompassing aim. In this speech for the first time he expresses his views on issues of the theory of science in more detail and summarizes science's overall objective:

> Whoever, in the pursuit of science, seeks after immediate practical utility, may generally rest assured that he will seek in vain. All that science can achieve is a complete knowledge [*Kenntnis*] and a complete understanding of the action of natural and mental forces.[429]

Helmholtz wants less to restrict the sciences to one theoretical objective, than to join the theoretical with the practical aspect in line with Bacon's dictum "knowledge is power", to which he specifically refers.[430] For the task of producing comprehensive knowledge about both nature and the mind can only be considered completed when in addition every challenge of the practical application of scientific knowledge has been resolved. Consequently, the aim is not the selective utilization of scientific findings elaborated specifically for that purpose but rather science's complete supremacy over the world.[431] From this it follows retrospectively that the validity of each individual finding cannot lie in its application and success. Neither Bacon nor Helmholtz is a pragmatist. Helmholtz is "convinced" that utility aside, "whatever contributes to the knowledge of the forces of nature or the powers of the human mind is worth cherishing".[432]

[428] This kind of commitment is already contained in **Helmholtz** (1847a).

[429] **Helmholtz** (1862), p. 182 (Germ.)/93 (Engl.).

[430] **Helmholtz** (1862), p. 180.

[431] **Helmholtz** (1862), p. 183.

[432] **Helmholtz** (1862), p. 182.

The text does not disclose whether science's overall aim is a realistic task or an ideal benchmark that science pursues in an unending progress process. However, in the program of reduction devised in *Conservation of Force,* there is almost no doubt that he is thinking of the possibility of a complete system of knowledge based on natural science. The problem seems rather to be how to determine the boundaries of such a system. For natural forces can only be conceived of as being causal. In *Conservation of Force* he left the range of validity for causality open, in order to account for spontaneity in nature.[433] In subsequent publications Helmholtz – I shall return to this later[434] – maintains that the range within which laws of nature hold can be determined empirically. He assumes that it is ultimately only known in cases where phenomena have been reduced to the effects of mechanically moved matter. Thus one must concede that he hardly does justice to his own position, when in the lecture of 1862 he says:

> Nature does not allow us for a moment to doubt that we have to do with a rigid chain of cause and effect, admitting of no exceptions.[435]

According to Helmholtz the moral sciences differ from natural science by the fact that the phenomena they elaborate are the exact opposite of causal nature. "The moral sciences deal directly with the nearest and dearest interest of the human mind and with the institutions it has brought into being".[436] We as humans have the "full power to act, without being subject to a stern inevitable law of causality".[437] This creates a tension between two extremes when classifying the subjects of science, namely causality on the one hand and free will on the other. Classification falters, however, when the relation between causal determination and spontaneity has not (yet) been settled, specifically regarding certain phenomena of human corporality and of nonhuman organisms. Helmholtz's remarks on topics of natural history are accordingly vague; he sees them as situated somewhere between the moral and the natural sciences.[438] A closer look reveals, however, that Helmholtz believes it impossible to scientifically investigate the "mental faculties" of either humans or animals.[439]

He tries to methodologically accommodate the basic difference between the two kinds of academic study by describing what he considers the "formal difference"[440] between the moral and the natural sciences as two opposite[441] kinds of procedures, namely the method of "aesthetic induction" and that of "logical induction".

---

433 See Section 6.1.1, Part β.

434 On Helmholtz's concept of nature see Section 6.3.2, especially Section 6.3.2.2.

435 **Helmholtz** (1862), p. 178 (Germ.)/90 (Engl.), cf. also p. 166.

436 **Helmholtz** (1862), p. 166.

437 **Helmholtz** (1862), p. 171, cf. also p. 174.

438 **Helmholtz** (1862), p. 175 (as Section 7.1, footnote 653).

439 Cf. **Helmholtz** (1868a), p. 362 (as Section 6.3.2, footnote 586).

440 **Helmholtz** (1862), p. 166.

441 **Helmholtz** (1862), p. 165 and 171.

He calls the one kind of induction aesthetic "because it is most conspicuous in the higher class of works of art".[442] Not only does it "play a leading part in the investigation of psychological processes", which explains why it is constitutive for the moral sciences but also the whole formation of "our perceptions depends upon it".[443] In contrast, physics, being the science that has progressed the farthest, has developed the method of logical scientific induction the furthest[444] and must abstract from the sensory appearances.[445] The distinct opposition of acquaintance through sensory perception on the one hand and knowledge derived by science on the other was thus cemented into the institutional classification of academic endeavor in a way that will mark Helmholtz's conception of science all his life.

In this speech it sets the stage for a definition of science, which, of course, only physical science can meet:

> It is not enough to be acquainted with the facts; scientific knowledge begins only when their laws and their causes are unveiled.[446]

Helmholtz defines a law, not to be confused with unobservable causes, as a "concept" that refers exclusively to phenomenal reality, i.e., to appearances accessible by natural science. It is in this sense of the word that I shall speak of the phenomenal meaning of the concept of law. According to Helmholtz, a law unites "a vast number of single instances and represents them in [...] memory".[447] But laws not only make thought more economical by facilitating a memory of all observed cases of a certain kind.[448] They may also allegedly be extended "to all similar cases that may be hereafter presented to us".[449] In other words, a law covers "an infinite number of cases".[450] Its defining feature is "exceptionalness"[451] or "unconditional validity".[452]

Naturally, such a concept of law is inapplicable to a sphere of the mind defined by free will (with "full power to act"). For "the experimental sciences to which mathematics are applied" though, this concept constitutes the crucial criterion:

> The essential differentia of these [the experimental sciences to which mathematics are applied – G.S.] sciences seems to me to consist in the comparative ease with which the individual results of observation and experiment are combined to render general laws of

---

[442] **Helmholtz** (1862), p. 171.

[443] Ibid.

[444] **Helmholtz** (1862), p. 175 ff.

[445] **Helmholtz** (1862), p. 165.

[446] **Helmholtz** (1862), p. 169 (Germ.)/83 (Engl.).

[447] Ibid.

[448] **König** (1968), p. 95.

[449] **Helmholtz** (1862), p. 170.

[450] **Helmholtz** (1862), p. 169.

[451] **Helmholtz** (1862), p. 171.

[452] **Helmholtz** (1862), p. 175. For Helmholtz's concept of law see also Section 6.1.1, Part α, and in general **Röseberg** (1994).

> unexceptionable validity and of an extraordinarily comprehensive character. In the moral sciences [...] this is just the point where insuperable difficulties are encountered. In mathematics the general propositions [...] are so few in number, so comprehensive and so immediately obvious, that no proof whatever is needed for them.[453]

The inductive method in the natural sciences yields laws that in terms of generality and logical stringency are no different than mathematical theorems. By calling the method used by science "logical induction", Helmholtz accommodates the smooth transition from induction to logic. Scientific induction, he says in another passage, can "be perfectly assimilated to forms of logical reasoning".[454] Logical structure gives laws an evidently avouched truth similar to that of mathematical theorems, which in turn expresses the sciences' classical claim to validity most clearly and declares it a privilege of natural research.

The secret of the ease with which the inductive method proceeds from individual cases to laws rests – as Helmholtz explicitly says – on consistently restricting research to the conducting of experiments:

> The experimental sciences have one great advantage over the observational sciences in the investigation of general laws of nature: they can change at pleasure the conditions under which a given result takes place and can thus confine themselves to a small number of characteristic instances, in order to discover the law. Of course its validity must then stand the test of application to more complex cases.[455]

Unity of induction and logic manifests itself in natural science's conduct of experiments in the Baconian tradition. Induction, which normally implies generalizing from a large number of individual cases, is reduced to the purposeful analysis of a few technically prepared incidents that quasi materialize the logic of the laws of nature. The infinite number of cases that are covered by a generally valid law is reduced to the variation of a single condition (variable) under otherwise constant, fixed circumstances. One single typical experiment set-up reflects the general law.

This ideal depiction of experimental practice, however, is not based on the notion that laws – like devised experimental equipment – are an invention of the human mind. For Helmholtz, the conducting of experiments does not generate laws, it simply "discovers" them and the chance of error in the search for laws lies in thinking that something is a law that in reality is not. In contrast to the modern understanding of science, experimental testing of the validity of laws served the sole purpose of eliminating such mistakes. Since testing is not essential to a law, it is only relevant for "complex cases" where the regularities that obtain are still unclear.

Although the "first general theorems" of mathematics cannot be clearly separated from inductive generalizations, he does not see the basis of their validity in experience but says they are "immediately evident" and therefore require no proof. The mathematical paradigm he has in mind appears to be arithmetic. Only arithmetic is "pure". He finds it worthy to mention that arithmetic is based on only three axioms

---

[453] **Helmholtz** (1862), p. 175 (Germ.)/88 (Engl.).

[454] **Helmholtz** (1862), p. 171.

[455] **Helmholtz** (1862), p. 177 (Germ.)/89 (Engl.).

(adding what is equal, unequal and transitivity). In a very superficial comparison, involving only the number of fundamental theorems, he includes geometry and theoretical mechanics in this scheme: neither are their axioms "more numerous" than those of arithmetic.[456]

Until the mid-1860s this is the only statement – in the sense of a theory of science – that Helmholtz makes on geometry and theoretical mechanics in public lectures and speaking. It makes one think that he wanted to locate the axioms of these two disciplines near the evidently recognizable and secured truth represented exemplarily by arithmetic. Does this mean his foundation for science and conception of nature erected thereupon was not scientistic after all but metaphysical?

In my opinion this question cannot be indubitably answered in the negative but there are good reasons for doing so. First of all, they could only be metaphysics in a very weak sense of the word, namely as reasoning independent of experience. Neither does the second meaning, that Helmholtz also uses, namely the derivation of fundamental principles of science from pure thinking, apply to his theory of science. He deduces no synthetic propositions a priori but rather presupposes only certain methodical and conceptual definitions that one can interpret as independent of experience if one neglects empiricist limitations. Besides the axioms of arithmetic, another example of this would be the assumptions contained in *Conservation of Force*.[457]

Regarding geometry and mechanics one cannot say that he was thinking of anything more than mere proximity to mathematics. He conspicuously does not call them "pure" as he does arithmetic. This may indicate that – in contrast to arithmetic – he considered these two disciplines to be empirical sciences. If this is so, then he would be demarcating them from arithmetic not by the nature of their validity but by what that validity is founded on. This may be another harbinger for the empirical foundation of geometry.[458] But one must remember that in private comments made in the 1850s Helmholtz still tended towards a metaphysical rationale. Thus in 1857 he wrote in a letter to his father, five years prior to the speech in question:

> I myself feel a vivid need to work through certain questions in more detail, questions with which no more recent philosopher has occupied himself and which lie entirely in the field of a priori concepts researched roughly by Kant, for example the deduction of geometrical and mechanical first principles [...]. I realize fully that things like this can only be solved by philosophical investigations.[459]

More clearly than seeing the tasks of deduction as belonging to the "field of a priori concepts", the fact that he calls them "philosophical investigations" indicates that here Helmholtz is obviously thinking of metaphysical reasoning. At any rate, in the letters to his father from this time he still commences from a thoroughly Kantian

[456] **Helmholtz** (1862), p. 175 f.

[457] See Section 6.1.1, Part α.

[458] See Section 6.2.

[459] **Koenigsberger** (1902 f.), Vol. 1, p. 292 f. (Letter dated March 4, 1857).

concept of philosophy that as the "doctrine of the sources and the activities of knowledge" is clearly distinguished from empirical science.[460] Nonetheless, his interest in working out an entirely new foundation reveals that the traditional assumption that Euclidean axioms are valid prior to experience could no longer be taken for granted.

If, then, one would also take Helmholtz's private correspondence into consideration, one might find that in his speech *Relation of the Natural Science* Helmholtz's remark on geometry and theoretical mechanics indicates that he is in a period of transition from a metaphysical to an empirical conception of geometrical and mechanical axioms. Here his private opinion is not significant for evaluating his conception of science but rather his publicly expressed standpoint. Publications from the first phase of the development of his conception of science, particularly the Introduction to *Conservation of Force,* contain neither a mention nor an implication of a metaphysical foundation for these axioms. Geometrical axioms go unmentioned and the absolute validity of mechanical propositions seems to have been confirmed by experiments in natural research. In order to obtain the special definitions required for dual mechanism, Helmholtz also took recourse to the concept of the chemical element.

Reviewing the entire first phase, one discovers a series of definitions that with sufficient clarity allows one to characterize Helmholtz's position as classical and scientistic: by defining the concept of a law he demands rigorous universality and unconditional necessity for every single finding. The term hypothesis does not appear once in those definitions. For him the aim of natural science is to obtain a complete system of knowledge that in a first approximation is delimited against mental phenomena. With the special status of certain axioms aside, experience alone supports the validity of scientific knowledge. That experience is clearly distinguished from the witness of sensory perception and characterized in line with Baconian tradition as being experimental.

6.3.1.3 Perceptions as Signs of the External World and Noumenal Causality

By delimiting the natural sciences from the moral sciences Helmholtz heightened the contrast between experimental findings and merely symbolical knowledge, bringing it to an institutionally confirmed level. Scientific logical-inductive procedure is characterized by its exclusive access to reality, the existence of which Helmholtz presupposes independent of sensory perception. Within this realistic principle it has hitherto remained unexplained how natural science gains that access

---

[460] **Koenigsberger** (1902 f.), Vol. 1, p. 284 (Letter dated December 31, 1856). This distinction becomes blurred in other private correspondence of later years, addressed not only to his father. See for instance **Koenigsberger** (1902 f.), Vol. 2, p. 162 (Letter to Ludwig dated March 28, 1869); Vol. 1, p. 243 (Letter to Fick, probably in the year 1875); additionally **Lipschitz** (1986), p. 131 f. (Letter to Lipschitz dated March 2, 1881). For the non-Kantian content of the two last letters see the text for Chapter 5, footnotes 239 and 241.

to reality experimentally, if it does not rely on sensory perception and furthermore, why sensory perception, although it has merely a symbolic nature, may be interpreted – like scientific findings – as appearances of an external world.

In the second phase of his "classical period" Helmholtz will try to solve both problems by mitigating the contrast between experimentally obtained knowledge and the direct witness of sensory perception. While everyday knowledge and scientific experience close ranks, they remain sufficiently separate that demarcation is identifiable. Later, the development of the relation between these two different kinds of knowledge will reveal how Helmholtz's conception of science changed. Findings gleaned from the physiological and psychological investigation of everyday sensory perception, which will increasingly guide his future work, no longer satisfy criteria determined for the classical claim to truth.

Very early on one finds signs in Helmholtz that he is aware of the problematic aspects of his own realism. First traces can be found in his lecture *On Human Vision.* They deal with the origin and the significance of perceptual insight in relation to the outside world. In this lecture delivered by Helmholtz in 1855 "as the best possible monument for Kant",[461] he basically reports on the state of the art in the sensory physiological and psychological research of vision, which in his opinion is "the point where philosophy and natural science come closest".[462] What he means of course is Kant's philosophy, which he here mentions publicly for the first time and by all means positively.[463] Nevertheless, the lecture does not confirm the common ground shared, as he claims, by Kant's epistemology and the physiological research of the senses.

Compared to his speech on Goethe from 1852, in this lecture he discusses the problem of the realistic assumptions made in the theory of perception and science by acuminating the contrast between scientific knowledge and the witness of immediate sensory perception. For the first time he separates sensations from the perceptions related to them. Using the example of light once more he explains that light sensations on the retina of the eye originate in the part of the nervous system located there[464] and "are passed onto the brain by the fibers of the optical nerve".[465] In the brain a "psychic process" then "transforms light sensation into perception of the external world".[466] Perception takes place when the subject "knows of objects of the external world" via sensations.[467] While in this depiction sensations designate a purely physiological activity and are entirely subject to empirically knowable natural laws, perceptions are part of an inner experience for which Helmholtz, as

461 **Helmholtz** (1884a), p. XIII.

462 **Helmholtz** (1855), p. 90.

463 For the context regarding the history of science and philosophy see Chapter 5.

464 **Helmholtz** (1855), p. 94.

465 **Helmholtz** (1855), p. 99.

466 **Helmholtz** (1855), p. 111. Later, Helmholtz will claim that this process continued without delay. Cf. Section 6.3.1.5.

467 **Helmholtz** (1855), p. 99f.

early as this lecture, already assumes that they are subject to the free will that is irreducible to causality.[468] According to his later conception of science the investigation of perception would have to be both a task for natural science, which would study the influence of sensations, as well as a task for the moral sciences, which would be responsible for examining mental activities and the "aesthetic induction" at work in them.

Compared to the inaugural speech, this clearly emphasizes the sheer subjective nature of perception. No longer does this nature result from the physical constitution of the nerves but also from the way that sensations are psychologically processed to become perceptions. Since psychological activities are not subject to the laws of nature, perception is in principle not suitable as empirical material for natural research. It will take ten years before Helmholtz in *Treatise* abandons this position and rates perception as capable of scientifically utilizable truth. But that it is difficult to fundamentally deny perception any sense of objectivity is already clear in the lecture from 1855. Namely, as long as one of the assumptions is that reality exists independent of being known, the question remains of how a perceptive consciousness with only signs of that reality to go on has any justification at all for relating its perceptions to external objects.

The effort alone to find an answer to this question prompts Helmholtz to include Kant in closing the lecture, although the problem that Helmholtz the realist suggests cannot even arise in Kant's transcendental philosophy:

> We must assume that the presence of external objects is the *cause* of nerve exquotation; for there can be no effect without a cause. We need [the law of causality] in order to even know that objects are in space around us, between which [then – G.S.] there can be [thus not: must be – G.S.] a relation of cause and effect [...]
>
> The study of sense perception also leads us to Kant's insight that the claim 'no effect without a cause' must be a law of thought that precedes all experience.[469]

Other than in *Conservation of Force* and the rectorate speech, here Helmholtz for the first time uses the law of causality in an entirely non-Kantian way. Previously he had viewed causality as solely concerning objects of natural research,[470] which he would have been able to think of as objects in an entirely incognizable reality, without that changing the contents of his statements. While he hitherto had not had to distinguish between an independently existing external world and the specific manner in which it is given, now he is interested in just that distinction. It is that difference that is to be bridged by applying Kant's principle of causality, such that the entirety of what is subjectively given can, prior to any definition, be interpreted as something that has been created causally. In this interpretation that lies beyond all spatio-temporal circumstances and for modern philosophy can be traced back to Descartes, causality serves neither as a guideline for empirical investigation nor does it provide reasons for the assumption of an external world.

[468] **Helmholtz** (1855), p. 116.

[469] Ibid.

[470] **Helmholtz** (1847a), p. 4 f., and **Helmholtz** (1862), p. 178 f.

Therefore it does not nullify realism by making it depend on a knowing subject who postulates the external world through causal reasoning. It merely uses causality to retrospectively justify a realistic attitude that cannot be based on conclusive argument.

When reconstituting Helmholtz's conception of science it is wise to clearly distinguish from one another the two functions of his concept of causality that he himself never parted. Following Kant's terminology I call the idea of causality applied to objects of natural research "phenomenal" causality. It demands that natural science think of phenomena as being unequivocally determined by preceding causes. Helmholtz spoke this way in the Introduction to *Conservation of Force*[471] and the rectorate lecture. In contrast, I call the notion of causality he used to legitimize realism "noumenal" causality.

While in *Conservation of Force* Helmholtz empirically restricted the scope of phenomenal causality by assuming that operative spontaneity is possible in nature and that ultimate causes exist, he now speaks in connection with noumenal causality much more widely as "a law [...] that precedes all experience". In contrast to Kant, Helmholtz once again does not establish apriority by arguing that causality, as a prerequisite of the possibility of experience, cannot itself be an object of experience. His argument is that the assumption of an external world is evident and this precedes any causal relations that (merely possibly) exist among phenomena.[472] He also indicates an empirical limit to the validity of noumenal causality: it is one of the "innate laws of thought".[473] Subsequently, Helmholtz extends this naturalistic definition, to which I shall return later, to become a "drive of human reason".

6.3.1.4 Laws as Causes and Limits of Phenomenal Causality

The crucial step towards retracting the opposition between scientific knowledge and the immediate witness of sense perception can be found in *§26 On Perceptions in General,* in the third part of the delivery of 1867 for the *Treatise*.[474] In theme, this paragraph takes up where the Kant lecture of 1855 left off. Helmholtz provides a first systematic account of his scientific theory of perception that still remains entirely within the framework of the opposition. Perception retains its principally subjective nature and the criteria of strict universality and completeness remain

[471] **Helmholtz** (1847a), p. 3 f., see Section 6.1.1, Part α.

[472] **Helmholtz** (1855), p. 116.

[473] **Helmholtz** (1855), p. 116.

[474] The *Treatise on Physiological Optics* appeared in three different shipments, in which the last pages of the delivery of 1860 already contained the first pages of §26 (**Helmholtz** (1856 ff.), p. 427–432). This means that the publication of this one paragraph, where Helmholtz provides his first comprehensive account of his theory of perception, respectively, signs, spanned a period of several years. The greater part of §26 was considerably revised for the second edition of the treatise (**Helmholtz** (1885 ff.), p. 576 ff.). From a total of 31 pages in the first edition, only approximately 12 were reprinted in the second. The continuities and differences existing between the two editions provide ample material for analyzing the development of Helmholtz's theory of perception and science (see Section 7.1.3.1, Excursus). For secondary literature see footnote 407.

valid for scientific knowledge, whose claim to objectivity has just been confirmed by the investigation of sensation and perception. What changes are not the criteria for science but the definitions of scientific method that are supposed to guarantee adherence to the criteria.

In order to retrace this correction I begin with the problem of induction as Helmholtz discusses it for the first time in the *Treatise*. It cannot be ruled out that his reflections were influenced by contemporary German and British debates in the philosophy of science. Here one finds his most important reference to John Stuart Mill, whose logic he considers "the best" and apparently novel "analysis" of the essence of inductive inferences.[475] Up to this point Helmholtz had considered "logical induction" a privilege of natural science. Now he explains that in sense perception the same thing happens unconsciously, as in inductive logic, where it is done "with thought and scrutiny": new experiences are accrued to a sum of previous experiences.[476] Exemplary of this is once more the unconscious processing of light sensation:

> Inasmuch as in an overwhelming majority of cases, whenever the parts of the retina in the outer corner of the eyes are stimulated, it has been found to be due to external light coming into the eye from the direction of the bridge of the nose, the inference we make is that it is so in every new case whenever this part of the retina is stimulated; just as we assert that every single individual now living will die, because all previous experience has shown that all men who were formerly alive have died.[477]

But to the extent that inductive logic does not fundamentally differ from natural perceptive activities, neither can it guarantee the claim that scientific knowledge about nature is true. This holds particularly for knowledge gained merely by passive observation and about which Helmholtz now, compared to his rectorate lecture from 1862, is much more skeptical. He points out that sense perception occasionally involves illusions and stresses that insofar as they rest on observation alone, inferences arrived at inductively have no incontestable claim to truth.[478] Now the experiment emerges as the decisive guarantee for the reliability of inductively gained knowledge. The experiment's two features – arbitrary manipulation of the conditions of observation and the reduction of instances observed[479] – permit the discovery of the causal relations at work in nature. In contrast to Mill, who sees causality as the characteristic outcome of applying the inductive method,[480] Helmholtz tries to establish induction based on causality:

> The peculiar ultimate basis, which gives convincing power to all our conscious inductions, is the law of causation. […] so long as we are limited to mere observations of such phenomena as occur by themselves without our help and without our being able to make

[475] **Helmholtz** (1856ff.), p. 447. On the influence of Mill's empiricism cf. Chapter 5, footnote 247 (contains an overview of all references to Mill by Helmholtz) and Chapter 8.

[476] **Helmholtz** (1856ff.), p. 448, **Helmholtz** (1868a), p. 358.

[477] **Helmholtz** (1856ff.), p. 430 (Germ.)/4 (Engl.). A more detailed characterization of these conclusions is given in Section 7.1.3.2, especially footnote 695.

[478] **Helmholtz** (1856ff.), p. 450f.

[479] **Helmholtz** (1856ff.), p. 452, cf. also **Helmholtz** (1862), p. 177, and here Section 6.3.1.1.

[480] **Mill** (1869ff.), Vol. 3, p. 299ff.

> experiments so as to vary complex causes, it is difficult to be sure that we have really ascertained all the factors that may have some influence on the result [of the experiment – G.S.].[481]

While in his treatise *Conservation of Force* Helmholtz still rigorously distinguished the conducting of experiments from the theoretical study of causes and then only later rudimentarily alluded to the relevance of experiments,[482] he is now convinced that causes, too, can be ascertained experimentally.[483] In 1847 he defined causes in two ways: on the one hand he understood all physical phenomena as the effects of forces, which ultimately rest on mechanical forces of motion. On the other, he considered elementary forces only conceivable as effects related to a material substrate. Experimentally ascertained laws of nature had to be explained in terms of matter and force, as he himself had done exemplarily for the theorem of the conservation of energy. If he now places the research of causes in the context of experimental method, then he no longer means matter and its associated forces when he speaks of causes. In fact, redefining the concept of cause, to which Helmholtz adhered all of his life, permits seeing laws as causes. The causal constitution of nature coincides with its lawfulness:

> But since we have to assume the laws of nature as being valid and as acting independently of our observation and thinking, whereas as generic notions they would concern at first only the order of our thinking, we call them *causes* and *forces*.[484]

This wording covers the possibility of forgoing material causes and with them a mechanistic conception of nature. Explanations of nature that forgo such causes are at least acknowledged, if they do no more than describe phenomena as lawfully constituted appearances. Specifying an empirical law and proving that it is applicable in itself constitutes a causal explanation of the corresponding phenomena.[485]

In terms of laws, this means that the concept of cause receives an additional meaning, one that is directed by the phenomenal nature of the concept of law and that follows the idea of "phenomenal causality". Conversely, however, the previous definition of the concept of law becomes enhanced. Causes are only called laws if they capture the structure of a reality that exists independent of being known and – "being noumenal"– causally evoke the world of appearances. Therefore, by

---

[481] **Helmholtz** (1856ff.), p. 450f. (Germ.)/29 (Engl.), cf. also p. 451f. Nevertheless Helmholtz interprets the concept of the experiment broadly to include not only scientific method but also any arbitrary manipulation of conditions (even performed by animals), under which an organism can perceive an object (cf. **Helmholtz** (1856ff.), p. 452ff.). For the pragmatic content of this broad concept see Section 6.3.1.5.

[482] For instance, in **Helmholtz** (1853), see here Section 6.3.1.1.

[483] Nevertheless according to Helmholtz, causes can be ascertained other than experimentally, namely through observation (as in planetary movement) (**Helmholtz** (1856ff.), p. 450f.).

[484] **Helmholtz** (1856ff.), p. 454f. (Germ.)/34 (Engl.) (italics in original).

[485] Thus the concept of explanation as defined in **Helmholtz** (1853), p. 40 (cf. footnote 426), has now undergone a transition that will lead to an explicit revision of the concept in **Helmholtz** (1878c), p. 187 (cf. Section 7.1.3.1).

being brought into a relation with causes, the concept of a law loses its previously exclusive phenomenal character.

Nevertheless, it is characteristic of §26's explanations in the theory of science that Helmholtz makes no distinction between the phenomenal and the noumenal meaning of his concepts of law and cause. I ascribe the fact that he acknowledges causality "as a law of thought preceding all experience" while simultaneously emphatically affirming the above mentioned absolute restrictions to its validity as being due to this lack of differentiation between the two meanings of causality.[486]

Compared to the apriority of the law of causality that he used particularly to legitimize his assumption of an external world (and which he claims here without mentioning Kant), the empiricist demarcation of validity for that law is much more important for evaluating his classical conception of science. What he previously mentioned in passing, he now says straightforwardly. It becomes evident that Helmholtz sees the scope of causal lawfulness as restricted by experience in two ways: first, he considers causal thought an innate property and in doing so makes the mistake that Kant had already criticized in a polemic paper addressing Eberhardt.[487] The "innate forms of intuition and laws of thought"[488] have now become an "urge":

> The law of sufficient reason is really nothing more than the *urge* of our intellect to bring all our perceptions under its own control. It is not a law of nature [...].
>
> Just as it is the characteristic function of the eye to have light-sensations, so the we can *see* the world only as a *luminous phenomenon,* so likewise it is the characteristic function of the intellect to form general conceptions, that is, to search for causes; and hence it can *conceive* [*begreifen*] of the world only as being *causal* connection.[489]

This analogy between the activity of the mind and light sensation, which has little to do with Carl Vogt's materialism, implies that while the contents of causal thought cannot be analyzed empirically, it can be done for the conditions of causal thought, just as it can be done for (e.g., electrically measurable) sensations. However, being a natural inclination that causal capacity will naturally vary among people. It almost looks as if Helmholtz considers it possible to empirically investigate varying intensities of the urge (although investigations would have to rest on non-causal prerequisites). The important thing is that within the framework of a classical conception of science, empiricist limitations do not (as in modern times) lead to relativizing validity but to affirming an absolute claim to knowledge.

Second, Helmholtz does not in fact subject empirical analysis without reservation to the necessity of a causal conception of nature. In 1847 he reaffirms his view[490] that natural research might also observe non-causal events:

---

[486] **Helmholtz** (1856 ff.), p. 453 f.

[487] **Kant** (1900 ff.), Vol. VIII, p. 221 f.

[488] **Helmholtz** (1855), p. 116, see footnote 473.

[489] **Helmholtz** (1856 ff.), p. 455 (Germ.)/34 f. (Engl.).

[490] Cf. Section 6.1.1, Part β, and here Section 6.3.1.2.

> In fact, by the assertions of our own consciousness, we positively assume both in beasts and in man a principle of free will, for which we most decidedly claim independence of the force of the law of causation. We could not positively convert the vitalistic physiologists who maintain that the law of causation is valid in inorganic nature; although in the organic world they relegate its action to a lower sphere.[491]

6.3.1.5 Correspondence of Perception and Reality: The Attributive Concept of Truth[492]

Since Helmholtz also considers non-causal natural activities a part of independently existing reality, for him phenomenal causality cannot be a criterion for the truth of knowledge. The decisive innovation of the *Treatise* then, is to define a criterion for truth that is independent of causality. It refers very generally to knowledge of reality or to what Helmholtz calls the "idea" *(Vorstellung)* of reality. He does not explicitly combine it with scientific knowledge until the late 1860s. By "idea" Helmholtz means not conceptually constituted knowledge but rather, in line with the theme from *§26 On Perceptions in General*, the "memory image of visual objects which is not accompanied by any present sensory sensations".[493] The definition, designed to fit investigations in the physiology of the senses and psychology, presupposes that sensations are processed to become perceptual experiences. Its purpose is to summarily express the various kinds of perception that a subject can have of an object.[494] But it does not prevent Helmholtz from also using the word "idea" in a very broad sense as a mental intuition or grasp of objects that includes both everyday meanings as well as the narrower interpretation of a scientifically produced definition.[495]

He answers the "much disputed question about the degree to which our ideas correspond at all to the objects they concern"[496] by partially retracting the mere symbolicalness of sensory perception. He postulates that in perception the time sequence of activities is represented just as those events happen in reality. This gives him – as Nietzsche would say – an "escape route" *(Schlupfweg)* from sensory perception to reality and breaks scientific knowledge's monopoly on the truth about reality.[497] Other than for Kant, for Helmholtz time is not a subjective form of intuition, whose objectivity must first be proven by applying causality but rather a dimension that creates absolute simultaneity for a world divided into inner and outer realms:

---

[491] **Helmholtz** (1856 ff.), p. 454 (Germ.)/33 (Engl.).

[492] On the distinction between an attributive and substantive concept of truth see Chapter 2.

[493] **Helmholtz** (1856 ff.), p. 435.

[494] **Helmholtz** (1856 ff.), p. 798.

[495] E.g. in **Helmholtz** (1856 ff.), p. 446 f., cf. quotation for footnote 507.

[496] **Helmholtz** (1856 ff.), p. 442.

[497] **Nietzsche** (1980), Vol. 2, p. 1092 f.

> The only respect in which there can be a real correspondence between our perceptions and the reality is the time-sequence of the events with their various peculiarities. Simultaneity, sequence, the regular recurrence of simultaneity or sequence, may occur likewise in the sensations as well as in the events. The external events, like their perceptions, proceed in time; and so the temporal relations of the latter may be faithful representation of the temporal relations of the former. [...] Yet here it certainly should be noted that the time-sequence of the sensations is not quite a faithful representation of the time-sequence of the external events, inasmuch as the transmission from the organs of sense to the brain takes time and in fact different time for different organs.[498]

In contrast to the *habilitation* lecture of 1852, Helmholtz no longer relates perception to isolated objects of reality but to a sequence of "events". What is given in reality is not individual objects, not their invariable identity in the face of change but rather the "temporal relations" with which various states are correlated. Here lies – apart from ranking sensory perception as capable of truth – the true significance of postulating temporal representation: the relation to objects is now replaced by the reproduction of relations. Helmholtz stresses explicitly:

> That in point of fact the *properties* of natural objects [...] do not denote something that is peculiar to the individual object by itself but invariably imply some relation to a second object (including our organs of sense).[499]

The *Treatise* does not yet demonstrate the consequences this statement will have for the classical concept of science. Helmholtz focuses entirely on characterizing sensory perception. Sensory perception can only experience anything about (realistically presupposed) objects when it is in a relation to them. But since the preconditions of bodily constitution also effect this relation, it is impossible for sensory perception to recognize the objective properties of things. That task remains for science, whose statements are valid "independently of the peculiar nature" of the sense organs.[500]

With the closing remark of the passage cited above[501] Helmholtz satisfies his scientism's demands by referring to research results he had obtained following his groundbreaking proof that the velocity of nerve impulses is finite. In those investigations he was able to show that in principle the velocity of sensation impressions can also be measured in humans and probably depends on extremely complex ancillary conditions.[502] Different values turned up for various organs and were also subject to temperature fluctuations.[503] Reliable results for the magnitude of the velocity were not available until 1867. Helmholtz knows that the first values he had

---

[498] **Helmholtz** (1856 ff.), p. 445 (Germ.)/22 (Engl.).

[499] **Helmholtz** (1856 ff.), p. 444 (Germ.)/21 (Engl.), and **Helmholtz** (1885 ff.), p. 589 (italics in original).

[500] **Helmholtz** (1856 ff.), p. 445, and **Helmholtz** (1885 ff.), p. 590.

[501] **Helmholtz** (1856 ff.), p. 445.

[502] **Helmholtz** (1850a), **Helmholtz** (1850b), **Helmholtz** (1852a), **Helmholtz** (1850b), p. 877 ff., **Helmholtz** (1867), p. 932.

[503] **Helmholtz** (1850a), p. 837 ff., **Helmholtz** (1856 ff.), p. 445, **Helmholtz** (1867), p. 932 f.

stated in 1850 deviate considerably from those found by other researchers, presumably due to experiment conditions and systematic errors that were not taken into consideration.[504] In addition, he only has measurements for the velocity of individual nerve excitations at his disposal but not the temporal pattern of a sequence of excitations. On the whole the results of his investigations therefore in no way provide sufficient material for empirically supporting the claim that a temporal "correspondence" exists between sensations or even perceptions and "the" real.

What Helmholtz passes off as an empirically provable relation of representation is only a poorly founded postulate for the purpose of at least partly eliminating the problem of the validity of sensory perception so typical of his realism. For no other feature of perception does he allow for reality – at least in the narrower context of his investigations in the physiology of the senses – any correspondence comparable to time.[505] Thus, spatial perception is only a sign of an absolutely postulated external world, whose spatial structure, as Helmholtz's work in geometry demonstrates, can only be objectively determined by empirical science.[506]

Retracting symbolicity for temporality is a necessary, but insufficient condition for determining the truth of an idea that refers to the external world:

> Thus the ideas of the external world are images of the regular sequence of natural events and if they are formed correctly according to the laws of our thinking and we are able by our actions to translate them back into reality again, the ideas we have are also the *only true* ones for our mental capacity. All others would be false.[507]

At this point a more precise determination of the concept of an idea would be desirable. On the one hand, the context in which Helmholtz uses the term "idea" when speaking of the lawfulness of time in sequences and "laws of human thinking" refers to scientific knowledge. On the other hand, when vaguely speaking of "the ideas we have", Helmholtz is alluding to an everyday concept. It is true that in terms of the theory of science, §26 of the *Treatise* obscures the distinction between scientific knowledge and quotidian perception. It would not be accurate, however, to say that Helmholtz says they are the same. According to his understanding of science, experimental and theoretical scientific research alone can judge the first of the two mentioned conditions for a true idea. For only experimental and theoretical research in science satisfies the "laws" of thought, i.e., meets the demands of logic and causality.[508] In contrast, the second condition is a pragmatic feature that should normally be irrelevant for the concept of truth in Helmholtz's classical conception of science.

---

[504] **Helmholtz** (1850c), p. 878, **Helmholtz** (1867), p. 933.

[505] He does make one exception in one of his popular reports on the physiology of the senses: cf. footnote 512.

[506] Helmholtz sees the two-dimensionality of visual and tactile perception as the reason why everyday perception lacks a spatial representation relation. See the text for Section 6.2, footnote 360.

[507] **Helmholtz** (1856ff.), p. 446f. (Germ.)/24 (Engl.) (italics in original). The expression "image"*(Bild)* is to be understood as meaning "representation" *(Abbild)*, cf. the quote for footnote 498.

[508] **Helmholtz** (1856ff.), p. 455.

Laws of nature that correspond to reality and the laws of thinking are true and require no further pragmatic examining authority. Helmholtz stated this in his rectorate lecture and will repeat it in the speech held for natural scientists in 1869.

Pragmatic elements gain significance for Helmholtz's concept of truth only inasmuch as scientific knowledge is not specifically discussed in §26 of the *Treatise*. These elements serve purposes of common sense, which, when navigating its way through the world, does not judge and evaluate the temporal correspondence of its perceptions with reality. In daily use the symbolic character of perceptions permits no other option than to evaluate the truth of ideas pragmatically:

> In my opinion [...] there can be no possible sense in speaking of any other truth of our ideas except of a practical truth. Our ideas of things cannot be anything but symbols, natural given signs for things which we learn how to use in order to regulate our movements and actions.[509]

Although Helmholtz speaks very generally of "ideas", he relates the pragmatic concept of truth only to the symbolic or sign nature of sensations, i.e., to the claim that they do not correspond to the external world. What this example reveals is also relevant for the other explanations in §26. Helmholtz does not consider his theory of perception a part of his theory of scientific knowledge. He discusses empirically researched everyday processes of perception without reflecting on the preconditions of experimental experience as required by the theory of perception. Neither in the speech *Relation of Natural Science* nor in the *Treatise* does Helmholtz develop a pragmatic conception of science, which he would have to do, if he were to apply to scientific knowledge the conditions of validity that hold for everyday perception. Conversely, based on his essentially non-pragmatically defined concept of science he interprets the psychological processing of sensory sensations in a way for which he aspires to achieve not hypothetical but absolute validity.[510] This will change in the 1870s when he develops his theory of perception to become epistemology and reverses the order of scientific knowledge and everyday perception. He will establish the relativity of the validity of the one based on mechanisms that are at work in the other.

Nonetheless, his explanations make it clear that the experimental method adds a pragmatic element to the classical-scientistic concept of science. The validity of experimental results depends significantly on the successful execution of the experiment, the extent to which the results are confirmed in technical contexts and

---

[509] **Helmholtz** (1856ff.), p. 443 (Germ.)/19 (Engl.) (italics in original).

[510] Helmholtz considers his interpretation of how sensations of perception are processed psychologically not a hypothesis but a "theory" (**Helmholtz** (1856ff.), p. 435, 819 and in other places) to which a "chain of facts [...] necessarily compels us" (**Helmholtz** (1868a), p. 352). Only because "the questions are not yet entirely ripe for deciding" does he believe that it is necessary to erect his "explanations [...] on the *smallest* possible number of hypotheses that are also *formulated specifically*" (**Helmholtz** (1856ff.), p. 819, and **Helmholtz** (1885ff.), p. 970 – italics in original). **Moulines** (1981), p. 70f., also stresses the claim to truth inherent in Helmholtz's theory of perception.

generally, whether the desired effect is achieved (among which one can also count being recognized as sound by the scientific community). By describing common sense's way of attaining truth as an experimental strategy, Helmholtz at least implicitly introduces a pragmatic aspect into the scientific experiment, which he had previously upgraded as an indispensable means for determining empirical laws and causes:

> The same great importance which experiment has for the certainty of our scientific convictions, it has also for [...our] perceptions. It is only by voluntarily bringing our organs of sense in various relations to the objects that we learn to be sure as to our judgments of the causes of our sensations.[511]

If the validity of scientific experimental findings were subject to the same criteria as everyday ideas, then the absolute criteria for the truth of natural law statements, namely universality and necessity, could no longer be fulfilled. But Helmholtz does not transfer to scientific knowledge the lack of clarity, the thoroughgoing ambiguity and lack of accuracy intrinsic to immediate sensory perception. By upholding the opposition between scientific knowledge and sensory perception it remains possible for him not to explicitly transfer the temporal postulate of representation to scientific knowledge.

Yet the elaborations of the *Treatise* prepared the way for a novel strategy for arguing the classical claim to truth: the reason that scientific knowledge does not have the nature of signs but instead meets the absolute criteria of objectivity is that science alone can represent temporal patterns of perception. The significance of the last two publications that will be used to reconstruct and evaluate Helmholtz's classical conception of science up until the late 1860s consists in the explicit formulation of that strategy. In line with his conception of nature it will be the equations of physical mechanics that satisfy the requirements for representing changes as desired.

6.3.1.6 Subjective Signs Versus Mechanical Images

The temporal correspondence of real events with the perceptions related to them, as postulated in the *Treatise*, subsequently – although not without fluctuation – becomes a central part of Helmholtz's concept of science. In the lecture *The Recent Progress in the Theory of Vision* [*Die neueren Fortschritte in der Theorie des Sehens*], published in 1868, he distinguishes more clearly the concept of a sign from that of an image:

> Nerve activity and the mental ideas which go with it may be 'images' of actual occurrences in the outer world, so far as the sequence of the former represents the sequence in time of the latter, so far as they represent sameness of objects by sameness of signs – that is, a lawful arrangement by a lawful arrangement.

[511] **Helmholtz** (1856ff.), p. 452 (Germ.)/30 (Engl.).

> This is obviously sufficient to enable the understanding to deduce what is constant from the varied changes of the external world and to formulate it as a notion or a law.[512]

Here Helmholtz uses the term "image" [*Bild*] to mean "representation" [*Abbild*]. The structure of the world, its temporal and not necessarily causal constitution, is pristinely reproduced by sensory sensation and perception. This concept of an image serves to redefine the tasks of science. The task of science now is to give adequate expression to the temporal structures embedded in perception. Scientific knowledge, which consists of causal law statements, becomes the pure depiction of the imagery part of sensory perception. Natural laws, Helmholtz soon says, are characterized by "the succession in time of natural phenomena".[513]

By reducing its relation to reality to that which is perceivable, science forfeits its unrestricted access to reality. The relational character of the representation penetrates, as a new element, the concept of natural law. Inasmuch as statements of natural law now only cover relations between the properties of objects, natural research can no longer claim to contact the objects themselves, the substantial causes of phenomena. Helmholtz appears for the first time to accept a limitation to the compass of truth as the price for securing truth based on sensory perception. Metaphorically speaking, using the analogy from his speech of 1852 once more[514]: science, inasmuch as it produces a relation to reality via sensory perception, is no longer in a position to describe the particular look of an "unfamiliar people".[515] What it is capable of registering are general changes over time, for example, the general phenomena of aging.

In the debate on Helmholtz's theory of perception, Ernst Cassirer insisted that the "notion of relation" embedded in the theory of signs does not imply the relativity of scientific knowledge but on the contrary "affirms" its absolute claim to validity.[516] According to Cassirer the theory of signs developed in the *Treatise* and subsequent lectures constitutes a "characteristic and typical development in the general scientific doctrine of knowledge" that can be used exemplarily to study the transition in the philosophy of science from orientation to a "concept of substance" to one of a "concept of function". Although with his realism Helmholtz still maintains a "concept of the absolute object", his theory of science phrased in terms of a theory of signs seems based on the conviction that is possible to know, without constraints, "not what is real [...] in its isolated, own, existing constitution but the rules that govern it and how it changes".[517]

---

[512] **Helmholtz** (1868a), p. 319 f. (Germ.)/167 (Engl.). The lecture is followed by an addendum that to my knowledge contains the only passage in which Helmholtz takes the correspondence between perception and reality to be more than a temporal relation (**Helmholtz** (1868a), p. 365). I assume that the claim of spatial correspondence rests on the notion of the absolute validity of Euclidean geometry. Cf. Section 6.2.

[513] **Helmholtz** (1869), p. 395.

[514] Cf. quotation for footnote 415.

[515] Ditto.

[516] **Cassirer** (1910), p. 404 f., cf. also **Cassirer** (1923 ff.), Vol. III, p. 378 f., and **Cassirer** (1957), p. 44 f., 198 f. and 276 f.

[517] **Cassirer** (1910), p. 404 f.

As law, scientific representation is neither characterized by intuition nor does it include an aspect of free depiction. Laws are time relationships described mathematically. They exist between causally linked events, can be clearly grasped, need no symbolization and as empirical statements, are also clearly distinct from substantial causes, i.e., material elements and the forces operating among them. But as before, the ambiguity of Helmholtz's concept of cause also permits conceiving of relationally understood laws as causes. Laws are causes, inasmuch as they are a part of reality that exists independent of being known.

In contrast to the representation relation, the acausal connection that Helmholtz sets up between signs and what they designate has no relation to the temporal structure of reality. From the first formulation of his theory of signs the only thing that held for signs was that each of them be assigned to an object. This reference to an object, trustworthy in everyday experience but entirely irrelevant for scientific knowledge, now becomes the precondition for its representational character.

Helmholtz uses the expression "sign" consistently to denote the real content of sensory sensations and related perceptions. His use of the concept of image, on the other hand, vacillates. Surveying his work on the theory of perception and epistemology one can say that he usually understands this concept ("image" [*Bild*]) in the sense just explained (i.e., as a representation [*Abbild*]) but sometimes also as meaning a sign.[518] When in some passages of his public speeches he attempts to illustrate the relationship of signs and images using examples from the arts one also gets the impression that he places the concept of an image with a third, more comprehensive meaning, above that of the sign. He says in the speech *On the Aim and Progress of Physical Science [Über das Ziel und die Fortschritte der Naturwissenschaft]*:

> An *image* must, in certain respects, be the *same* as the original object; a statue, for instance, has the same corporeal form as the human being after which it is made; a picture the same color and perspective projection. For a *sign* it is sufficient that it become apparent as often as the occurrence to be depicted makes its appearance, the conformity between them being restricted to their presenting themselves simultaneously.[519]

What suffices for making a sign a sign is the property that Helmholtz uses to establish the representation relation. Since the sign exists simultaneous to what it designates, the succession of events in both agree.[520] Since therefore the sign is of one of a kind with what it designates in a relation, it belongs to the encompassing set of images that includes statues and paintings as well as letters of the alphabet. When the concept of an image is adopted from artwork it is bestowed with a content extending beyond a

[518] Meaning a representation: **Helmholtz** (1856 ff.), p. 446 (cf. footnote 507), **Helmholtz** (1878a), p. 222, **Helmholtz** (1892), p. 358; meaning a sign: **Helmholtz** (1878a), p. 222, **Helmholtz** (1885 ff.), p. 590 and 599.

[519] **Helmholtz** (1869), p. 393 (Germ.)/221 f. (Engl.) (italics in original), the same but without mention of simultaneity: **Helmholtz** (1878a), p. 222.

[520] **Helmholtz** (1856 ff.), p. 445.

mere relation of sameness or similarity and which can be devised intentionally in numerous ways, making is useless for science.

The expansion of the concept of law marks a shift in the development of Helmholtz's conception of science. It will remain characteristic of Helmholtz and in 1892, two years before his death, he will once again call it the fundament of the inductive method:

> As soon as it came to be accepted that correct perceptions could be obtained through the senses, the subsequent path of investigation, that is, the inductive method of the natural sciences, was prescribed.[521]

In terms of the theory of science, the upshot of Helmholtz's theory of signs is that perception evidently involves elements that are not only of a sign but also of a representational nature and which therefore, according to Helmholtz, may be a part of the foundation of science. This conclusion puts Helmholtz at the onset of a trend towards a theory of representation that will come to have considerable consequences. In the famous *Principles of Mechanics* published in 1894, Helmholtz's follower Heinrich Hertz develops the concept of an image further. It plays a central part in his philosophy of science[522] which, in turn, is said to have heavily influenced Ludwig Wittgenstein's theory of the representation of meaning.[523]

While later Helmholtz no longer thinks that the relation of representation is always temporal, its characterization as such is the focus of his final arguments for the classical conception of science, as expressed in the opening speech delivered in Innsbruck in 1869 to the Association of German Natural Researchers and Doctors [*Versammlung Deutscher Naturforscher und Ärzte*] and titled *On the Aim and Progress of Physical Science*. Therein the postulate provides him with an absolute criterion for the truth of perceptions and scientific statements and creates the link to his mechanistic aims. Summarizing this speech to the members of one of the nineteenth century's most important associations in experimental science in Germany and thus in my opinion also summarizing the entire previous development of his conception of science, Helmholtz says:

> We found at the beginning, that what our science strives after is the knowledge of laws, in other words, the knowledge how at different times under the same conditions the same results are brought about; and we found in the last instance how all laws can be reduced to laws of motion. We now find, in conclusion, that our sensations are merely signs of changes taking place in the external world and can only be regarded as pictures in that they depict succession in time. But, for this very reason they are in a position to represent lawfulness directly, in regard to succession in time, of natural phenomena. [...] That which our organs of sense perform is just sufficient to meet the demands of science.[524]

---

[521] **Helmholtz** (1892), p. 338f. (Germ.)/480 (Engl.).

[522] Cf. Excursus for Section 7.1.3.5.

[523] While Wittgenstein himself only mentions Hertz a few times (cf. *Tractatus logico-philosophicus*, No. 4.04, in: **Wittgenstein** (1970), Vol. 1, and *Blue Book*, in: loc. cit. Vol. 5, p. 50), several authors see Wittgenstein's theory of representation of meaning as guided by Hertz's philosophy of science. See e.g. **Pitcher** (1967), p. 227, or **Hacker** (1978), p. 19ff.

[524] **Helmholtz** (1869), p. 394f. (Germ.)/223 (Engl.) (italics in original).

The ambiguity of the concept of law that since the publication of the *Treatise* can be understood both phenomenally and noumenally is here now enhanced to include yet another component: here the "law(s) of motion" to which "all laws [...] are reducible" means submicroscopic laws of motion that are not accessible through perception and are bound to the existence of elementary matter. As Helmholtz hinted in the Introduction to *Conservation of Force,* he – but now with unmistakable resolve – identifies mechanically moved matter with chemical elements, in which he is convinced to have found the immutable components of all ponderable matter.[525] In the Introduction the "ultimate causes" of all phenomena were the elements and the central forces operating among them[526] and this view also supports the statements of his speech to the association of scientists. The connection with mechanically moved matter distinguishes the fundamental laws of motion from the purely relationally defined phenomenal concept of law. One might be tempted to apply the idea Helmholtz demonstrably has of a noumenal cause that evokes all phenomena to the ultimate elements of matter. However, since for him these elements are potentially an object of empirical science, they belong just as much to the world of appearances as the phenomena they produce.

The way Helmholtz, in the passage cited, creates the aberrant impression that the activity of the sensory organs could suffice for establishing the mechanistic program of reduction, is by no means characteristic of the entire speech. He himself seems to know that reducing all empirical laws to mechanical ones cannot be accomplished by a theory of perception but rather in principle only either a priori or with reference to the state of research in the empirical sciences. This speech follows the second, namely scientistic strategy. He addresses "the progress of physical science as a whole"[527] and deduces mechanism from previously obtained research findings as the conception of nature that suits them. Describing the contents of that legitimating connection to natural research, however, is not part of reconstructing his conception of science and will be dealt with in the following section with an outline of his conception of nature.

Before closing this reconstruction I would like to point out how the speech to the scientists' association adapts the concept of force to the altered concepts of cause and law.[528] Likewise, taking up immediately where the *Treatise* left off, Helmholtz no longer uses this term with a merely mechanical or energetics meaning but rather as a means for stressing the importance of laws as causes.[529] For example, if one were to call the laws of chemical affinity or refraction forces (affinity forces or refraction forces), it would do justice to the fact that these laws have "not been invented at random" but are "essential laws of phenomena".[530] In this way the

---

[525] **Helmholtz** (1869), p. 378 f., cf. quotation for Section 6.3.2.1, footnote 553.

[526] See Section 6.1.1, Part β.

[527] **Helmholtz** (1869), p. 377.

[528] **Heidelberger** (1994a) sees the changes in the concept evident in **Helmholtz** (1869) as "first signs of a significant and deep change in his view of force" (loc. cit. p. 474), in which his orientation towards mechanistic ontology is partially replaced by phenomenological objectives in physics (loc. cit. p. 474 ff.).

[529] Cf. **Helmholtz** (1856 ff.), p. 454 f. (as footnote 484).

[530] **Helmholtz** (1869), p. 376.

concept of force goes beyond having a mere mechanical meaning and stands for the equivocal concept of law: on the one hand forces generally designate phenomenal causality or, as Helmholtz says, "the lawfulness of nature [...] as a causal relation".[531] On the other, they stand for a noumenally effectual reality that exists independent of being known.

The "simplest form of expression of force" is a "mechanical force acting on a point of the mass".[532] As in the Introduction to *Conservation of Force,* here Helmholtz does not metaphysically deduce the determination of this central concept of his conception of nature. Once again he is referring to contents of statements in natural research, although he does not explicitly mention the concept of the chemical element in this passage. Instead, he limits himself to a putatively empirically imparted knowledge of nature that can be obtained "by trying to find from the phenomena attending large visible masses laws for the effect of the infinitely small material particles".[533] These he calls "smallest elementary particles" and thinks of the law governing their effect as being the same as force's "purest form". Here the term 'force' not only means that submicroscopic regularities are the "ultimate" causes of perceptible natural events. The term also designates the mechanical, clearly defined laws of motion governing fundamental particles, the most primordial of all changes in mechanism: "Force [...] is made equivalent to the acceleration of the mass on which it takes effect, multiplied by this mass".[534]

6.3.1.7 Summary

When reconstructing the formal aspects of Helmholtz's mechanism up to the late 1860s I found it particularly important to isolate, as much as possible, the determinations of his theory of science from those of his theory of perception in order to do justice to the opposition between scientific knowledge and the witness of immediate sensory perception. This opposition so characteristic of Helmholtz's early conception of science is particularly suitable for deciphering the classical features of that conception.

From this standpoint of reconstruction one sees that in the second phase Helmholtz does not entirely dissolve the opposition, he only retracts part of it. The classical fundamental determinations for science, given by the concept of law and the worthwhile aim of a complete system of knowledge, persist and are not applied to the witness of sensory perception. Stuck with the symbolical nature of sensations, everyday knowledge based on sensory perception can only claim to be true in a pragmatic sense. However, Helmholtz demands that the law statements of science have exclusive validity for the whole period and can be derived in its necessity

[531] **Helmholtz** (1869), p. 377.

[532] **Helmholtz** (1869), p. 376.

[533] Ibid.

[534] Ibid.

no different from logical deduction. Towards the late 1860s he has secured the absolute claim to validity in two ways: in one, a temporal correspondence exists between empirical law statements and the structure of the real world; in the other, Helmholtz assumes that empirical laws are clearly reducible to mechanical laws and that the system of scientific knowledge will – inasmuch as it is possible – someday be finished by a complete mechanistic reduction.

While the continuous elements of Helmholtz's conception result from upholding the distinction between science and sensory perception, each mitigation of that difference is attended by a change in his grasp of the problem. This relates less to the classical criteria of science and more to the justification of the claim to truth that they express. The pressure for justification mounts once sensory perception, due to its psychological characterization, has been contrasted even more decisively with natural regularity. It then seems necessary to make a substantial correction, if the absolute validity of the inductive method has been relativized. Helmholtz's partial retraction of the opposition between sensory perception and scientific knowledge is in the end well suited for establishing the previously inductively guaranteed claim to truth for classical science in a new way.

At this point it is wise to once again remind the reader that the present study is a reconstruction. Just as the emphasis on the opposition between sensory perception and scientific knowledge served to characterize Helmholtz's classical conception of science, the portrayal of the partial retraction of that contrast served the same purpose. All that can be said is that this is probably linked to the problem of justification in the theory of science. Furthermore, one can ask whether the problem of justification is among the influential factors that led Helmholtz to correct his theories of perception, respectively, signs or whether it was the other way around. The latter seems improbable, since the theory of perception arguments that Helmholtz himself provides are not very convincing. He never did supply proof of the alleged simultaneity of the occurrence of perception and the perceived. But the weakness of the argument is far from sufficient for evaluating the possible motives for developing a theory of science and perception. At this point I do not want to get ahead of myself by prematurely coming to the reasons for the transition (see Chapter 8) but only make two remarks: first, it must be taken into consideration that in the course of the 1860s Helmholtz increasingly preoccupied himself with questions in the theory of perception. Even though he did this while entertaining a certain conception of science, it can by no means be ruled out that the preoccupation initiated reasons for the change in that conception. Second, while it is possible by studying the texts to demonstrate how the legitimation problem developed, it cannot be sufficiently explained. Questions about why (in the mid-1850s) he heightens the opposition between sensory perception and scientific knowledge or (in the 1860s) relativizes the previously asserted absolute validity of "logical induction" indicate that the transition process rests on a complex network of conditions that we have for the most part not yet considered.

The essential result of the classical correction lies in presenting the first argumentation for the validity of empirical law statements based on a theory of perception and in determining their scope in a world whose structure is perhaps not completely causal. The image character of perception related to time is one condition for meeting the classical claim to truth, insofar as it does not refer to knowledge about individual objects but to the relations existing among properties. It is clear that this argumentation strategy cannot touch the mechanistic program of reduction for the simple reason that mechanism is based on structural determinations that Helmholtz obtains independent of considerations in the theory of perception by referring to the state of research in physics and chemistry. The link between a mechanistic conception of nature, aimed at discovering ultimate causes and the knowledge of laws, aimed at compiling the temporal relations in the changes of observable processes, on the one hand represents the further development, in terms of the theory of science, of the relationship between his mechanistic program and the conservation of energy, as postulated in the paper of 1847. Empirical laws are probably no longer merely pure heuristics for mechanism but remain as a goal of the complete explanation of nature subordinate to it. On the other hand, a tension exists between the argumentation for validity based on the theory of perception and the program of mechanistic research, which can be found right down to the determinations of the fundamental concepts of law, cause and force. Although both the argumentation for validity based on the theory of perception and the classical mechanism aim for the complete knowledge of nature, in contrast to the latter the former is limited to what is perceptible and requires no further specific objective in terms of content.

6.3.2 Helmholtz's Classical-Mechanistic Conception of Nature

> Helmholtz was the last great representative of the mechanical view of nature and though he by no means presupposed it dogmatically, it stands as the background of all his critical epistemological considerations and gives them their characteristic coloring. From this standpoint matter becomes what is fundamentally real and must be described by the concepts of natural science
>
> (Ernst Cassirer, *Determinism and Indeterminism in Modern Physics*).

The difficulties discussed thus far do not concern the main endeavor of the lectures and speeches Helmholtz delivered up until the late 1860s, where he was usually less interested in offering a theory of nature foundation for his conception of nature than in explaining natural phenomena, technical constructs and the scientific basis of painting and music. He elucidates, among other things, the physiological conditions for vision and harmony,[535] the optical prerequisites of painting,[536] the physical and technical aspects of the theorem of the conservation of energy,[537] scientific explanations for

[535] **Helmholtz** (1855), first publ. Koenigsberg 1855; **Helmholtz** (1868a), first publ. Berlin 1868; **Helmholtz** (1857), first publ. **Helmholtz** (1865).

[536] **Helmholtz** (1871 ff.), first publ. **Helmholtz** (1876).

[537] **Helmholtz** (1854a), first publ. Koenigsberg 1854; **Helmholtz** (1861), first publ. in: Proc. Roy. Inst. (London), Vol. III (1861); **Helmholtz** (1862/3), first publ. **Helmholtz** (1871).

the origin of ice, glaciers and storms[538] and outlines a mechanical theory of the origin of the planetary system.[539]

Nowhere does Helmholtz expound a coherent all-encompassing interpretation of nature. The majority of this writing comprises phenomenon-oriented individual investigations that involve the application of specific empirical laws. It is the phenomena of life that particularly get in the way of formulating a unified view, just the ones that had been of considerable importance for reconstructing his classical conception of science up to this point. In both the Introduction to *Conservation of Force* and in the *Treatise* Helmholtz had empirically limited the validity of causal explanation by pointing out that the phenomena of life seem to involve exceptions. This may be one reason why he needed to bring a non-causal criterion for truth into his conception of science. Helmholtz's doubt about someday being able to exhaustively explain the phenomena of life by the mechanical structure of matter led, as it were, to halved mechanism. In range, mechanism was to apply only – but then with certainty – to the material side of natural phenomena but not to psychological, respectively, mental kinds of appearances, such as those that are – in contrast to lifeless matter – characteristic of life. According to Helmholtz that would be accomplishment enough, because it would mean that mechanistic explanation covers everything that can be explained causally.

The pervading and most obvious flaw in the previously scant systematic attention given to Helmholtz's conception of nature has been not to see the doubt about explaining life mechanistically that lingers behind the comprehensive program of reduction he formulated for all of natural science.[540] The fault is understandable, since Helmholtz himself very rarely expressed his skepticism and he was significantly

---

[538] **Helmholtz** (1865a), first publ. **Helmholtz** (1865); **Helmholtz** (1875a), first publ. in: *Deutsche Rundschau* (1876).

[539] **Helmholtz** (1854a), first publ. see above; **Helmholtz** (1871a), first publ. in **Helmholtz** (1876).

[540] Except for his theory of matter (**Heimann** (1974), cf. Section 6.1.1) neither the history of science nor that of the philosophy of nature has made Helmholtz's conception of nature an isolated object of study. Anthologies commemorating the 100th anniversary of his death (**Cahan (Ed.)** (1994) and **Krüger (Ed.)** (1994)) contain only one essay on it (**Schiemann** (1994)). **Krüger** (1994) presumably speaks for many when he says: "Everyone knows and therefore it need not [...] be stated again, that Helmholtz saw the ideal of complete comprehensibility of nature realized in the mechanistic model of the world" (loc. cit. p. 206). But the "world model" Helmholtz pursued remains exponible. Individual aspects of his conception of nature have been discussed within the context of his work on mechanics (**Koenigsberger** (1895) and **Breger** (1982)), thermodynamics (**Klein** (1972), **Bierhalter** (1981) and **Bierhalter** (1987)), physiology (esp. **Mendelsohn** (1964) and **Lenoir** (1982)), electrodynamics (**Buchwald** (1994a), **Darrigol** (1994)) and epistemology (**Heidelberger** (1994a) and (1994b) and **Krüger** (1994)). Although the distinction between mental and causal phenomena in his conception of science has been stressed (e.g., **Cahan** (1994b) and **Hatfield** (1994)), little attention has been given to the fact that Helmholtz never denied animal mental capacities. In addition, often the erroneous impression has been created that his mechanism applied to all of nature, e.g., in **Koenigsberger** (1895), p. 3, **Erdmann** (1921), p. 13, **Wise** (1983), p. 8ff., **Mendelsohn** (1964), **Herneck** (1973) and **Breger** (1982). An exception to this general tendency are **Hörz and Wollgast** (1971) and **Wollgast** (1973), who consider Helmholtz's conception of the world "pantheistic", also regarding his speculation concerning the origin of life (cf. Section 6.3.1.4 of this book).

involved in introducing and implementing mechanistic explanations to countervail the vitalistic interpretations of life still widespread in physiology throughout the first half of the nineteenth century. Particularly his enunciation of the theorem of the conservation of energy was taken for a convincing refutation of the vitalistic postulate of inner warmth [*calor innatus*].[541] According to Helmholtz the source of organic body warmth is not a purposeful force peculiar to life but the mechanical motion of dead matter.[542] His resolutely iterated opinion that life forces are either idle explanations or violate the theorem of energy was meant to apply strictly to the physical and chemical metabolic processes in living organisms. Although he believed that research could generally do without life forces, he did not definitely say so.[543] Future developments in experimental research would probably shift the demarcations between the mind and nature but in principle never eliminate them.

In contrast to his statements in the theory of science during the same period, his explanations in popular science exhibit no comparable developmental phases.[544] While a delayed recognition of Darwin's theory of evolution is salient, it does not reflect a change in Helmholtz's basic attitude. More striking are the consistent elements of his mechanistic theory of matter, which, alongside the status of phenomena of life, constitute the trait characteristic of his mechanism in terms of content. How the *Conservation of Energy* in 1847 and his speech to scientists in 1869 are interrelated is of particular interest. In the 1869 speech to the association of scientists he once again affirms the program already formulated in 1847. He stresses his conviction that substantial success has already been achieved in reducing natural phenomena to matter in motion, expounds the dual character of his mechanism once more and challenges his audience to pursue the program to completion, to merge all natural science "into mechanics".[545]

In terms of content, these two publications can be seen as marking the beginning and the close of a fairly homogenous period. Together they create the framework for my portrayal of Helmholtz's fundamental statements in the philosophy of

---

541 Representative of many are **Du Bois-Reymond** (1848), p. 19, **Koenigsberger** (1902 f.), Vol. 1, p. 58 f., **Lenoir** (1982), p. 195 ff., and **Eckart and Gradmann** (1994), p. 106.

542 **Helmholtz** (1861), p. 563, **Helmholtz** (1869), p. 385 ff. et passim. As early as in his first papers in science Helmholtz had dealt with the particularly physiological aspects of body warmth in organisms and critically discussed the concept of life forces. In this work that can also be considered preparatory for formulating the theorem of the conservation of energy, he concludes that the phenomenon of organic warmth cannot be explained by life forces and instead develops his own explanation for it based on physics and chemistry (**Helmholtz** (1845), **Helmholtz** (1846), **Helmholtz** (1847b)).

543 **Helmholtz** (1856 ff.), p. 796 f. and 804 f. (**Helmholtz** (1885 ff.), p. 946 f. and 955 f.), **Helmholtz** (1861), p. 579 f., **Helmholtz** (1869), p. 385 ff. and 391, and later: **Helmholtz** (1878c), p. 178 ff., **Helmholtz** (1892), p. 351, **Helmholtz** (1897 ff.), Vol. I.1, p. 16.

544 See first publications in footnote 535 ff.: except for the speeches on painting (**Helmholtz** (1871 ff.)) and on tornadoes and storms (**Helmholtz** (1875a)) they all took place between 1855–1871.

545 **Helmholtz** (1869), p. 379 (as Section 1.1, footnote 31).

nature, which now follows. First I shall finally evaluate whether viewing his mechanism as belonging to the dual tradition is correct for the period ending in the late 1860s. To do so I shall discuss the justification for the theory of matter that he repeats in his speech to the association of scientists because up to now reconstructing it was based on important but few passages from the Introduction of 1847 (Section 6.3.2.1). I shall then elaborate his philosophy of nature, which thus far has only been mentioned in passing, by demonstrating the exceptional position he awards life in contrast to matter (Section 6.3.2.2) and list four invariant determinations that he uses to characterize the fundamental traits of nature: duality (Section 6.3.2.3) (i), universality of the principles of mechanics (ii), the finiteness of natural history (iii) and the exploitability of natural energies (iv).

6.3.2.1 Dual Mechanism: Matter and Force

Helmholtz arranged the central passages[546] of his speech for the association of scientists similar to the classical-scientistic layout used for the Introduction to *Conservation of Force*: fairly detached from the determinant contents of the speech he first asserts science's claim to truth. Although the conceptual specifications involved are themselves already aligned to the principles of mechanics, they afford no specific conclusions.[547] Not until Helmholtz includes experimental and theoretical findings from physical mechanics and chemistry to guarantee the classical conditions of validity does he arrive at a determination that permits viewing his conception of nature as belonging to one of the three traditions of mechanism expounded above.[548]

With its axioms (the theorem of inertia and definition of force) and mathematical methods (differential calculus), early modern physical mechanics had erected the framework for a set of lucid concepts.[549] As mentioned earlier, Helmholtz applies the mechanical term 'force', which he calls "the purest form of expression of force" to the "smallest elementary particles", thus defining force as the cause, respectively, law of change for uniform straight inert motion. Instead of considering though that the duality of force and matter generates a theoretical conception that must not necessarily be real, he considers it an empirical finding that "attend[s] large physical masses".[550]

Helmholtz adopts the more detailed physical determinations for the material aspect of his dual conception of nature not from physics but from chemistry,[551]

---

[546] **Helmholtz** (1869), p. 374–379.

[547] **Helmholtz** (1869), p. 374–376.

[548] **Helmholtz** (1869), p. 376–379.

[549] **Helmholtz** (1869), p. 377.

[550] **Helmholtz** (1869), p. 376, cf. Section 6.3.1.6.

[551] **Helmholtz** (1869), p. 377.

which plays a key part in this speech just as it did in the Introduction from 1847. At that time he had not assumed that matter could be split atomistically until he had incorporated the concept of the chemical element and that – besides the mechanical principles borrowed from physics – provided the crucial argument for establishing his dual mechanism. Recourse to chemistry, however, was expressed only in one sentence explicitly and also added a hypothetical aspect to the arguments.[552] He now devotes two detailed passages to the ideas of the atom, which in his opinion follow with necessity from research in chemistry. The atomistically interpreted concept of the element has meanwhile lost its hypothetical connotations and Helmholtz believes that the epochal and scientifically secured advance of empirical knowledge supports the claim to validity bound to it:

> The great progress of chemistry [has] conclusively solved the ancient problem of discovering the elementary substances [...] in place of the four primitive metaphysical elements – fire, water, air and earth – we now have the sixty-five elements of modern chemistry. Science has shown that these elements are really indestructible, inalterable in their mass, unalterable also in their properties.[553]

Helmholtz's evocation of early modern natural research is typical for how the classical conception of science saw itself in relation to the "metaphysical", i.e., non-experimentally "proven" doctrine of the four elements from the ancient and medieval philosophy of nature. Not only does he adopt the claim to truth asserted from the days of pre-Socratic philosophy of nature but suggests that it is justified for the first time and indubitably by a new basis for validity as guaranteed by research. What had always been thought to be true, prior to all scientific experience, namely the existence of elements, now seems verified by recent findings in chemistry.

But the material aspect of natural phenomena has not been completely characterized until it has been verified that no other matter exists in nature besides the elements. Without specifically mentioning that this condition is tied to the assumption that there must be another substance, namely ether, which differs from chemical substances (cf. Section 6.3.2.3(i)), Helmholtz explains that this piece of evidence can also be found in chemistry. In doing so, he proceeds directly to the foundation for his mechanism:

> From this invariability of the elements follows another and wider consequence. Chemistry shows by actual investigation that all matter is made up of the elements which have been already isolated. These elements may exhibit many differences as regards combination or mixture, the mode of aggregation or molecular structure – that is to say, they may vary the mode of their *distribution in space*.[554]

This last statement increases the amount of interpretation in Helmholtz's report on research in chemistry considerably compared to the amount of description it contains. From the phenomenally detectable combination and separation of chemical

[552] **Helmholtz** (1847a), p. 5, cf. Section 6.1.1, Part γ.

[553] **Helmholtz** (1869), p. 378 (Germ.)/210 (Engl.).

[554] **Helmholtz** (1869), p. 378 f. (Germ.)/211 (Engl.) (as Section 1.1, footnote 31).

substances, from which he had derived the concept of the element, he moves effortlessly to more fundamental processes that he reduces to purely spatial changes of the elements. Here, however, he creates the impression that it is not about interpretation, not about an approach for reducing phenomena to matter in motion but about a fact revealed by "actual investigation" in chemistry. However, even he does not want to claim final certainty for it yet. Still, that is not the reason why the passage following the last quote, where he proceeds to formulate his mechanism, is phrased conditionally. Instead, it underscores the fact that the mechanistic program will follow with necessity from research in chemistry when chemistry has reached its already distinctly foreseeable goal – a goal that paradigmatically meets the demand of classical science to erect a complete system of true knowledge:

> If, then, all elementary substances are unchangeable in respect to their properties and only changeable in respect to their distribution in space – it follows that all changes in the world are changes in the local distribution of elementary matter and are eventually brought about through *motion*.
>
> If, however, motion be the primordial change which lies at the root of all other changes occurring in the world, every elementary force is a force of motion and the ultimate aim of physical science must be to determine the movements which are the real causes of all other phenomena and discover the motive powers on which they depend; in other words, to merge itself into mechanics.[555]

Consequently, Helmholtz's investigation into the duality of matter and force is divided interdisciplinarily. On the one hand, in phenomenological investigations chemistry demonstrates the (discrete) composition of matter that combines qualitatively different elements. It also confirms the notion that those elements only change by spatial movement. Thus beyond all physical mechanics there exists the reason for applying mechanical principles to nature's elementary structures. On the other hand mechanics – particularly due to its concept of force developed without assistance from chemistry – provides the theoretical framework for performing the reduction of phenomena to matter in motion. If elements are the ultimate components of matter, then the mechanical forces related to them are the ultimate causes of natural events.

Compared to the Introduction from 1847 the relation of matter to force has thus now been transposed and lets the dual character of Helmholtz's conception of nature throughout the entire period between 1847 and 1869 emerge once more. While in the Introduction matter represented the quantitative and force the qualitative aspect of scientific knowledge, now certain qualitative prerequisites are tied to the concept of matter as defined by chemistry and must be reproduced by the forces as quantitative relations. While the materialistic and dynamic traditions of mechanism each select one of the two concepts as its basic principle, in dual mechanism matter and force are not reducible to one another.

---

[555] **Helmholtz** (1869), p. 379 (Germ.)/211 (Engl.).

6.3.2.2 Halved Mechanism: Matter and Life

When Helmholtz says that the goal of all the natural sciences is to "merge itself into mechanics" based on the assumption that motion is the root of all change "in the world", he does not mean that natural science should also reduce the thought and action elaborated in the moral sciences to mechanics. In the speech *Relation of Natural Science* he held a scientific, i.e., causal-law-guided compilation of the phenomena of the moral sciences for the impossible and argued explicitly against them being reduced to natural laws:

> In fact, in ascribing to ourselves free-will, that is, full power to act, without being subject to a stern inevitable law of causality, we deny *in toto* the possibility of referring at least one of the ways in which our mental activity expresses itself to a rigorous law.[556]

The distinction between moral and natural sciences generated by the difference of free will and causality corresponds to an antagonism existing between mind and nature in his conception of the world, respectively, nature. For Helmholtz to have a mind means more than simply being aware of a difference between an internal and an external world and grasping the laws of nature, it means especially to have "the full power to act" and the capacity to master the world, respectively, nature.[557] In later speeches he will affirm this opinion and, in spite of his increasingly reductionist stance on thinking,[558] stresses repeatedly that natural science is the crucial means for subordinating nature to the command of the mind.[559]

In this view, marked by two incompatible extremes, nature is essentially what is causally determined and contains no phenomena generated by purposefully active forces or created by the interference of transcendental powers.[560] In this context one can also say that nature is that which is knowable through concepts, respectively, abstractions of moving, indestructible and inert matter on the one hand and a motion-altering mechanical force on the other. Helmholtz directs this mechanistic concept of nature particularly at contemporary idealistic and romantic explanations of nature, which disputably assume that mind and nature are identical and yearn to "make the laws of the mind also the laws of reality".[561]

Here the issue is not how far the scope of his concept of nature reaches[562] but instead, of where to position life between causality and freedom of will. My thesis is that in Helmholtz's conception of nature, human and nonhuman organisms together

---

556 **Helmholtz** (1862), p. 171 (Germ.)/85 (Engl.) (as Section 6.3.1.2, footnote 437), cf. also **Helmholtz** (1855), p. 161.

557 **Helmholtz** (1862), p. 171 and 180 ff.

558 On the reduced conception of logical thinking developed towards the late 1860s cf. Section 6.3.1.4 and on the later development Section 7.1.3.2.

559 **Helmholtz** (1871b), p. 43 and 50, **Helmholtz** (1878a), p. 243, **Helmholtz** (1897 ff.), Vol. I.1, p. 21, and other places.

560 **Helmholtz** (1862), p. 179, **Helmholtz** (1856 ff.), p. 443, and other places.

561 **Helmholtz** (1855), p. 99. Helmholtz does not explicitly mention idealistic and romantic explanations of nature but instead speaks of "more recent philosophy" (Ibid.).

562 Cf. Sections 6.1.1, Part α, and 6.3.1.2.

take up an intermediate position, which he sees as in harmony with his findings from research into perception and physics. Life is, as far as the metabolism of chemical substances is concerned, part of causally determined nature. But humans and also to a lesser degree animals are further characterized by their mental capacities that allow them to act as they will. Since for Helmholtz the capacity to act cannot be reduced to causality, it is understandable that he cannot applaud any theory that would view an entirely mechanically determined world as being the origin of (animal and human) life. My thesis is based on his theory of perception, his standpoint on the origin of life and the Darwinian theory of evolution and on his interpretation of the second law of thermodynamics. Three factors may have moved Helmholtz to view life ontologically as an independent entity alongside matter: first, his empiricist theory of perception suggests it. Second, evolutionary theory that relies on variability may offer a causal explanation for the development of everything organic but it cannot causally explain how life originated from matter. Third, the (inadmissible) application of thermodynamic regularities makes the existence of complex structures, such as those of organisms, an inexplicable phenomenon. Here I shall restrict my discussion to his theory of perception.[563]

The first clues in Helmholtz's public lectures and speeches[564] indicating that phenomena of life have a different place than that of material nature can be found in his *habilitation* lecture *On the Nature of Human Sensory Perception* from 1852, where he establishes the symbolic character of sensations using a law stemming from the vitalistic tradition – "the law of the specific nerve energies" formulated by his mentor Johannes Müller.[565] For Müller physical sensations and the perceptions based on them were caused by a life force to which he traced all traits characteristic of life. Life force generates, as the "ultimate cause" *(Endursache)* of every creature, the distinct, holistic and functional essence of every organism. It is also responsible for the inner transmission (and partially the production) of movements and the capacity for sensation concentrated in the nerves.[566] To the latter Müller traced the subjectivity of the perception characteristic of both humans and animals.[567] His vitalistic interpretation of sensations and perceptions stood in explicit

---

[563] For his standpoint on the origin of life, on the Darwinian theory of evolution and on the second theorem of thermodynamics see **Schiemann** (1997), p. 295 ff.

[564] Helmholtz's papers in science contain an even earlier mention that the phenomenon of life takes a special position: **Helmholtz** (1843), his first publication following his dissertation, contains a vitalistic interpretation of fermentation.

[565] Cf. Section 6.3.1.1. Müller understood the law of the specific nerve energies as a fundamental law for the physiology of the senses and made it one of the primary "necessary pre-concepts" for relevant investigations: "Sensation is not the transmission of a quality or a state of the external bodies to consciousness, it is the transmission of a quality or state of a sensory nerve to consciousness, instigated by an external cause and these qualities are different for each type of sensory nerve, they are energies of the senses" (**Müller** (1833 ff.), Vol. 2, p. 254). Another version of the law is stated in the programmatic preface to **Müller** (1826), p. XVI, where he also discusses mechanistic conceptions.

[566] **Müller** (1833 ff.), Vol. 1, p. 18, 24 f. and 47.

[567] Müller's ideas in physiology underwent considerable change from a romantic attitude to one that was rather positivistic and experimental. He stresses the subjectivity of sensation particularly in his early writing **Müller** (1826), p. 39 ff.

opposition to a long tradition of mechanistic theories of perception that had described sensations as the transmission of stimuli through which consciousness received a direct report on the state of an outer body.[568] In contrast to a mechanism that directly represents objects in the perceptive mind he stressed that in interaction with the outside world a person only perceives himself and the ideas of the constitution of the outward body that are transmitted by the senses as only "relatively correct".[569]

In his lecture Helmholtz describes the objective properties of light as understood by Newton's experimental physics in order to show how sensations are processed. He emphasizes the intrinsic difference from vitalist physiology by explicitly rejecting Goethe's theory of color, which Müller favored.[570] In spite of this critical initial position and although in the context of his investigation on animal warmth Helmholtz also clearly distanced himself from his mentor's vitalism, Müller remains the discoverer of the "merely subjective value"[571] of sensations not in name only. What Müller's theory and Helmholtz's own theory of signs have in common is obvious, even though Helmholtz never discusses the former in detail: he speaks of "the special activity of the optical nerve".[572] Through their capacity for sensation humans are confronted with an entirely insensible, causally determined external world. By saying that sensations have a sign character, Helmholtz makes irreducible aspects of subjectivity a part of perception. For Helmholtz, similar to Müller's theory, without scientific assistance subjects are incapable of objective knowledge: when they perceive an object, they are in reality only sensing themselves.

In the following years Helmholtz begins to rule out any vitalistic interpretation of his theory of signs, based on elaborating where he stands on life forces in physiology. Once mid-century research in physiology had proven that the chemical and physical composition of different identifiable nerves is identical,[573] Helmholtz gives a new meaning to Müller's "law of the specific nerve energies", which he continues to use for establishing his theory of signs.[574] On this new interpretation, the subjectivity of perceptive experience is no longer significant. Instead of providing an explanation that would permit a vitalistic premise, Helmholtz offers what "sounds trivial", namely, that "*under different conditions the same causes can produce different effects*"[575]:

---

[568] **Müller** (1826), p. XI ff. On interpreting the "process of perception as transmission in the meaning understood by the mechanistic conception of nature" cf. **Post** (1905), p. 9 ff. and 29, and on traditions that go back further, for example John Locke, see **Maier** (1938).

[569] **Müller** (1833 ff.), Vol. 2, p. 254 ff., where Müller attempts to also empirically prove that there are limits to the truth of ideas.

[570] **Helmholtz** (1852), p. 601.

[571] **Helmholtz** (1852), p. 607 (as Section 6.3.1.1, footnote 408).

[572] **Helmholtz** (1852), p. 605 (as Section 6.3.1.1, footnote 413).

[573] On himself he reports: **Helmholtz** (1868a), p. 296 ff.

[574] **Helmholtz** (1869), p. 392, and later: **Helmholtz** (1878a), p. 219 f. (however, without direct reference), and **Helmholtz** (1892), p. 357.

[575] **Helmholtz** (1868a), p. 298 (wide letter spacing in the original has been italicized).

> All the difference which is seen in the excitation of different nerves depends only upon the difference of the organs to which the nerve is united and to which it transmits the state of excitation.[576]

But it would be a mistake to think that by stressing his distance to vitalistic interpretation Helmholtz is paving the way for the mechanistic interpretation of perception of which Müller disapproved and in doing so, is also making way for a mechanistic theory of vital phenomena. The opposite is the case. Helmholtz supplements his particularly vision-based theory of perception with just that element of action theory that I already pointed out for the context of the theory of science: similar to experimental procedure in science, perceptive individuals navigate their way through the world – a world for which they only have signs to go on–by inductively evaluating tentative actions. They adjust their bodies in various spatial positions in order to obtain various views of an object, which they then generalize to become an idea. The truth of such an idea can only be decided pragmatically, i.e., depending on the success of the action.[577] The theory's "empiricist" content and potential application to non-human living beings is what makes it important in terms of the philosophy of nature. While Müller thought that specific nerve energies are innate capacities,[578] Helmholtz now claims that perception only truly represents the external world if one assumes that the individual perceiving subject can adjust flexibly:

> The empiricist theory of vision assumes that none of our sensations give us anything more than 'signs' for external objects and processes and that we can only learn how to interpret these signs by means of experience and practice.[579]

Here "learning by means of experience" basically implies the opposite of innate, respectively, instinctive mechanisms that causally determine certain behavior. For Helmholtz "experience", designating an action-imparting learning process, can only occur where a certain degree of freedom enables the perceiving individual to move in space at will.[580] However, that does not mean that learning processes are not subject to any kind of regularity. Normally the inductive processing of individual perceptions occurs unconsciously and the ideas of external objects that result present themselves to the subject with "compelling necessity".[581] Thus one can no longer recall how one in early childhood learned to estimate distances for everyday

---

576 **Helmholtz** (1868a), p. 297 f. (Germ.)/150 (Engl.).

577 **Helmholtz** (1856 ff.), p. 445 ff., cf. Sections 6.3.1.4, 6.3.1.5 and 7.1.3.2.

578 Cf. Helmholtz's critique on Müller in **Helmholtz** (1856 ff.), p. 805. Helmholtz assumes that not only Müller but also Kant presupposed that the processing of sensations is an innate capacity. Müller allegedly "enhanced" Kant's transcendental philosophy with his own theory of perception. Nevertheless, Müller's writing contains no such reference to Kant.

579 **Helmholtz** (1868a), p. 332 (Germ.)/177 (Engl.), cf. Section 6.3.1, footnote 407.

580 **Helmholtz** (1856 ff.), p. 431 ff., and later in more detail in **Helmholtz** (1878a), p. 223 ff.

581 **Helmholtz** (1856 ff.), p. 430, cf. also **Helmholtz** (1855), p. 112 ff., and **Helmholtz** (1868a), p. 358 f. Sometimes Helmholtz confuses unconsciousness with absence of will: **Helmholtz** (1856 ff.), p. 804.

practical purposes, yet under altered circumstances (continuous use of spectacles) one could learn to do so again.[582]

Understandably, the philosophically interesting distinction of how instinct relates to experience is one of the central issues in Helmholtz's theory. He seems to be aware that particularly investigations into sensation processing in infants and animals might afford insight into the matter.[583] However, since a sufficient quantity of research material is not available, Helmholtz permits himself only to speculate. As fragmentary as his rare statements from the 1850s and1860s are, they clearly share the assumption that the transition from animal to human sensation processing is smooth. By also reviewing his later remarks on the topic, one finds that Helmholtz thought that even in animal sensation processing is not entirely innate.[584] In my opinion this is why, in the *Treatise* he resolutely asserts that "both animals and humans [possess] a principle of free will", which he claims is "decidedly detached from the rigorousness of causal law".[585]

In *The Recent Progress in the Theory of Vision*, a speech consisting of a more popularly phrased version of a few of the *Treatise* chapters, Helmholtz mentions the "mental capacities" of animals.[586] His farsighted example from behavioral physiology is (interpreted slightly differently) still current. Opposed to seeing a parallel between the allegedly instinct-guided behavior of animals and small children he provides a familiar example for the learning patterns of both, which today is known as imprinting:

> Just as a child that has learned to drink from a bottle afterwards no longer desires the breast, so also do ducklings that have grown up in the kitchen eschew the pond and a chick that has known no hen before becoming five days old will follow a person who has cared for it and not a hen. This seems to demonstrate that in comparison to experienced facts, instincts that operate initially as long as memory is a tabula rasa, quickly lose their impact. [...] precisely this distinguishes humans from animals, namely that [...] in humans innate instincts are reduced to the smallest possible order.[587]

By closing ranks between humans and animals the mental phenomena that they characteristically have in common are ostracized from causally determined natural phenomena. The fact that Helmholtz in various passages speaks of "life" as opposed to "lifeless nature" or "matter" indicates what he considers to be the independence of life-related

[582] **Helmholtz** (1855), p. 114.

[583] **Helmholtz** (1855), p. 115, **Helmholtz** (1856 ff.), p. 431, **Helmholtz** (1868a), p. 362 f.

[584] **Helmholtz** (1885 ff.), p. 599 ff., esp. p. 600 ("Mental Activity in Animals"; "Seelenthätigkeiten der Thiere") and p. 602 (cf. Section 7.1, footnote 696). Cf. also the undated paper on the history of mechanics in **Koenigsberger** (1902 f.), Vol. 2, p. 31 ff.

[585] **Helmholtz** (1856 ff.), p. 454 (as Section 6.3.1.4, footnote 491).

[586] **Helmholtz** (1868a), p. 362.

[587] **Helmholtz** (1868a), p. 363. The translation by David Cahan does not contain this passage. Like today's term 'imprinting', Helmholtz's example expresses less an ability for learning than the fact that a certain inalterable fixation is inborn.

phenomena.[588] It culminates in the assumption that life cannot originate in matter. According to Helmholtz the material conditions of the earth can be explained by a theory of the origin of the planetary system but the same cannot be done for the origin of species.[589]

6.3.2.3 Mechanistic Invariants in Nature

As a program of reduction, mechanism undertakes to causally explain individual phenomena; as a philosophy of nature it goes beyond the science-immanent function of reducing and declares that basic traits are inherent to the entirety of nature.[590] The fundamental traits or invariants of nature found in Helmholtz's popular lectures and speeches generalize specific successful scientific explanations, thereby making them the conditions underlying all natural events. Within the context of the philosophy of nature that makes natural science a science of conditions. The traits also concern human technological interaction with nature and mankind's fundamental attitude towards nature. A science of conditions thus acquires another meaning: it is subject to the normative demand to contribute to altering the material conditions of nature such that natural forces become exploitable for mankind's purposes.

(i) The invariants in Helmholtz's philosophy of nature include first of all the elements of a duality in principles already mentioned:

- The opposition between nature and mind corresponding to the distinction he makes between natural and moral sciences.
- The disjoining of lifeless nature (matter) and life, which is meant to distinguish both the mental capacities that people and perhaps also animals have, as well as the origin of life.
- The dual constitution of (non-mental) nature, where elementary substances and forces are mutually referential. Generalized, this means that Helmholtz allows no interaction without the presence of a material substrate. For this reason, since he denies the corpuscular theory of light, he finds it inconceivable that light waves traverse a space void of matter. Space in which no matter is detectable but which is nonetheless traversed by light, must, according to Helmholtz, contain a different

[588] **Helmholtz** (1871a), p. 90: "lifeless nature", and **Helmholtz** (1874a), p. 419: "whether [...] life [...] is perhaps not just as old as matter"; cf. the corresponding distinctions in **Helmholtz** (1854a), p. 66 and 75f., **Helmholtz** (1862/3), p. 226f., **Helmholtz** (1862), p. 180, **Helmholtz** (1871a), p. 89ff., and later in **Helmholtz** (1878a), p. 246, and **Helmholtz** (1892), p. 360.

[589] **Helmholtz** (1854a) and (1871a).

[590] Cf. Section 1.1.

substance, namely ether.[591] The result is that he subdivides matter into ethereal and chemically (already) detectable substances.

Although Helmholtz also elaborates mechanical equations for ether, being a medium of electromagnetic phenomena it also has properties that cannot be explained mechanically.[592] Helmholtz holds the knowledge of the structure of chemical substances alone as sufficient for establishing (in line with his scientism) a dual mechanism, according to which (non-mental) natural phenomena should be traceable to mechanical movements of elementary particles.

(ii) Second, Helmholtz acknowledges no laws of nature that might contradict the mechanical principles and the theorem of energy derived (under boundary conditions from his theory of matter) therefrom. Thus he rejects electromagnetic laws of force, which – like Weber's law,[593] that in German physics was accepted until the 1870s – postulate effects of velocity-dependent forces or he criticizes principles that – like his mentor Johannes Müller's vitalistic life force – are supposed to be the cause of work but do not consume energy.[594]

Helmholtz accepts regularities and assumptions that seem to be compatible with mechanical principles although they are not (yet) explainable in terms of mechanical premises and integrates them into his conception of nature. The most important

---

[591] Initially for Helmholtz the existence of ether follows directly from the mechanistic interpretation of optical appearances, which he – in terms of energy – identifies with heat rays (e.g., **Helmholtz** (1854a), p. 79, **Helmholtz** (1855), p. 98 ff., and **Helmholtz** (1871a), p. 71 ff.). Later the mathematical theory from his electrodynamics, where he links the separate equations for matter and ether by a joint term, will become the central argument (esp. **Helmholtz** (1875b), **Helmholtz** (1881), p. 256 ff., **Helmholtz** (1892b) and **Helmholtz** (1897 ff.), Vol. IV, p. 1 ff., and Vol. V, p. 43). Cf. also: **Woodruff** (1968) and **Buchwald** (1985).

[592] The non-mechanical properties of electromagnetic substrate substances had been for the most part uncontested after recognition of the undulation theory of light at the onset of the century and consisted in the straight-line dispersion of transversal waves.

[593] On the predominance of Weber's theory of electrons in the third quarter of the nineteenth century cf. **Helmholtz** (1894b), p. XII, and **Boltzmann** (1899), p. 204 f. In this theory all electrodynamic phenomena were to be traced back to the effects of charges in motion. Weber followed up Gustav Theodor Fechner's hypothesis that electrical current is based on paired movement of positive and negative charge carriers, that move with the same velocity but in opposite directions. The result was a term for the force existing between two electric charges, which in contrast to the time-independent force of gravitation depended on the velocity and the acceleration of the charge carriers. Weber's law of force for two resting and moving charges (Q_1, Q_2) at distance r reads with a constant c:

$$F = [(Q_1 \cdot Q_2)/r^2] \cdot [1 - (1/c^2) \cdot (dr/dt)^2 + (2r/c) \cdot (d^2r/dt^2)]$$

Helmholtz was able to demonstrate that under certain conditions Weber's law was incompatible with the theorem of the conservation of energy, which in his opinion confirmed that one of the fundamental forces of nature is to be immutable over time. Cf. overall: **Whittaker** (1958), Vol. 1, p. 203 ff., **Kaiser** (1981), p. 92 ff., and specifically on the relation between Helmholtz and Weber: **Archibald** (1989).

[594] As footnote 543.

example from physics is the second law of thermodynamics.[595] It also includes though perhaps non-mechanically structured ether or the "mental functions" of animals.

Finally, laws of force that satisfy the form of mechanical (central) forces are universally valid. As early as the rectorate speech he presents a version of the Newtonian law of gravity that allows for no principle difference between the various dimensions:

> Every bit of ponderable matter in the universe attracts every other bit with a force varying inversely as the square of the distance.[596]

The force of gravity is effective among the planets just as it is between double stars that are light years away from planets. Based on his conviction that the conservation of energy holds without restriction throughout all of nature Helmholtz concludes that all material interaction must be based on central mechanical forces. He is convinced that differences in interactions can be reduced to the algebraic sign and the quantity of the absolute value of the underlying forces.[597] Although interactions in atomic dimensions are not yet measurable, he therefore believes that their essential features can already be concluded from rudimentary observation. Thus he concludes from combustion phenomena that an "attractive force" exists between the carbon and the oxygen atom, which "only acts through them with extraordinary power, if the smallest particles of the two substances are in closest proximity to each other".[598]

Central forces that depend solely on distance not only span the entire cosmos but are also responsible for its evolutionary genesis. Helmholtz lectures on Kant's and Laplace's "hypotheses" about the origin of the planetary system, making minor corrections[599] and philosophically asserts that cosmology is mechanistic, comparing obtained knowledge to useful everyday navigation instruments:

> Physical-mechanical laws are, as it were, the telescopes of our spiritual eye; which can penetrate into the deepest night of time, past and to come.[600]

The notion that the scientific study of nature can limitlessly transcend the present rests on a rigorously deterministic view, which nonetheless does not claim to know the "origins of heat and light"[601] or life. Helmholtz uses the phrase "physical-mechanical" to reflect the fact that thus far very few of the known natural laws have

---

595 Cf. **Schiemann** (1997), p. 298 ff.

596 **Helmholtz** (1862), p. 176 (Germ.)/89 (Engl.).

597 Cf. Section 6.1.1, Part γ.

598 **Helmholtz** (1862/3), p. 219, cf. also **Helmholtz** (1861), p. 569 f.

599 **Helmholtz** (1854a), p. 68 ff. Helmholtz calls the assumption that the masses of the planetary system were "originally distributed nebularly in space" the only hypothesis in the speech (loc. cit. p. 72). Nonetheless, he must also hypothetically presuppose a "slow rotating motion" for this "nebular ball" (loc. cit, p. 69). He calls Kant's and Laplace's hypotheses "one of the fortunate moments in science" (**Helmholtz** (1871a), p. 84).

600 **Helmholtz** (1854a), p. 80 (Germ.)/40 (Engl.).

601 **Helmholtz** (1854a), p. 69.

been reduced to mechanics. It is also reflected paradigmatically in both principles that in his opinion contain the fundamental basic conditions for natural processes: the principle of the conservation of energy derived from mechanical theorems and the quasi-parallel principle of entropy (second law of thermodynamics).

Aided by these two invariant principles "we can now at pleasure regard this or the other side of the surrounding world".[602] Seen through the spectacles of energetics and thermodynamics, nature becomes a reservoir of potential work effects and energy conversion. Helmholtz splits the "total force store of the universe" into one part consisting of eternal heat and another consisting of the part convertible into work, where the latter "constitutes the whole wealth of change which takes place in nature".[603]

(iii) This view contains the theory that the universe will end in heat.[604] It is another of the invariant features of Helmholtz's conception of nature – namely, his explicit claim that the development of nature is finite. Under the increasing decline of religious-cosmological worldviews, Helmholtz's calculations for the earth's life expectancy and the maximally expected continuation of life on earth[605] met with great public interest and also drew the attention of some of his colleagues.[606] No longer from the pulpit but instead now from the podium of a lecture hall for the association of physics and economics in Königsberg one could hearken when and how "the last day of the human race" draws near.[607] But instead of opening the gates to a new kingdom, the scientifically predicted demise of earth's history means that the book will be closed forever. While yet distant, still the discernibly harrowing future is fitting for mankind's negligible place in the cosmos:

> that we, who like to consider ourselves the center and purpose of creation, are only motes of dust on earth and the earth is a mote of dust in the fathomless universe.[608]

Cloaked as the scientific discussion of physical conditions a project unfolds that was already implicit in the theory of science's claim to complete explanation: to relieve theology of interpreting the world and the future.

(iv) While Helmholtz's scientifically objectionable interpretation of the second theorem permits an outlook on the distant physical future, the law of the conservation of energy guides the look back to the present, where mankind need not yet be concerned with its termination but instead finds itself on the way towards mastering

---

[602] **Helmholtz** (1854a), p. 67.

[603] **Helmholtz** (1854a), p. 66.

[604] **Helmholtz** (1854a), p. 66f. Cf. **Schiemann** (1997), p. 298ff.

[605] **Helmholtz** (1854a), p. 80ff., corresp. to **Helmholtz** (1871a), p. 87f.

[606] In 1884 Helmholtz himself reports that his talk *On the Origin of the Planetary System* [*Ueber die Entstehung des Planetensystems*] has "recently been a favorite topic in reviews on popular science and philosophy" (**Helmholtz** (1884a), Vol. II, p. VI), of which that of **Lange** (1866), p. 666ff., is probably most well-known today. On how Helmholtz was received in British physics cf. **Smith and Wise** (1989), p. 498ff., 520ff. and 560f.

[607] **Helmholtz** (1854a), p. 83, and similarly, **Helmholtz** (1871a), p. 88ff.

[608] **Helmholtz** (1871a), p. 88.

nature. In a metaphorical manner, abstaining from mechanistic interpretation and considering energetics aspects, Helmholtz describes nature as a kind of steam engine. This metaphor addresses a fourth invariant of his conception of nature: external nature is essentially a donor of usable energy.

In cosmic dimensions "the sun runs on earth a kind of steam engine whose performance is far greater than that of artificially constructed machines".[609] As examples he mentions the hydrological cycle, tides[610] and the metabolism of organic beings. As early as the 1850s he points out:

> The animal body therefore does not differ from the steam engine as regards the manner in which it obtains heat and force but does differ from it in terms of the purposes for which and in the manner in which the force gained is to be made use of [. . .].[611]

While definitions of purpose distinguish creatures from machines, mankind is distinguished from the animal kingdom by the capacity to generate purposes. Still, man's possible actions are limited by the basic energetic conditions:

> We cannot create [...] mechanical force but we may help ourselves from the general storehouse of Nature.[612]

Of course, the fact that Helmholtz mentions the finiteness of reserves has little to do with the insight into the idea that mankind might exhaust resources in the foreseeable future; it has more to do with the idea that the conservation of energy implies that the total amount of all energy remains constant.

How unaware he is of the necessity of sparingly consuming natural carriers of energy is demonstrated by the great attention he gives the concept of work [*Arbeit*] in his remarks on energetics and that is paradigmatic for technical applications.[613] With respect to the exploitability of natural energies the concept of work obtains a remarkable double meaning. On the one hand he means physical value (of the product of trajectory and directed force),[614] which Helmholtz considers the general unit of measurement for natural energy converted in technical processes.[615] On the other hand, he also speaks of "human labor" [*menschliche Arbeit*] in a wider sense that can only partly be defined in terms of "the force which is expended in it" and partly "by the skill, which is brought into action".[616] The distinction being made here

---

609 **Helmholtz** (1871a), p. 79.

610 Ibid.

611 **Helmholtz** (1854a), p. 75 (Germ.)/37 (Engl.).

612 **Helmholtz** (1854a), p. 65 (Germ.)/29 (Engl.).

613 **Helmholtz** (1854a), p. 54 ff., **Helmholtz** (1862/3), p. 192 ff. Here technical applications are clearly distinguished from observation in the theory of science because for Helmholtz the universality of a natural law is first proven in the validity it has independent of all application contexts (**Helmholtz** (1854a), p. 63, and **Helmholtz** (1862/3), p. 227).

614 Or the product of weight and height of fall (**Helmholtz** (1862/3), p. 196). Note on translation: the meaning in physics of the English term 'work'.

615 **Helmholtz** (1854a), p. 54 and p. 58: and as a general unit of measurement: **Helmholtz** (1862/3), p. 196, cf. Section 6.1.2, Part γ.

616 **Helmholtz** (1854a), p. 54.

is one of labor seen as a qualitatively human capacity in contrast to the quantitative unit of measurement for work that is entirely sufficient for thinking of machines in terms of energy but Helmholtz fails to keep the two meanings clearly apart. Indeed, there exist domains of action where human skill "cannot be supplied by machines"[617] but conversely, often machines are able to "work skillfully"[618] and are therefore qualitatively commensurable with human performance. Inasmuch as human labor can be reduced to the physical concept of work, it is comparable with machine work[619] and indeed, "by the magnitude of their effects" machines can surpass "the power of men and animals".[620] The steam engine is thus not merely a metaphor from the philosophy of nature's energetics approach. It is the symbol for the true purpose of natural forces in Helmholtz's worldview: to provide energy for ever spreading industrialization. Aware of the epochal commencement he declares:

> You all know how powerful and varied are the effects of which steam engines are capable; with them has really begun the great development of industry which has characterized our century before all others. [...] By means of these machines we can develop motive power to almost an indefinite extent at any place on the earth's surface [. . .].[621]

---

[617] Ibid.

[618] Ibid. **Breger** (1982), p. 224 f., points out this passage.

[619] **Helmholtz** (1862/3), p. 193.

[620] **Helmholtz** (1854a), p. 54, and corresp. p. 52.

[621] **Helmholtz** (1862/3), p. 211 (Germ.)/112 f. (Engl.).

Chapter 7
The Hypothetization of Helmholtz's Mechanism

7.1 Helmholtz's Conception of Science from the Early 1870s on

Unmistakably, in the 1870s Helmholtz's conception of science and nature deviates fundamentally from his former conception at least from time to time. While his 1869 speech held for scientists had proclaimed that the "ultimate aim of physical science must be [...] to merge itself into mechanics,"[622] less than ten years later he distances himself in the speech *The Facts in Perception* [*Die Thatsachen in der Wahrnehmung*] in principle from absolute reductionist claims:

> Every reduction of the phenomena to the underlying substances and forces claims to have found something unchangeable and definitive. We are never justified in making an unconditional claim of this type; for it grants neither the fragmentary nature of our knowledge nor the nature of inductive conclusions, upon which rests, from the first step on, all our perception of the real.[623]

Such a statement no longer fits a classical conception of science offhand and can only be accommodated by an operose argumentative effort. While the "fragmentary nature" of knowledge may be minimized as scientific progress advances, the methodological considerations Helmholtz mentions are of a fundamental nature and directly contradict the conception of science he had propagated during the 1860s: elementary processes of perception, which in the 1860s were supposed to guarantee the possibility of the reduction to the "ultimate forces" by occurring simultaneously to the events they represented, now decisively contribute to the fact that the reduction to just these forces will "never" be claimed unconditionally.

Here Helmholtz is not denying reduction in general. By – in my opinion purposefully – using the definite article and speaking of "the underlying substances and forces" he presupposes fundamental entities that include the atoms and central forces

[622] **Helmholtz** (1869), p. 379 (as Sections 1.1, footnote 31, and 6.3.2, footnote 555).

[623] **Helmholtz** (1878a), p. 243 (Germ.)/362 f. (Engl.) .

of the mechanistic conception of nature and serve the purpose of a reductionist explanation of phenomena. The deviation from his conception of science throughout the 1850–1860s thus appears to be less an alteration of the methodical definition of science's objective than much more the overall abandonment of the absolute claim to validity previously bound to that task.

Its rejection is precisely the feature characteristic of the modern conception of science. In modern times the program of science intended to create an exclusively valid system of knowledge has been replaced by a procedural conception of science shaped by the belief that scientific statements are conditional and their truth is open. While the classical system of knowledge would have no use for scientific hypotheses, in modern times hypotheses are the epitome of what is considered scientific. Hypotheticity means neither arbitrariness nor insufficient empirical verifiability of statements; it means attributing unquestionable relativity to scientific statements, no matter how well they have been corroborated or proven.[624]

At the same time, modern science can acknowledge poorly corroborated and barely proven hypotheses as being scientific. To the extent that in terms of content research is no longer bound to one uniform aim, on the theoretical level science is characterized by diversity. This gives scientistic conceptions of nature more options for scientifically justifying their statements. Thus not until modern times could mechanism, as well, follow-up on its reductionist conception of nature, without emphatically claiming to be true, as the classical conception had required of it.

In the following I would like to use Helmholtz's lectures and speeches to explore whether and how much in the course of the 1870s his conception of science underwent a process of transition – headed towards a modern interpretation. Do these lectures and speeches contain proof of a fundamental correction that would allow severing his conception from the classical and associating it with the modern understanding? Or can one find merely occasional isolated statements that contradict a classical framework that he otherwise further entertained? If a transition process can be detected, can one also discover when it began to emerge?

I would like to propose a thesis in answer to the last question only: in his public lectures and speeches, the changes in Helmholtz's conception of science are accompanied by a break in the development of his mechanism. In 1871 he delivers and publishes a speech in which he for the first time distances himself from the previously uncontested foundations of his conception of nature. If a process of transition can be detected, then this thesis accesses a further-reaching understanding stating that in the 1870s the development of his conception of science correlates to the emergence of a difficulty in his theory of nature.

---

[624] See Section 4.1.

The changes in Helmholtz's conception of science and nature were not given greater attention until the most recent secondary literature.[625] Probably the most frequently mentioned indication of a possible transitional process, the change in his publicly attested attitude regarding Kant's philosophy, relates above all to the theme of causality, which I also find very significant.[626] In the 1870s, however, Helmholtz's understanding of causality is so far removed from the Kantian position that his relation to transcendental philosophy hardly provides an appropriate standard for adequately evaluating the changes in question.

Using material from popular publications I shall now first trace the signs for a possible transition, beginning with the mechanism-critical speech *Gustav Magnus. In Memoriam.* [*Zum Gedächtniss an Gustav Magnus*]. To do justice to the transition's inner dynamics I distinguish two levels: (1) distancing himself from atomism, followed by (2) a reflection on the scientific method. I find that in the last two speeches from this period, namely *On Thought in Medicine* [*Das Denken in der Medicin*] (1877) and *The Facts in Perception* (1878), Helmholtz formulates a position to which in the subsequent period he will continue to adhere without modification. In reconstructing it, aspects of the contents are of lesser importance, because the interesting issue is where to place Helmholtz's conception of science within the tension between the classical and the modern conceptions, particularly in terms of how they are founded (3). The question of Helmholtz's continued relation to mechanism will be expounded in the next section.[627]

7.1.1 Emerging Critique of Atomistic Hypotheses (1871)

In 1871 Helmholtz abruptly begins to scrutinize atomistic conceptions with a resoluteness that contrasts remarkably with the fact that two years earlier he had paid

---

[625] See the literature listed for his conception of science in the introduction, footnote 24, on the early changes in that conception in Section 6.3.1, footnote 403, and for his concept of nature in Section 6.3.2, footnote 540. That his conception of science and nature was subject to general change was noted early in the history of receiving his works, although only in passing (see the introduction, footnote 24, and later: **Heimann** (1974), **Elkana** (1974), **Winters** (1985)). Not until the most recent literature has more thoroughgoing significance been attached to the elements of change in his philosophy of science and nature (**Buchwald** (1994a), **Darrigol** (1994), **Hatfield** (1990), **Hatfield** (1994), **Heidelberger** (1994a), **Heidelberger** (1994b), **Krüger** (1994), **Röseberg** (1994) and **Schiemann** (1994)). It is uncontroversial that **Helmholtz** (1847a) marks the beginning of a position and **Helmholtz** (1878a) marks one that clearly deviates from that. The essays between these two poles are sometimes associated with the one, sometimes with the other and the choice of one or the other can depend on which aspect of his development is in focus.

[626] **Riehl** (1904), **Erdmann** (1921), **Hörz and Wollgast** (1971), **Heimann** (1974), **Darrigol** (1994), **Heidelberger** (1994a), **Heidelberger** (1994b) and **Krüger** (1994). **Buchwald** (1994a), p. 366f., stresses the anti-metaphysical character of Helmholtz's early conception of nature. **Hatfield** (1994) calls Helmholtz's position in the philosophy of science overall "naturalistic" (loc. cit. p. 524 and other places) and places it opposite of Kant's idealistic position, as he had already done in **Hatfield** (1990).

[627] See the introduction to this book.

homage to it in the most prominent place, from the podium before assembled scientists.[628] At that time he had concluded from the proven quantitative and qualitative immutability of chemical elements that the elements constitute a real structure of discrete entities and are subject exclusively to the mechanical laws of motion.[629] As positive examples from natural research that had already rudimentarily succeeded in reducing phenomena to elementary forces of motion, Helmholtz mentions besides astronomy and the "purely mechanical part of physics" also acoustics, optics and the study of electricity.[630] As if he had never endorsed attempting a mechanistic reduction, he now explains in his speech *Gustav Magnus. In Memoriam:*

> In the reference to atoms in theoretical physics, Sir W. Thomson says, with much weight, that their assumption can explain no property of the body which has not previously been attributed to the atoms. Whilst assenting to this opinion, I would in no way express myself against the existence of atoms but only against the endeavor to deduce the principles of theoretical physics from purely hypothetical assumptions as to the atomic structure of bodies. We know now that many of these hypotheses, which found favor in their day, far overshot the mark.[631]

In its generality this critique is addressed not only to the research approaches he had unhesitatingly approved two years earlier. In an elementary way it also relates to the existing foundation of mechanism, which was based on just that assumption of the atomic constitution of bodies, i.e., of their being configurations of mechanically moved elements. In contrast to this conception there now emerged the possibility of setting up a theory without assuming substances, a theory limited to the mathematical reconstruction of observable phenomena. The existence of atoms, which continues to be assumed, thus loses its relevance for defining the tasks of theoretical research. What remains is mathematical elementarianism, according to which the world consists of volume elements whose actions accrue to become natural phenomena. With reference to mathematicians and physicists C.F. Gauss, F.E. Neumann, G.G. Stokes, J.C. Maxwell and W. Thomson, Helmholtz continues:

> We now know that even mathematical physics is pure empirical science [...] In immediate experience we find only extended diversely configured and composite bodies; only on these can we make our observations and perform experiments. Their actions are composed of the actions which all parts contribute to the whole, such that if we want to know the simplest and most general laws of action [...] we must return to the laws governing the smallest [...] elements of volume. But these are not, like the atoms, disparate and diverse, they are continuous and uniform.[632]

If the elementarianism advocated here was to remain within the general framework of mechanics, then now turning to continuous quantities would only be significant

---

[628] **Buchwald** (1994a), p. 370 f., has also stressed that this speech marks a break in the development of Helmholtz's conception of science. Cf. also **Heidelberger** (1994a), p. 479.

[629] **Helmholtz** (1869), p. 378 f. Cf. Section 6.3.2.1.

[630] **Helmholtz** (1869), p. 379.

[631] **Helmholtz** (1871b), p. 45 (Germ.)/17 (Engl.). The context surrounding this text indicates that Helmholtz is probably referring generally to notions of the atom from the first half of the nineteenth century.

[632] **Helmholtz** (1871b), p. 45 f. (Germ.)/17 (Engl.).

if the dynamics of discrete mass points were in principle to be distinguished from continuous distributions of masses. However, such a distinction is not compelling and Helmholtz was able, without a radical revision, to formulate his concept of central force using constantly changing mechanical quantities. But here this inner-mechanical problem of theory production to which Helmholtz is alluding is not what is most important. It is a result of making the concept of the atom an issue and indicates upgrading the field of observable physical objects.[633]

That this change does not deflect from the classical conception of science is revealed in the fact that it evaluates hypotheses negatively. Helmholtz disapproves of assumptions that "far overshot the mark", calling them "purely hypothetical". In this emphatic speech (with nationalistic undertones) Helmholtz raises the banner of "eager, ruthless and unselfish love of truth" to ward off the hypothetical. This love of truth, on which one could, in his opinion particularly in Germany, pride oneself, is what impels scientists to "pursue the questions of principle to their ultimate sources", without overly troubling themselves with "practical consequences and [...] useful applications".[634] Science legitimated by such bold motives must clearly distinguish "what is empirical matter of fact" from what is "mere verbal definition" and what is "only hypothesis".[635] The task of science is to discover "*the laws of facts*".[636]

Nonetheless, in one remark he does disclose that the "characteristic properties" of elementary volumes, to which he would now like to reduce phenomena, must be presupposed "hypothetically", whenever knowledge of the observable total effect is insufficient for determining those properties.[637] Hypothetical are therefore assumptions concerning conditions under which regularities operate in the submicroscopic range and which as middle terms enter into the deductive explanation of phenomena. The meaning of the concept of hypothesis used here is still related to that found in Helmholtz's essay of 1847, used then to refer to the atomic composition of bodies.[638] Both meanings can be summarized as saying that hypotheses designate objects or properties that are presupposed as causes or conditions for the purpose of experimental result interpretation. These objects may be either substances, which one hopes to prove beyond a doubt in the future (e.g., atoms), or non-existing,

---

[633] Helmholtz's reassessment of physic's descriptive tasks or rudimentary characterization of physics overall as being a descriptive science, conspicuous from when the speech *In Memory of Gustav Magnus* was delivered, is closely tied to the attention he draws to Kirchhoff's dictum after 1875 (**Helmholtz** (1875), p. 46f., **Helmholtz** (1878a), p. 242, **Helmholtz** (1892), p. 352, and **Helmholtz** (1897ff.), Vol. 1.1, p. 13), namely that the task of mechanics is not to discover the cause of the phenomena but to "describe the movements that occur in nature [...] completely and in the simplest manner" (**Kirchhoff** (1876), p. V, cf. Chapter 8, footnote 974 and the related text).

[634] **Helmholtz** (1871b), p. 41, Helmholtz does not explicitly mention "scientists" but speaks rather of "us", meaning the participants of the Leibniz meeting at the Academy of Sciences in Berlin.

[635] **Helmholtz** (1871b), p. 45.

[636] **Helmholtz** (1871b), p. 47 (wide letter spacing in original has been italicized).

[637] **Helmholtz** (1871b), p. 46.

[638] **Helmholtz** (1847a), p. 24 (as Section 6.1.1, Part γ, footnote 295) and a similar use of the term in **Helmholtz** (1852), p. 593ff.

fictitious entities, introduced for the sole purpose of explanation (e.g., volume elements). Compared to what is observable, hypotheses appear to be an inevitable evil tolerated for the needs of scientific explanation.

7.1.2 Re-Evaluating Hypotheses in Scientific Procedure (1874)

Equally as abrupt as Helmholtz's critique of atomist assumptions in theoretical physics, in popular publications three years later one finds he has reassessed hypotheses in terms of the philosophy of science. The concept of hypotheses he had previously used rather in passing and as meaning what has just been explained above now broadens in meaning, moves to the center of his description of scientific procedure and for the first time is seen as systematically advancing knowledge. Helmholtz introduces the new definitions, I presume, deliberately in a preface to a German rendition of a book by two representatives of British physics, William Thomson and Peter Guthrie Tait's *Treatise on Natural Philosophy* [*Handbuch der theoretischen Physik*], which he helped translate.[639] He defends their inductive method against objections raised by Carl Friedrich Zöllner. In contrast to Zöllner, who in his book *The Nature of Comets* [*Ueber die Natur der Kometen*] precedes empirical science with a series of a priori natural principles (including gravitation and generatio aequivoca),[640] Helmholtz demonstratively takes up the stance of the empiricist, acknowledging no a priori "conceptual necessity" [*Denknotwendigkeit*].[641] In doing so he sets the stage for some methodological remarks that clearly deviate from the notions he had advocated up until the late 1860s. He focuses particularly on those paragraphs of Thomson and Tait's *Treatise* that discuss the central importance and various usage of the term 'hypothesis' in physics.[642]

Now Helmholtz sees the inductive method no longer indistinguishably tied to logical deduction[643] but instead clearly distinct from deductions subordinated to the purposes of verification:

> We have all [...] used the inductive method to find new laws, or hypotheses, respectively and the deductive [method] to develop the consequences they have for the purpose of verification.[644]

In this preface Helmholtz provides various examples of hypotheses. One of them illustrates well how induction and the production of hypotheses are connected.

---

[639] **Helmholtz** (1874a).

[640] **Zöllner** (1872). On the controversy between Helmholtz and Zöllner see Chapter 8.

[641] **Helmholtz** (1874a), p. 420.

[642] **Helmholtz** (1874a), p. 415. On Thomson and Tait's concept of the hypothesis cf. **Schiemann** (1997), p. 317, footnote 24.

[643] Cf. Section 6.3.1.2.

[644] **Helmholtz** (1874a), p. 414.

Helmholtz defends the hypothesis that life on earth cannot be of a terrestrial origin because all efforts "to generate organisms out of lifeless substance"[645] have failed. His hypothesis generalizes a large number of findings in one sentence, the truth of which he is not (yet) asserting.[646] Now hypotheses are understood as – a phrase Helmholtz will use later – a "preliminary stage to a law"[647] and subordinate to the concept of it:

> Every legitimate hypothesis is an attempt to establish a new, more general law that covers more facts than have been observed. Verifying it then means that we try to develop all consequences it will have, particularly those that can be compared to observable facts.[648]

By extending what is grasped of nature in natural laws, hypotheses gain a positive function as part of the scientific process for obtaining knowledge. In order to cover a larger segment of nature, hypotheses rest on previously known laws and make prognoses (develop "consequences") that extend beyond already observed facts. They add a prognostic component to the inductive component of tentative generalization. If the prognoses are confirmed, the hypotheses themselves become more general laws that in turn act as the starting point for producing new hypotheses. Thus the hypotheses take an intermediate position, mediating the transition between the old and new natural laws.

One can therefore distinguish two meanings to Helmholtz's concept of the hypothesis. The first refers to entities, the existence of which is either only assumed as fictitious or is yet with certainty unproven. These may be middle terms of deductive explanation. Besides the examples of atoms and volume elements mentioned above, the preface to the German version of Thomson and Tait's *Treatise* includes two others: Wilhelm Eduard Weber's postulated electric charge particles[649] and Newton's postulated corpuscular light particles.[650] The second meaning of the concept of the hypothesis refers to laws that enter deductive explanations as premises but are not yet empirically verified. While the assumption of hypothetical entities serves the purpose of theoretical explanation exclusively and is thus grouped with deduction, the tentative establishment of laws is the result of applying a research method that proceeds inductively.

Another preface, also published in 1874, namely the preface to a German rendition of John Tyndall's *Fragments of Science*,[651] which Helmholtz also edited, shows that the positive assessment of hypotheses now becomes part of reorienting the definition of scientific procedure. In the 1860s Helmholtz's characterization of experimental

---

645 **Helmholtz** (1874a), p. 419. Helmholtz does not explicitly state the hypothesis that a terrestrial origin of life is impossible. It precedes his idea that life-germs might have been carried by meteorites to the earth (ibid. and **Helmholtz** (1871a), p. 89).

646 On Helmholtz's analogous foundation for the law of the conservation of energy see Section 6.1.2.

647 **Helmholtz** (1897 ff.), Vol. I.1, p. 18.

648 **Helmholtz** (1874a), p. 416 f.

649 Cf. Section 6.3.2.3(i), footnote 594.

650 **Newton** (1704).

651 **Tyndall** (1874).

science included a contrast to "aesthetic" procedures in the moral sciences based on immediate witness through sensory perception.[652] While at that time he wished to restrict overlaps between methods in the natural and the moral sciences to "a few areas" of natural science such as "subjects in natural history,"[653] he now introduces elements of an aesthetic approach to nature directly into the methods of physics.[654] What formerly separated natural from moral science, now serves to distinguish theory from practice within physics:

> There are two ways of searching for lawful interrelations in nature: one by the use of abstract concepts, the other by carrying out thoroughgoing experimental research. The first method, involving the use of mathematical analysis, leads ultimately to a precise quantitative knowledge of phenomena. It can only advance, however, where the second method has already at least partially opened up a region – that is, where experimentation has already resulted in an *inductive knowledge of the laws* [...] and where the problem is now only to test and refine the laws already found, to pass from them to the most general and ultimate principles of the region in question [...] The second method leads to a rich knowledge of the behavior of natural substances and forces; but by this method the laws or regularities are understood at first only in *the way artists grasp them*, that is, by means of vivid sensuous intuitions of types of action or behavior. [...] These two sides of the physicist's work are never quite separate.[655]

The experimental method no longer merely derives the representation of the time sequence of real events from sensuous intuition. Instead, it so fully exhibits the qualitative totality and immediate presence of sensuous intuition that Helmholtz considers it analogous to the work of the artist, without, however, denying that the experimental "way" of grasping lawful relations is legitimate on its own. The experiment separated from permeation by concepts and mathematics leads to knowledge of natural laws that "now only" become the object of "testing and refining" in order to become laws. Helmholtz could also have called the empirical probing of laws "experimental hypothesis" and distinguish it from theoretical study. For the latter the metamorphosis into laws would happen by investigation through "experimental research".

Thus the pursuit of "completed knowledge"[656] of natural laws continues to be the aim of natural science and laws are not hypothetical. In subject matter law hypotheses

---

[652] Cf. my remarks on **Helmholtz** (1862) in Section 6.3.1.2.

[653] **Helmholtz** (1862), p. 175.

[654] **Hatfield** (1994), p. 546 f., also stresses that within the development of Helmholtz's thought the relation of scientific conclusions to aesthetic induction underwent "a complete turnabout". He sees the origin of this tendency in how Helmholtz formulates his theory of perception in **Helmholtz** (1856 ff.) and **Helmholtz** (1868a), where Helmholtz, however, does not explicitly discuss knowledge gained by aesthetic perception.

[655] **Helmholtz** (1874b), p. 430 (Germ.)/336 (Engl.) (my italics). Cf. also **Helmholtz** (1871b), p. 47. From the start Helmholtz rigorously distinguished theory from practice in physics, cf. **Helmholtz** (1847a), p. 3 f. (as Section 6.1.1, Part α, footnote 266). In the course of the 1850s he retracted the difference between (empirical) laws and causes that he had associated with that difference; cf. Section 6.3.1.4.

[656] **Helmholtz** (1874b), p. 425.

are totally identical to laws. What they need to become laws is either empirical confirmation that they are true or to be stated in a universally valid mathematical form. This contrast of law and hypothesis demonstrates how the dominant classical conception of science still holds. The hypothetical is precisely that which does not yet satisfy the claim to truth demanded of science. The acknowledgment of hypotheses will not lead to a modernization in the understanding of science until the concept – that was already accepted anyway – erodes the claim to truth. But as long as hypotheses remain simply "quasi questions raised for further research,"[657] that will not be the case.

7.1.3 Approximating the Modern Concept of Science (1877 et seq.)

Up to this point I have tried to identify two breaks: the first (emerging in 1871) concerns Helmholtz's critique of atomistic "hypotheses", i.e., an initial questioning of what comprises the foundations underlying the mechanistic program. The second (alluded to in 1874) concerns amending the definition of the methodical procedure in the natural sciences. Whether these two revisions are merely episodic or mark the onset of a fundamental transformation in Helmholtz's conception of science, inasmuch as can be said judged by his popular publications, can now be investigated by examining two speeches he held a few years later on different topics. One is a speech given in 1877 titled *On Thought in Medicine* that became known particularly because of the dissociation from materialism added to the second edition.[658] The other is Helmholtz's probably most important lecture *The Facts in Perception* from 1878, that he supplements with an appendix (partially also written in 1878) and includes essentially in the reworked §26 of the second edition of the *Treatise*.[659] Helmholtz published this part of the new edition in 1894, the year he died, which suggests that he still considered the contents of the lecture correct.

Both speeches so intensify changes that hitherto had not left the framework of the classical conception of science that former criteria of true science start to unhinge and the first signs appear unmistakably pointing in the direction of a modern conception of science. My reconstruction takes clues from the distinction already made in 1874 regarding induction and deduction. Whenever Helmholtz's later conception of science deviates from that of the two speeches I shall indicate the subsequent publications (usually in remarks) and discuss them in two excurses. The criteria for evaluating the transitional process are the aspects of hypothetization used in Section 4.1 to characterize the modern concept of science: hypothetisizing

[657] **Helmholtz** (1874b), p. 427.

[658] **Helmholtz** (1877a).

[659] **Helmholtz** (1878a).

statements of natural law (criterion no. 1), mathematic-logical propositions (criterion no. 2) and the empirically given (criterion no. 3).

I begin with Helmholtz's continued critique of atomistic assumptions and with the dematerialization of the concept of law it implied. The definitions of this concept so central to his conception of science clearly indicate that Helmholtz now substitutes the former mechanical bent of his endeavors at establishing a theoretical foundation for science using a paradigm from the theory of perception. I consider it the first consequence of his change in approach that the postulate of temporal representation loses significance. That postulate had been his most recent suggestion for founding truth (Section 7.1.3.1).

Helmholtz's problematization of the difficulties of inductive procedure, which he still finds fundamental for producing and justifying scientific knowledge, continues right up to his rejection of unconditional claims to validity, which in itself can already be considered modern. The importance of the loss of truth is reflected in his discovery of a non-inductive, intuitive source of knowledge about the laws of nature (Section 7.1.3.2).

Weakening the inductive guarantee for validity puts more emphasis on the empirical investigation of the formulations of natural laws and on the deductions made in that context. But the process of verification can still never be completed. Helmholtz's position on induction and deduction can be seen in how the relation between law and hypotheses changes (criterion no. 1) (Section 7.1.3.3).

His conception of the empirical foundations of geometry remains basically intact. Signs of a transitional process arise solely from re-weighting formerly articulated arguments. Although not elaborated in depth, they still indicate increased emphasis on those modern elements already contained in Helmholtz's earlier writing. It at least implicitly subjects mathematic-logical statements and the empirically given (criteria nos. 2 and 3) to hypothetization (Section 7.1.3.4).

The gravest revision is probably the hypothetization of his own standpoint in epistemology. By calling his version of realism a "metaphysical hypothesis" Helmholtz on the one hand acknowledges that irrefutable alternative standpoints are possible, which can be understood as part of the diversity of arguments characteristic of modern times. On the other hand, this relativization can be interpreted as expressing difficulties arising within his own standpoint, which Helmholtz tries to evade in an appendix to the speech *Facts of Perception* by reformulating his position on classical realism. He attenuates the theory of perception reference dominant in the speech and justifies a remainder of uncontested assumptions about reality by reviving the noumenal understanding of causality from previous years. Helmholtz pays a stiff price to save this classical core of the foundation of science: even empirical concepts' reference to the given can only be established hypothetically (criterion no. 3). In my opinion this is the reason why, in a later publication, he also calls scientific theories *images*, meaning models that do not correspond to reality (Section 7.1.3.5).

While Helmholtz's treatment of realism basically does not answer the question of his position between the classical and the modern conception of science, his idea of the aim of scientific progress provides further clues. The search for "final causes" remains the "regulative principle" and theoretical major task of natural research (Section 7.1.3.6).

Helmholtz's reasons for this aim show that by using unclear concepts he can override the contradictions arising between classical and modern elements that lie so close in his arguments. One explanation that begins with a classical commitment to the a priori validity of (noumenal) causality ends by subjecting the concept of causality to a pragmatic conception of truth that makes the validity of epistemological prerequisites depend on the individual success of applications. Even further reaching than this publicly expressed relativization of causality's validity is the insight recorded in a note found posthumously that the "law of causality is [...] only a hypothesis" (Section 7.1.3.7).

I shall illustrate my concluding assessment of his conception of science with his attempt to demarcate new characteristics of science from non-scientific knowledge (Section 7.1.4).

7.1.3.1 The Concept of the Atom and Dematerializing the Concept of Natural Law

With respect to the transitional process in classical mechanism both speeches share, in terms of content, further poignant criticism of the concept of the atom, consisting of two brief, almost casual, remarks that are strategically placed. In his 1877 speech Helmholtz explains the task characteristic of scientific research, namely "to search for the law of facts" as follows:

> In recognizing the *law* found, as a power which rules the processes in nature, we conceive it objectively as a *force* and such a reference of individual cases to a force which under given conditions produces a definite result, that we designate as a causal explanation of phenomena. We cannot always refer to the forces of atoms [. . .].[660]

This first admission of difficulties in reducing all phenomena to atomic forces occurs entirely abruptly alongside a new definition of the concept of explanation that no longer includes final causes – mechanical central forces and displaced matter.[661] Against the backdrop of his previous mechanistic program this new standpoint appears to have arisen out of problems related to the quest for mechanical causes. Reduction aimed at finding mechanical causes is now replaced by explanations that must only indicate conditions under which phenomenally detectable laws operate. This definition implies that laws are only valid for a restricted field that normally depends on specific experimental conditions from which individual phenomena can only be deduced, when the relevant initial or boundary conditions are known. Thus explanations are no longer supposed to deduce empirical laws from other, more

[660] **Helmholtz** (1878c), p. 187 (Germ.)/324 (Engl.) (italics in original).

[661] For the classical definition of the mechanistic concept of explanation cf. the quotation for Section 6.3.1, footnote 426. On the preparation of the transition cf. the quotation in the same section, footnote 484.

fundamental laws but are exhausted by naming the prerequisites of their applicability. Subordinating the validity of propositions to the conditions of their applicability is what I have called conditionalizing.[662] Here it reflects the problematization involved with the classical foundations for science and already excludes absolute validity if it be bound to the claim of being unconditional. In Helmholtz it is attended by a noticeable tendency towards the production of theories free of assumptions about substance.[663]

One year later, in the *Facts of Perception,* Helmholtz gives a reason for the apparently problematic situation in microphysics. Once again in a central passage, he makes a discussion of the difficulties of the contents of mechanistic reduction an occasion for considering the concept of substance for the first time:

> The concept of substance can only be attained through exhaustive testing and it always remains problematic insofar as further testing is always held in reserve. In previous times, light and heat were thought to be substances, until it later turned out that they may be transient forms of motion. Moreover, we must always be prepared for new decompositions of what are today considered to be chemical elements.[664]

While in 1871 – despite his emerging skepticism about atomic hypotheses – Helmholtz was still hesitant to "declare that atoms do not exist,"[665] this apparently lapidary comment asserts that the end of atomism is possible, which for him also implies the onset of a period in which the concept of substance will "always remain problematic".[666] In a vague phrase he now separates the concept of substance from what is perceptible, which is now surprisingly upgraded, by being directly associated with laws:

> We call *substance* that which, without dependence on other things, remains the same over time; and we call the constant relationship between changeable quantities the *law* that binds them. It is only the latter that we perceive directly. [...] The first product of the reflective understanding of a phenomenon is the *law-like.*[667]

This phrasing is ambiguous because while it contrasts the concept of natural law with substances, it could also designate a substance itself, if the "constant relationship" recognized as a law were to require nothing else to define it. But since here Helmholtz's concept of substance is obviously related to material entities and not to law-like definitions of relation, I prefer to call it dematerializing the concept of law, instead of desubstantializing it. The previous distinction of submicroscopic matter, namely being unobservably small and accessible solely through science, is now

---

662 See Section 4.1.

663 See Section 7.1.1.

664 **Helmholtz** (1878a), p. 240 (Germ.)/361 (Engl.). Helmholtz used the quoted passage word for word in **Helmholtz** (1885 ff.), p. 592.

665 **Helmholtz** (1871b), p. 45 (as footnote 631).

666 **Helmholtz** (1878a), p. 240.

667 **Helmholtz** (1878a), p. 240 (Germ.)/361 (Engl.) (emphasis in the German original). Also in **Helmholtz** (1885 ff.), p. 591.

seen as a flaw that should not mar statements of law. The validity of laws can no longer be proven by tracing them back to mechanical laws of motion but exclusively by being applicable to observable phenomena.

Thus reference to final mechanical causes is no longer relevant for the concept of natural law, just as it was dropped for reformulating the concept of explanation. The meaning is restricted to the relational character of relations that Helmholtz had introduced when applying the concept of the picture to scientific statements.[668] As quantitative relations, statements of natural laws still need no intuitive demonstration.[669] Helmholtz calls "conceptualization of natural law the final goal" of natural research, compared to which "knowledge by intuition is only a facilitating aid, a false appearance to be overcome."[670] Furthermore, laws continue to fulfill the classical claims to validity. Their statements must continue to satisfy criteria of logical necessity and rigorous universality. In both speeches and throughout the entire subsequent period Helmholtz especially upholds the distinction between hypothesis and law.[671] What is lost by dematerializing the concept of law is initially only the mechanistic foundation for the claim to validity that since the 1860s Helmholtz had also founded on the theory of perception by postulating the relation of simultaneous representation.

But while up to this point Helmholtz's recourse to the theory of perception had been limited to this postulate, he now begins to establish the basic concepts of his conception of science using the theory of perception in a way that indicates an entirely new direction for the foundation of science. He begins with the lingering ambiguity of phenomenal and noumenal meanings of the concept of natural law and its continued relationship to the concepts of cause and force. As was already

---

668 Cf. Section 6.3.1.6.

669 As the transition in his conception of science gets underway, Helmholtz stresses this requirement (that had presented no problem before the 1860s) at first explicitly: "completion and purification of the findings is subsequently preferably a job for *mathematical analysis*, since it always deals with the equality of quantity" (**Helmholtz** (1874b), p. 431 – emphasis in original). In the *Introduction to Lectures on Theoretical Physics* he still solves the problem of "freeing [previously postulated – G.S.] natural laws from contingencies" by tracing them back to "completely general spatial relations". This makes "theoretical physics predominantly mathematical" (**Helmholtz** (1897ff.), Vol. I.1, p. 22f.). According to Helmholtz mathematical physics can arrive at propositions independent of intuition (**Helmholtz** (1856ff.), p. 444, and **Helmholtz** (1885ff.), p. 586).

In contrast to Kant, Helmholtz thinks of intuition directly in terms of perception: we restrict "the term [...] intuition to perception attended by physical sensations" (**Helmholtz** (1856ff.), p. 435, and **Helmholtz** (1885ff.), p. 609, cf. also **Helmholtz** (1885ff.), p. 600). In Section 7.1.3.4 I shall discuss the enhanced concept of intuition that he introduced later (**Helmholtz** (1870), p. 30f., and **Helmholtz** (1878a), p. 231f.) and that eliminates the evident validity of geometrical axioms.

670 **Helmholtz** (1878b), p. 655.

671 On the distinction between law and hypothesis in both speeches cf. **Helmholtz** (1878c), p. 184f., and **Helmholtz** (1878a), p. 240. While in the earlier speech (**Helmholtz** (1878c) he uses the term 'hypothesis' almost exclusively negatively, in **Helmholtz** (1878a) he explicitly emphasizes its heuristic value (loc. cit. p. 235f. and 242, cf. with **Helmholtz** (1874a), p. 417). He also calls epistemological prerequisites of scientific work (realism and idealism) hypotheses (**Helmholtz** (1878a), p. 238f., cf. Section 7.1.3.5).

the case in the *Treatise,* Helmholtz also calls law causes as a way of expressing his ontological interpretation of regularities. As causes, laws belong to a world that originally "exists" and "abides" "behind the change" (of phenomena – G.S.).[672] Once more starting with former definitions, he anchors the concept in realism and reinforces it by using the term 'force' to mean causes that operate according to laws and which present themselves to man in the form of natural powers.[673] According to Helmholtz, the objective effectiveness of what is lawful indicates that there exists a reality independent of being perceived, a "noumenal" cause of all appearances. But while previously he had taken the assumption of reality to be evident,[674] he now offers a foundation for it from the theory of perception:

> This concept of a power opposing us is directly conditioned by the ways and means that our simplest perceptions occur. From the very start, the changes which we ourselves make by our acts of will are separated off from those which are not made by our will and which cannot be overcome by our will.[675]

The concept of force is no longer defined as the "purest form" of mechanical force[676] but by its reference to what is perceptible. The phenomena he is talking about are not – as they would be for natural laws – data from experimental practice. What he has in mind are rather elementary experiences of perception that are supposed to guarantee the existence of an external world existing independently of the perceiving subject by being separate from willful action. The 'Real' is what the subject has not generated himself; it is resistant to action, it changes without influence by man. This idea that comes close to Aristotle's conception of nature is meant to explain why reality presents itself to the subject as "an objective Given".[677] The realistically understood world of reality, which Helmholtz now explicitly equates with "the thing in itself" [*das Ding an sich*],[678] remains the basis of his foundation for science.[679]

By deriving his assumption of an external world from elementary experiences of perception in which the human will to change becomes aware of its own limits in a very primal way, Helmholtz relocates the principles of his foundations for science in the theory of perception. Thus his formerly constitutive contrast between scientific knowledge and subjective witness through perception begins to fade. The consequences are severe: everyday perception only has access to reality-representing contents, which natural laws are supposed to present in pure form, only as a changing

---

[672] **Helmholtz** (1878a), p. 240.

[673] **Helmholtz** (1856ff.), p. 454f. (as Section 6.3.1.4, footnote 484), and **Helmholtz** (1869), p. 376 (as in Section 6.3.1.6, footnote 530f.).

[674] Cf. Sections 6.3.1.3 and 6.3.1.4.

[675] **Helmholtz** (1878a), p. 241 (Germ.)/361 (Engl.), cf. also p. 225ff., 237 and 239.

[676] **Helmholtz** (1869), p. 376 (as Section 6.3.1.6, footnote 532).

[677] **Helmholtz** (1878a), p. 226.

[678] **Helmholtz** (1878a), p. 223.

[679] **Helmholtz** (1878a), p. 241.

of the signs. Hence, for Helmholtz everyday perception can only be true in a pragmatic sense.[680] Now the relative character of validity finds its way into his foundation of science. Elementary processes of perception simultaneously guarantee and limit – almost as if it were a modern life world a priori – the conditions of validity for scientific knowledge of nature.

The concept of law demonstrates how this new foundation is related to his earlier reference to the theory of perception. Now the existence of a *temporal* representation relation between perception and the perceived[681] that formerly had been the condition for the possibility of absolute true statements of natural law, is no longer an explicit criterion:

> Every natural law says that, given preconditions which are alike in certain respects, consequences which are alike in certain other respects will always follow. Since likeness in our world of sensation is shown by like signs, then there will also correspond to the natural-law consequence of like effects upon like causes a regular consequence in the field of our sensations.
>
> Hence, even though our sensations are, in their quality, only *signs* [...] they are nonetheless certainly not to be dismissed as empty appearance; rather they are precisely signs of *something* [...] and, what is most important, they can delineate for us the *law* of this occurring.[682]

By claiming the existence of a representational relation Helmholtz is once more affirming that natural law statements can claim to be true.[683] However, without an explicit reference to an objective structural feature of the external world, for instance its temporal constitution, the assertion remains oddly vacuous. If characterizing the sequence of sensory signs as "just as regular" is meant as indicating the representation of the time sequence of real events, Helmholtz should have explained that differently. The correspondence between one sign and its reference to something real (that he introduced as early as his *habilitation* speech in 1852)[684] produces no such relation. It follows merely that something corresponds to an immutable sign. The temporal development of the perceived sign might be determined by factors other than the things of the external world, even if they appeared regularly with respect to causality. By no longer connecting his postulate of representation to temporal relations, when he mentions it at all, he readied its devaluation.[685]

---

680 Cf. Section 6.3.1.4.

681 **Helmholtz** (1869), p. 394f. (as Section 6.3.1.5, footnote 524).

682 **Helmholtz** (1878a), p. 222f. (Germ.)/348 (Engl.), (emphasis in German original) and somewhat abridged in **Helmholtz** (1885ff.), p. 586, cf. **Helmholtz** (1852), p. 608, and **Helmholtz** (1892), p. 357.

683 The speech begins by asserting that "science in contrast [to philosophy – G.S.] tries to clip what constitutes a definition, a designation, a conceptual form, a *hypothesis*" (**Helmholtz** (1878a), p. 218 (my emphasis), cf. also p. 240 and p. 245).

684 **Helmholtz** (1852), p. 608 (as Section 6.3.1.1, footnote 415).

685 **Schiemann** (1997), p. 331ff.

7.1.3.2 Discovering Hypothetical Laws Inductively and Intuitively

Dematerializing the concept of natural law signalizes more than just a reversal of the relationship between science and the theory of perception. It sheds new light on the reductionist aim of Helmholtz's previous conception of science. It is no longer the standard for evaluating quotidian theoretical work in scientific research, which now means for Helmholtz: to explain and predict individual phenomena.[686] The fact that statements of natural law that enter into explanations in the form of premises, can only prove their value by being applied to experimentally testable phenomena, does not mean that they themselves no longer need to be established in terms of causality. Reducing phenomena described by empirical laws to non-observable causes should not be entirely eliminated from natural research. Like phenomena in general, Helmholtz also interprets regular changes in nature as events that are brought forth by immutable causes embedded in a comprehensive context. Following a repeated affirmation, made in the speech from 1877, that the goal of science is the "*completed* insight into [the] causal connection of natural processes,"[687] a year later he specifies this overall task of science as the search for "a final unchangeable something [that is the] *cause* of the observed changes".[688]

In spite of difficulties, which he meanwhile admits, in realizing the mechanistic program, Helmholtz's conception of science remains reductionist in terms of content. This makes his re-evaluation of the claims to validity involved with that program all the more significant, as expressed in the passage quoted above from the speech of 1878:

> Every reduction of the phenomena to the underlying substances and forces claims to have found something unchangeable and definitive. We are never justified in making an unconditional claim of this type; for it grants neither the fragmentary nature of our knowledge nor the nature of inductive conclusions, upon which rests, from the first step on, all our perception of the real.[689]

After first introducing applicability conditions for the explanation of individual phenomena, here Helmholtz signalizes further conditionalizing. Empirical laws and their reductionistic foundation, as well, are subject to inevitable conditions of validity. Besides the impossibility of eliminating the imperfection of knowledge, once again elementary conditions of perception are at fault for making it impossible to find absolute "final causes". Helmholtz takes recourse to his work in the theory of perception from the 1860s, when he had developed the concept of "conclusion by induction" [*Induktionsschluss*]. A closer look at the relevant writings shows, however, as Friedrich Conrat demonstrated in his work on Helmholtz's views in

---

686 See Section 7.1.3.1.

687 **Helmholtz** (1878c), p. 170 (emphasis in German original).

688 **Helmholtz** (1878a), p. 243 (emphasis in German original). For more on this goal and the concept of "final cause" it involves see Section 7.1.3.6.

689 **Helmholtz** (1878a), p. 243 (Germ.)/362 f. (Engl.) (as footnote 623).

psychology[690] that the concept of conclusion by induction is permeated by an ambiguity that – if it were to be disregarded – would lead to misconceptions when applied to scientific statements.

On the one hand Helmholtz uses the expression (and synonyms)[691] to designate the epistemological condition for achieving the realistic interpretation of ideas. In a conclusion the perceiving individual recognizes that the external world, as it appears to him, exists independent of his own existence.[692] The premise of this conclusion by induction (the first kind), that is in no way relevant for action and therefore cannot be subject to pragmatic criteria, is a priori presupposed noumenal causality according to which the external world is conceived as the cause of all phenomena. Inasmuch as Helmholtz in his speech of 1878 does not doubt the absolute validity of this causality,[693] the realistic interpretation of the external world is no argument for the limitation of validity claims.

But he also uses the expression to designate what he considers the fundamental process of sensation processing in living organisms[694]: ideas and actions related to sensations are produced inductively by an indigenous [*naturwüchsig*] learning process. Effects of sensation that are similar and successful processing of sensations are constantly stored in memory, without the individual in question normally being aware of it and generalized to become rules for thought and action. If a comparable sensation happens again, those rules – now without exception operating unconsciously – determine ideas and actions that will be reactions to it. The major principle for this conclusion (the second kind) is not the law of causality but the rule obtained by

---

690 **Conrat** (1904), p. 92 ff.

691 Instead of "conclusion by induction" (e.g., **Helmholtz** (1856 ff.), p. 449, **Helmholtz** (1878a), p. 226 and 243 f., **Helmholtz** (1885 ff.), p. 602) Helmholtz uses "subconscious conclusion" (for the first time in **Helmholtz** (1856 ff.), p. 430 and 447 ff., later in **Helmholtz** (1868a), p. 358 ff., and **Helmholtz** (1878a), p. 233) or, in earlier work, simply of "conclusion" (**Helmholtz** (1855), p. 116).

692 **Helmholtz** (1855), p. 112 and 115 f., **Helmholtz** (1856 ff.), p. 430, 449 and 453 f., **Helmholtz** (1878a), p. 243 f. This meaning can be discovered in **Helmholtz** (1856 ff.) where Helmholtz writes of "ideas of a cause" (loc. cit. p. 430) or how one concludes "from sensations that an external cause for them exists" (loc. cit. p. 449).

693 **Helmholtz** (1878a), p. 243 f., cf. Section 7.1.3.7.

694 **Helmholtz** (1855), p. 122, **Helmholtz** (1856 ff.), p. 447 and 449, **Helmholtz** (1868a), p. 358 ff. This meaning can be found in **Helmholtz** (1856 ff.), where he writes of the "development" of ideas and perception (loc. cit.).

In a posthumously published essay planned for the second edition of the *Treatise* (but which for unknown reasons was not entirely printed in it, see **Erdmann** (1921), p. 25), Helmholtz explains expressly that all empirical material provided by sensory perception probably rests on conclusions by induction: "Inductive conclusions play an important part in generating intuitions produced by the unconscious work of memory [...] It seems doubtful whether an adult can have knowledge by acquaintance [*Kenntnis*] from any other source (**Helmholtz** (1894a), p. 553). Helmholtz uses the term *Kenntnis* to designate "the pictorial intuition of an object as it is produced only by sense impressions" (**Helmholtz** (1894a), p. 540, and **Helmholtz** (1885 ff.), p. 598). See footnote 273.

induction.[695] This constitutes the formation and application of a psychological schema, the results of which are apparently not unconditionally valid.

- There is no guarantee that conclusions by induction are correct because they rest on averaging a number of similar experiences and their range of application is not clearly demarcated. Sensations may act as a middle term in a conclusion, although they only approximately match the premise. Helmholtz attributes sensory illusions to this sort of false conclusion.[696]
- The learning process that is never entirely finished in the course of a finite lifetime results in a continuous modification of inductive methods, which subsequently can never be the foundation for something that is "unchangeable and definitive".

Inasmuch as the function of perception is to provide the empirical material for inductively deriving laws, one can assume that Helmholtz basically relates the limitation to validity that follows from the elementary conditions of perception to scientific inductions, as well. In an appendix to the speech *The Facts in Perception* [*Die Thatsachen in der Wahrnehmung*], to which we shall return when discussing questions in geometry, he writes of "the only approximate proof of all natural laws by induction".[697] Later (1892) he states more resolutely:

> All knowledge of natural laws is inductive; no induction is ever absolutely complete.[698]

But why does Helmholtz believe that the validity of scientific induction and reduction based on it can be measured by an elementary method of perception processing, for which he stresses that while it is practical for everyday purposes, it is also imprecise and not universally valid? I see his reason for doing so in the way he equivocally extends his original concept of the theory of perception to become an epistemological approach that subordinates all thought to the rule of inductive reasoning:

> We are here [regarding conclusions by induction – G.S.] obviously concerned with an elementary process which underlies all actual, so-called thinking, even if the critical sighting

---

[695] **Helmholtz** (1856 ff.), p. 447 f. Helmholtz thinks that the major premise of these inductively obtained conclusions, whose logical circularity he finds "best studied" by John Stuart Mill (**Helmholtz** (1856 ff.), p. 447, as Section 6.3.1.4, footnote 475), consists of a summary processing of experiences that have occurred in interaction with a certain kind of sensation (e.g., a reaction to a stimulus on the right side of the retina); the minor premise is the occurrence of an actual affection and the conclusion is the interpretation of that affection or the behavior that corresponds to it (e.g., glance to the left, loc. cit. p. 447 ff.). Cf. Sections 6.3.1.4, 6.3.1.3 and 7.1.3.5.

[696] Helmholtz calls "false inductions", that occur in animals "much more frequently", sensory illusions. He attributes their occurrence to the fact that they rest on an "insufficient number of observed cases" (**Helmholtz** (1885 ff.), p. 602). See also the corresponding passages (not included in the second edition) in **Helmholtz** (1856 ff.), p. 438 ff., where the problem of validity involved in sensory illusions is obvious: "Here it must be explained [...] how experience counteracts experience" (loc. cit. p. 438). Cf. also **Helmholtz** (1868a), p. 357 f.

[697] **Helmholtz** (1878a), p. 393.

[698] **Helmholtz** (1892), p. 358.

> and completeness of the individual steps which enter into the scientific formation of concepts and conclusions is still missing.[699]

Helmholtz now views thinking as being pre-determined right down to its "individual steps", so that he can actually no longer use the term "thinking" in its actual sense to designate the individual activity of a mind. Thinking can no longer transcend the horizon of validity drawn by the theory of perception. This also includes the production of scientific terms, which Helmholtz explicitly mentions and understands as a part of the inductive procedure.

Thus the denial of unconditional validity for a special class of scientific statements, namely that of the reductions typical of mechanism, leads to the limitation of validity for the entire human capacity for knowledge. This argumentation proceeds similar to that used for redefining the concepts of explanation and law: the problematization of fixing, in terms of contents, definitions for the formerly approved mechanistic program appears to be the starting point of a change in his conception of science. But while the classical claim to truth was not questioned when reformulating the concepts of explanation and law, distancing himself now from the unconditional character of validity for the mechanistic goal of comprehensive reduction leads him to take up a position that can already be considered *modern*. The hypotheticity of knowledge claimed by the modern conception of science consists basically of a general denial of classical science's emphatic claim to truth, be it founded metaphysically or scientistically.

At this point it should be noted that only now does Helmholtz begin to take up the epistemological claim[700] he had made early within the context of his theory of perception. Up to now his studies were confined to physiological and psychological aspects of perception in living, preferably human organisms and were based on a scientific self-image that itself is not subject to the conditions of perception.[701] But as epistemology, the theory of perception becomes reflexive because it also includes the scientific knowledge from which it claims to arise. Helmholtz formulates this self-reference so typical of modern conceptions of science for the first time explicitly in the speech of 1877 and he emphasizes it later frequently.[702] In an after-dinner speech at the celebration of his seventieth birthday (1891) he recalls:

> I took up another field to which my studies on sensation and sensory perception had led me, namely that of epistemology. Just as a physicist must examine the telescope and galvanometer he will work with and must clarify what he intends to achieve with them and how they may mislead him, so it seemed necessary to me to also investigate the performance of our capacity to think.[703]

---

[699] **Helmholtz** (1878a), p. 233 (Germ.)/355 f. (Engl.), cf. also p. 232. Early forms of this approach can be found where he first took up the problem of inductive method in **Helmholtz** (1856 ff.), p. 448, and **Helmholtz** (1868a), p. 358 (quoted in Section 6.3.1.4, footnote 476).

[700] A claim to be working in epistemology is raised as early as **Helmholtz** (1855) (cf. Section 6.3.1.3) and in 1857 in a letter to his father (quoted in Section 6.3.1.2, footnote 459).

[701] Cf. Section 6.3.1.4.

[702] **Helmholtz** (1878c), p. 188, **Helmholtz** (1891), p. 16, and **Helmholtz** (1892), p. 338.

[703] **Helmholtz** (1891), p. 16.

Helmholtz's naturalistic epistemology that aspires to empirically determine the boundaries of "the capacity to think" brings him into a circle that has continued to be typical for similar investigations until today – one may think of the evolutionary epistemology of Konrad Lorenz, Gerhard Vollmer and others – who have continued to pursue it. The limits of knowledge are to be determined scientifically. But scientific knowledge itself is subject to those limits to knowledge, such that it is impossible to draw borders, without first presupposing them. Helmholtz tries to escape this circle by consequently excluding the statements in the theory of perception that he uses to assert the relative conditions of the validity of the "capacity to think" from precisely those conditions. That must not necessarily be interpreted as an inconsistency in an epistemology insufficiently worked out in terms of the philosophy of science. For in my opinion we can find in Helmholtz's writings at least rudimentary forms of one of the most significant arguments given until today for justifying the special status of statements from the theory of perception. Namely, the familiarity of everyday experience which he can repeatedly use to refer to elementary experiences of perception and which award his statements a great deal of immediate plausibility. Exaggerated, one could say that science gains a life world foundation to the extent that it loses objectively distinguished certainty of its validity. However, this solution for the problem of providing a foundation for science is only hinted at in Helmholtz's writing. He does not explicitly discuss the validity character of his epistemology and theory of perception, so that although they can be called reflexive, they cannot be called self-reflexive.

By denying autonomy to pure thinking, Helmholtz underlines the anti-metaphysical character of his epistemology. For his conception of science this empiricism amounts to expelling even the last metaphysical elements from his foundation of science. Although his earlier foundation for science already stood within the scientistic tradition, it still contained elements that originated not in scientific experience but in pure thinking. For example, he founded the truth of arithmetical axioms in evidence, he then did not clearly demarcate the validity of statements in geometry from that and justified a realistic presupposition of the external world based on a priori valid (noumenal) causality.[704] Regarding geometry, I have already discussed how its unempirical foundation was destroyed. As I shall show below, in the 1870s Helmholtz will explicitly formulate (Section 7.1.3.4) the anti-metaphysical line of attack implied by it. His wavering position on the relation of causality to experience[705] will also emerge more clearly and finally lead to entirely relinquishing apriority (Section 7.1.3.7).

Not until modern times does the critique of metaphysics coincide with the relativization of claims to validity. It rests on change in the scientistic evaluation of experience, for which Helmholtz's altered evaluation of inductively obtained statements is an example. For him induction still remains the most important procedure

[704] Cf. Sections 6.3.1.2 and 6.3.1.3.

[705] Cf. Sections 6.3.1.3 and 6.3.1.4.

for obtaining and justifying statements of natural law.[706] But it can no longer be presupposed that the method guarantees certain truth. A harbinger for this transformation was the incompleteness of inductions that he had noted in the first edition of the *Treatise,* which – without applying the experimental-causal method – rests entirely on observation.[707] But since Helmholtz continues to uphold classical criteria for statements of natural law, the result is a tension between the maintained claim to truth and the inability to found it in terms of theory of science. Once the representation relation claimed to be valid of natural law propositions has – due to the fact that it is no longer characterized by the simultaneousness of events and representations – become in need of justification, continued evaluation of the inductive method is the first step towards hypothetisizing the propositions of natural law.

The significance of this loss of certainty of truth is reflected in Helmholtz's effort to evade the progressive succession of inductively gained scientific statements by further mitigating the contrast between science and art. In the course of the 1870s he says that "finding laws" can also be done in a way analogous to aesthetic intuition. The "first discovery of previously unknown laws of nature", he had already written in a preface to Tyndall's book, is a "matter of *wit* (taking the word in its broadest meaning)".[708] In the speech *Thought in Medicine* he elaborates on this idea:

> The first discovery of a new law is the discovery of a similarity which has hitherto been concealed in the course of natural processes. It is a manifestation of that which our forefathers in a serious sense described as 'wit'[*Witz*]; it is of the same quality as the highest performances of artistic intuition in the discovery of novel types of expressive appearance. It is something which cannot be forced and which cannot be acquired by any known method.[709]

Being due to "wit" makes the discovery of something new essentially dependent on a specifically human ability. In the broad meaning of the term as accepted before Helmholtz's time, "wit" [*Witz*] meant not cleverness but also ingenuity.[710] While it is true that the discoverer's wit gets its material from what is given in intuition and from experimentally accessed "natural processes", the "similarity" he discovers

---

[706] On the importance of induction for natural science see **Helmholtz** (1878c), p. 175, **Helmholtz** (1892), p. 338 and 358, and **Helmholtz** (1897 ff.), Vol. I.1, p. 7 ff.

[707] Cf. Section 6.3.1.4.

[708] **Helmholtz** (1874b), p. 431, where I believe Helmholtz expresses this idea for the first time.

[709] **Helmholtz** (1878c), p. 184 (Germ.)/322 (Engl.). Similar characterizations can be found in later publications. In **Helmholtz** (1878a), p. 232 f., he does not expressly mention "wit"; but what conforms to law here also originates directly in intuition. In **Helmholtz** (1892), his second speech on Goethe, he says that "the first inventive thought that must precede wording" can "always only happen in a way similar to aesthetic intuition, as a hunch of a new regularity" (see footnote 711). He also writes in the *Introduction to Lectures on Theoretical Physics* (**Helmholtz** (1897 ff.), Vol. I.1), that trying to generate new laws always takes "a certain inventive spirit": "it takes quasi guessing and generally only one will be successful at it, who knows how to make good guesses". The assumptions found in this way he also calls hypotheses that "serve the sole purpose of providing starting points for later formulating the law". This means that "necessarily the road of investigation in natural science leads through hypotheses" (loc. cit. p. 18 f.).

[710] Cf. **Grimm** (1854 ff.), Vol. 14-II (1960), Col. 861 ff.

may not be obvious, if revealing it is to be a question of wit. The thought that recognizes or produces the law must be capable of linking disparate and reluctant material in a novel view of the whole thing.

By viewing science as parallel to art, Helmholtz adds a creative element to his philosophy of science. Just as "new types of expressive appearance" are created, so also do laws appear as new creations that transcend existing explanatory approaches in natural science. Not the confirmation of familiar conceptions and theories but their keen extension, breaking through their boundaries, is what distinguishes this form of intuitive scientific production of knowledge. It differs from the inductive method in several ways: while induction starts with particulars, "wit" works with insight into the whole; while induction claims to be guided by available cases, "wit" plays with those cases; while inductively obtained laws appear to be generalizations from what is already known, those found through "wit" produce something genuinely novel; while inductive method proceeds slowly by nature, "witty" knowledge happens abruptly. Helmholtz was aware of the difference between inductive and intuitive insight. In 1892 in a second speech on Goethe he calls Goethe's discovery of the vertebral structure of the cranium "typical" for the kind of discovery consisting of a "hunch about a new regularity" and happening "analogously" to "scientific work".[711] Otherwise the latter obtains its "insight into the complicated workings of nature" in a different, namely inductive way.[712]

As long as intuitively grasped natural laws still lack empirical confirmation and proof of universal general validity, Helmholtz considers them merely hypothetical.[713] Of course, being rough drafts for laws they present a special sort of hypothesis, the analogy to highly esteemed artistic productivity leaves them in the realm of perennial truths: an idea gets "produced by an in-depth look into the connection of the whole"[714] and in 1892 Helmholtz calls it "a sudden insight" [...]: "*divination*" [...] a kind of godlike inspiration".[715] Just when findings in physiology question whether truth can be guaranteed by induction, Helmholtz discovers "wit" as an alternative source of knowledge, whose equally hypothetical character he tries to compensate with an aura of being transcendent.

7.1.3.3 Deductive Justification for Natural Law Propositions

If an unrestricted claim to validity cannot be guaranteed for laws by applying the inductive method and one nonetheless desires not to forego that claim, then the

711 **Helmholtz** (1892), p. 348.

712 **Helmholtz** (1892), p. 338 f.

713 **Helmholtz** (1874a), p. 415 f., **Helmholtz** (1878c), p. 184 f., and later: "The first yet insufficiently tested attempts to formulate a natural law can only be called *hypotheses*" (**Helmholtz** (1892), p. 339 – emphasis in German original).

714 **Helmholtz** (1878c), p. 185.

715 **Helmholtz** (1892), p. 348 (emphasis in German original).

entire burden of founding that claim rests on subsequent empirical verification. Experimental investigation of hypothetically assumed laws is then no longer only – as Helmholtz had said in the 1860s – indispensable for the "more complicated cases,"[716] it becomes constitutive for the scientistic foundation of science, such that any limitation of its validity inevitably leads to a succession of the classical conception of science. While the hitherto upgrading of hypotheses to mean empirically not verified law propositions still intended to create a systematic link to non-hypothetical statements, now the problem of verification, which we have yet to discuss, undermines the distinction between hypotheses and laws. Laws themselves acquire a hypothetical character and can thus no longer satisfy classical criteria.

For Helmholtz, experience as the basis of validity – in this he remains consistent – is determined experimentally.[717] The prerequisite for applying hypotheses to experience is the "business of deduction"[718]:

> When we fancy that we have arrived at a law, the business of deduction commences. It is then our duty to develop the consequences of our law as completely as may be but in the first place only to apply to them the test of experience, so far as they can be tested and then to decide by this test whether the law holds and to what extent. *This is a test, which really never ceases.* The true natural philosopher reflects at each new phenomenon, whether the *best established* laws of the best known forces may not experience a change; it can of course only be a question of a change which does not contradict the whole store of our previously collected experiences. It *never thus attains unconditional truth* but such a high degree of probability that it is practically equal to certainty.[719]

Helmholtz immediately retracts this waiver for absolute knowledge (which – going by the speeches' chronology – he here publicly expresses for the first time), by using the known and previously employed[720] rather meaningless formula of probability bordering on certainty. More important than the assertion itself is the reason he offers as to why it is impossible to determine with finality the validity of a law: this time unrelated to the incompleteness of the inductive procedure, empirical verifiability is limited by experience. Even should the truth of a natural law proposition be considered proven for a certain range of application, it seems difficult to determine more precisely the exact extent of that range. The unlimited number of individual cases subsumable under a law is not the greatest challenge but instead the possibility that novel kinds of classes of phenomena may infringe on the validity of known laws. The process of empirical examination should permit every amendment that does not "contradict the whole store of our previously collected experiences". It would be a mistake to think that Helmholtz considered this criterion a measure taken to avoid a fundamental change in the foundation of physics, which in current terms would imply a shift of paradigm. On the contrary, it seems as if he now holds no law in physics to be sacred.

[716] **Helmholtz** (1862), p. 177 (as Section 6.3.1.2, footnote 455).

[717] **Helmholtz** (1878c), p. 180, and in other places.

[718] **Helmholtz** (1878c), p. 183.

[719] **Helmholtz** (1878c), p. 183 (Germ.)/323 (Engl.) (my italics).

[720] Cf. **Helmholtz** (1862), p. 170.

Thus he claims in the same speech: "Astronomers have never believed that Newton's force excluded the simultaneous action of other forces".[721] Indeed, even if post-Newtonian astronomers had always entertained that position,[722] in Helmholtz's own mechanism there had hitherto been no room for forces of that sort.[723] Their operating, namely, would only have been detectable as deviating from gravity, making gravitation no longer a universal law and model for elementary physical interaction. This concession renders a cornerstone of Helmholtz's scientifically founded conception of nature vulnerable.

Once natural law propositions become assailable to this extent, in contrast to hypotheses the certainty of the validity demanded of every single finding begins to dwindle. In both speeches mentioned here Helmholtz does clearly separate law propositions that have not yet been empirically tested – being hypotheses – from laws proper.[724] But the difficulty in formulating a foundation shows that laws cannot be kept entirely free of hypothetical validity. The hypothetical character a law statement's origin can at the most be reduced as empirical investigation proceeds but not entirely eliminated. As negligible as the hypothetical residue may seem, its very existence suffices to rule out, in principle, that a particular statement has ultimate validity. Even the most tried and proven physical laws are no longer protected from a loss of validity.

In *Introduction to Lectures on Theoretical Physics* [*Einleitung zu den Vorlesungen über theoretische Physik*], held in 1893, Helmholtz mitigates the distinction between law and hypothesis even further. Scientific work can only minimize the hypothetical character of law propositions to a certain degree. Selective experimenting can:

> Increasingly remove the hypothetical element [from law propositions – G.S.], the more pertinent investigations into specifics are done.[725]

The first step towards hypothesizing law propositions consisted in relativizing the inductive guarantee of validity. Doing that shifted the burden of establishing a foundation for science to deductive-experimental justification, which, in a second step, reveals the extent of hypothetization de facto already completed when Helmholtz admits the possibility – put in modern terms – of a shift in paradigm.

In this loss of certainty for natural law statements I see an element of Helmholtz's conception of science that is related to the first aforementioned criterion for a modern understanding of science, according to which laws of nature can definitely be neither verified nor immunized against falsification brought about by contrary experience.[726] To the extent to which Helmholtz makes the validity of laws depend

---

[721] **Helmholtz** (1878c), p. 175.

[722] Helmholtz is probably referring to the efforts of astronomers to explain observable planetary movements based on calculations involving only two stars (the so-called "two body problem").

[723] **Helmholtz** (1862), p. 176; **Helmholtz** (1869), p. 378, cf. Section 6.3.2.3(ii).

[724] Cf. footnote 671.

[725] **Helmholtz** (1897 ff.), Vol. I.1, p. 19, cf. also **Helmholtz** (1892), p. 338 f.

[726] Cf. Section 4.1.

on empirical testing, they are no longer eternal and invariably valid statements. For Helmholtz this hypothetization relates not to the concept of a law but only to its manifestation in scientific work. The concept, with its claim to universality and necessity, designates an ideal that science should continue to pursue. However, the fact that this ideal now becomes unattainable indicates an incompleteness that runs counter to the classical demand for a closed system of knowledge. Whether or not this circumstance is already part of a modern conception of science is something I shall return to below when discussing the aims of progress.

Nonetheless, the justification of laws based solely on empiricism need by no means lead to hypothetization. Classical scientism, as advocated by Helmholtz, founds the claim to truth for scientific statements precisely in recourse to experimental practice. The dwindling certainty of validity for statements of natural law is thus an expression of a transition process (that Helmholtz did not explicitly discuss) in evaluating scientific experience. It is no longer seen as being the indubitable guarantor for absolute validity but simply one case in which claims must stand the test to varying degrees.

7.1.3.4 Empiricizing Geometry

How much does the re-evaluation of experience concern Helmholtz's conception of science? If it were also to include the hitherto undisputed validity of the formal structures of logic and mathematics it would have meant another and indeed an extremely consequential step in the direction of modernity. In the later 1860s Helmholtz held formal truths to be the epitome of the emphatic claim to truth that he considered evident for mathematical axioms[727] and established for geometry by reduction to the immutability of solid bodies.[728] This gives geometry a key function: together with "pure mathematics", geometry is paradigmatic for absolute validity and because of its empirical fundament it is effected by any change to the concept of experience.

At this point I would like to recall that Helmholtz's empiricizing of geometry in the 1860s already contained two elements of a modern conception of science: first, the deduction and applicability of axioms in geometry depends on the validity of empirically testable properties of solid bodies; if the former were merely relative, then so would the latter. Second, his approach directly produces varying equivalent geometrical representations for physical space, such that it loses its uniqueness. Helmholtz himself had considered his approach classical as it appeared in the writings developing his conception of geometry, which were completed by the speech from 1870, *On the Origin and Significance of Geometrical Axioms*. Because the certainty of geometrical axioms was supposed to be on par with absolute validity traditionally guaranteed for by evidence, he took for granted that the assumed

[727] Cf. Section 6.3.1.2.

[728] Cf. Section 6.2.

properties of solid bodies could be completely manifested. Behind his recognition of the various geometrical systems of axioms he put his conviction that only one of them is manifested in space, is accurately measurable and deviates merely negligibly from Euclidian axioms. This classical position was directed programmatically against Riemann's almost modern claim that the foundations of geometry are hypothetical in character.

In remarks on geometry that can be found in Helmholtz's public lectures and speeches after 1870, he did not revise this fundamental conception. It is noticeable, however, that the elements of modernity already ensconced in his foundation for geometry do emerge more clearly. It is open to debate whether he was ever aware of the consequences of possibly losing validity for geometrical propositions, or whether he only suspected them, or largely misconceived them.

To begin with, Helmholtz modifies the significance of his foundation for geometry by presenting it with a differently weighted motivation. He is primarily no longer defending the absolute claim to truth against Riemann's conception that geometry's axioms are hypothetical but instead, defending it as opposed to a metaphysical foundation allegedly traceable to Kant.[729] This altered background pervading *Facts in Perception* had a year earlier already shaped his remarks on geometry made in *Thought in Medicine:*

> All metaphysicians united to fight against any attempt to resolve the intuitions into their rational elements; whether the so-called pure or the empirical, the axioms of geometry, the principles of mechanics, or the perceptions of vision. For this reason, therefore, the newer mathematical investigations of Lobatschewsky, Gauss and Riemann on the alterations which are logically possible in the axioms of geometry; and the proof that the axioms are principles which are to be confirmed or perhaps even refuted by experience and *can accordingly be acquired from experience* – these I consider to be very important steps.[730]

---

[729] Critique of Kant plays no role in Helmholtz's first essays on empirical foundations (**Helmholtz** (1868b), **Helmholtz** (1868c) and **Helmholtz** (1869a)). It is also not the focus of his closing exposition (**Helmholtz** (1870), p. 4, 22 and 30). Later it seems as if dealing with Kantian conceptions of geometry had become pivotal for clarifying his relationship to Kant's philosophy as a whole. In a report that **Koenigsberger** (1902 f.), Vol. 2, p. 141 f., printed and dated as being from 1888, Helmholtz writes: "The nub of these errors [caused by the incomplete development of individual sciences during Kant's lifetime – G.S.] comprises the axioms of geometry [...] In my opinion, one may preserve Kant's great achievements only by disregarding his error about the purely transcendental meaning of geometrical and mechanical axioms".

[730] **Helmholtz** (1878c), p. 186 (Germ.)/324 (Engl.) (my italics). The last sentence was taken almost word for word from **Helmholtz** (1870), p. 30, and slightly altered, from p. 22, (cf. Section 6.2, footnote 382) and occurs again (slightly modified) in **Helmholtz** (1878a), p. 233.

Including the "principles of mechanics" among the "intuitions" to be decomposed "into their rational elements" is probably related to the changes in "mechanical principles" in non-Euclidean geometry that Helmholtz had already stated in **Helmholtz** (1870), p. 29. See also the corresponding statement in **Helmholtz** (1878b), p. 643: non-Euclidean "systems [...] and the system of mechanics belonging to it". Helmholtz probably assumed that each kind of geometry has its own corresponding (classical) mechanics, such that refutability of the "principles of mechanics" would mean that one such system of mechanics could not be applied to a system of geometry to which it did not belong. At least this is the way it is understood in **Helmholtz** (1870), p. 22.

Physiological explanation for intuitive spatial forms and an empirical fundament for geometry become the paradigm for a scientistic critique of non-empirical and therefore metaphysical foundations for science.[731] In thus examining Kant, Helmholtz now stresses one of the elements from the classical framework, which he does not leave, that already indicates a modern interpretation of geometry. According to Helmholtz, Kant had claimed, "that spatial relations which might contradict Euclid's axioms cannot even be imagined".[732] If contrary to Kant one agrees with Helmholtz that a notion designates not foremost conceptually processed knowledge but instead a perception caused by external physical stimulation,[733] then this statement makes sense. In a transcendental philosophical system, non-Euclidean geometries would be entirely conceivable as objects of an intellectual intuition that man cannot grasp conceptually[734] but as the specification of real space and the perceptions and notions related to it, they are inconceivable. The one follows directly from transcendental aesthetics, the other corresponds to the entire structure of Kant's critical philosophy focused on the ways of human knowing, which, as Kant wrote in hindsight, can only be understood as specific if one has "other possible intuition in mind".[735]

What Kant considered beyond human forms of intuition now becomes for Helmholtz a possible object of external perception. Directly following his earlier explanations he demonstrates in *Facts in Perception* that non-Euclidean geometries are "intuitively imaginable, because [the] series of sense impressions" that results when an Euclidean object is studied from the perspective of a curved world "can be given completely and clearly".[736] While in his closing essay from 1870 he compensates the abandonment of truth evoked by the proof of equivalent systems of axioms for geometry by making the (erroneous) claim that only one system can be manifested in space, here that argument moves to the background and the emphasis is on the possibility of its intuitive demonstration.[737] The idle question of the real

---

[731] Particularly in **Helmholtz** (1878a), p. 218f. and 223ff.

[732] **Helmholtz** (1878a), p. 230, and correspondingly **Helmholtz** (1878b), p. 642f. A comparable statement can be found in Helmholtz's writings on geometry prior to 1878 only in **Helmholtz** (1870), p. 22, where he writes that if "spaces of this kind [...] are imaginable, this itself would refute that the axioms of geometry are necessary forms of an a priori given transcendental form of intuition in Kant's meaning of the word". But in this writing it is not his main concern, only a secondary issue; cf. also loc. cit. p. 28.

[733] On Helmholtz's concept of a notion cf. Section 6.3.1.5.

[734] **Kant** (1781), B XL. Of course, Kant does not yet speak of non-Euclidean geometries but that they are conceivable is implied by his thought that there could exist "intellectual beings, to which our sensual capacity for intuition does not apply" (**Kant** (1781), B 309, see also B 72). Kant explicitly states that empiricist foundations for geometry do not constitute arguments against the reality of space having more than three dimensions; see loc. cit. A 24.

[735] **Kant** (1790), B 346. **Loh** (1990) drew my attention to this passage.

[736] **Helmholtz** (1878a), p. 230f.

[737] In publications after 1870 Helmholtz no longer explicitly mentions his former claim that spatial curvature is constant.

structure of space also seems to have lost significance.[738] Helmholtz is now interested in using an altered concept of non-immediately experienced intuition as a foundation for the mathematically already proven equivalence of various systems:

> The task of imagining the spatial relations in meta mathematical spaces requires, in fact, some exercise [...]
>
> This, however, contradicts the older concept of intuition whose representation comes to consciousness immediately with the sense impression and without recollection and effort. Our attempts to imagine mathematical spaces do not, in fact, have [this] ease, rapidity and lightning-like self-evidence.[739]

Claiming that forms of intuition can be learned (which follows the theory of the second kind of conclusions by induction), pulls the ground out from under evidence, the typical fundament of validity for the classical conception of science and to which Helmholtz had taken recourse in the 1860s for establishing the "first general principles" of mathematics.[740] An in principle illimitable number of differing systems of theorems that all equally well define the spatial representation of something empirically given replace the one evident system of theorems on which geometry was previously based. If the choice of a system of theorems alone were to determine the access to what is empirically given, then the evaluation of it would need to be subjected to fundamental change. No spatial property could be represented in a nonambiguous way, it could only be shown relative to the chosen system of geometric axioms. While none of these presentations can claim to be more adequate to what is empirically given than any other, special (e.g., conventionally selected or life worldly established) criteria would need to be introduced in order to settle the question of presentational form. As a result, the assumption of the existence of space, if it were discarded altogether, would have the status of being, in principle, open in terms of truth; in other words, it would be a hypothetical statement.

The hypothetization that emerges in the context of the diversity of descriptions negates the classical precondition of truth as stated by empiricism, namely that what is empirically given is available as an authority against which propositions and theories can be checked, regardless of how it may be presented. This constitutes the third criterion that I suggest for a modern concept of science.[741] The fundament of science characteristic for modern times rests on the claim that access to experience is indubitably of a linguistic nature, consisting in scientific research not only of informal communication regarding objects of scientific study but also in the definition of (both empirical and theoretical) concepts and in the design of theories. Experience is said to be permeated by "theoreticity" or to be "theory-laden". The systems of geometrical axioms that Helmholtz discusses imply that the access to

---

[738] When discussing geometrical axioms in **Helmholtz** (1878a) the "investigation of empirical facts" is mentioned only in passing, on p. 233.

[739] **Helmholtz** (1878a), p. 231 f. (Germ.)/354 (Engl.).

[740] **Helmholtz** (1862), p. 175, cf. Section 6.3.1.2.

[741] See Section 4.1.

experience in terms of space is theory-laden in just this way and thus contribute to the hypothetization of what is empirically given.

One might underestimate this shift in evaluating the empirical foundation for validity if one were to draw it only from Helmholtz's remarks on geometry found in the speech *Facts in Perception*. Hence, at this point I would like to note that in other passages of the same speech, which I shall discuss in the next part of this section, Helmholtz explicitly attributes hypothetical status to the realistic origins of his foundations for science as a whole and thereby relativizes experience's capacity to access truth, which he had hitherto presupposed without question. Helmholtz's occupation with Kant's understanding of geometry shows that scientistic foundations of science lean towards modernity as soon as they take up the kind of skepticism regarding empirical knowledge that is characteristic of metaphysical underpinnings and that Kant with respect to geometry expressed by saying: "What is derived from experience has only comparative universality, namely that which is obtained through induction".[742]

But once again Helmholtz labors to invalidate the modern component of his conception of geometry by claiming that reality has immutable properties. While in 1870 this effort was part of his following the principles of mechanics, he now primarily names considerations from the theory of perception that allow making a statement about space to which there is no alternative. Similar to what he had said for the concept of natural law and for the inductive method, meanwhile here also his remarks on geometry are subordinated to the physiological-psychological context of his theory of perception.

As mentioned above, Helmholtz traces the assumption of an external world, one that presents itself to human beings as a "power", back to the fact that external perceptions that cannot be altered by an act of will might be isolable on an elementary level. The claim that space has properties invariant to how they may be presented allows him to assume that on the side of the subject there exist natural conditions that must be given to assure that in perception anything at all is resistant to action. This concerns the presumed constitutive possibility of bodily movements:

> And space would be an *innate form of intuition prior to all experience* insofar as its perception would be tied to the possibility of the will's motor[ic] impulses and for which the mental and corporeal ability must be given us through our organization before we can have spatial intuition.[743]

Of course, what Helmholtz calls non-empirical lies wholly within the range of what is empirically accessible: the conversion of impulses of will is the object of inner perception[744]; detecting how the human body is organized is the task of empirical

---

[742] **Kant** (1781), A 24.

[743] **Helmholtz** (1878a), p. 225 (Germ.)/349 (Engl.) (emphasis in German original). According to Helmholtz the impulses of will converted into bodily movements do not lead to acknowledgement of the external world as the cause of all appearances, but to the more elementary distinction of inner and outer.

[744] **Helmholtz** (1878a), p. 223.

science. Nonetheless, the knowledge it procures inductively is subject to limitations of validity that cannot be cancelled by subsequent empirical verification.[745] This is perhaps the reason why, in the passage quoted above, Helmholtz speaks of what should actually be certain a priori, in the conditional. It is valid only under the provision that it can never be completely scientifically corroborated.

Although the theory of perception clearly takes center stage in Helmholtz's comments, physical mechanics maintains its distinguished place among the empirical sciences. Particularly his remarks on geometry in the 1870s show that referring to mechanics still serves to guarantee that the validity of scientific statements is certain. Helmholtz continues to assert that axioms of geometry present a system of statements that can be "acquired from experience"[746] and whose disproof is merely an abstract possibility. The special character of the underlying mechanical experience is expressed in his continued claim of real existing immutability of the shape and the free mobility of solid bodies.[747] Helmholtz still measures the validity of mechanically derived geometrical statements by metaphysical fundaments of science and regards geometry "as the first and most completed of the natural sciences".[748]

Once more it is within the context of examining Kant's conception of geometry that compels Helmholtz, at least in one passage, to relinquish the special status of geometry. In the heat of debate with "strictly observant Kantians,"[749] he confronts it as an integral component of scientific research with a loss of certainty that clearly reaches beyond the classical conception of science: some claim that:

> Even if the axioms may be theorems of experience, we cannot be absolutely convinced of their correctness, as we indeed surely may be. The contention turns on precisely this very point. [...] If the axioms are natural laws, then it is also correct that they naturally play a part (through induction) in the (only approximate) provability of all natural laws. However, the wish to want to know exact laws is still not itself proof that there may be any such.[750]

Now the validity of applying a specific geometric system no longer depends on different degrees of measurable accuracy, it is the validity of the axioms of geometry themselves, i.e., in particular the validity of the properties of solid bodies, that does so. It cannot yet be ruled out that the merely approximate validation of natural law statements may improve progressively and towards an ideal state. Yet the reference made here to inductive method alludes to the principle limitations to validity that

---

745 See Sections 7.1.3.2 and 7.1.3.3.

746 **Helmholtz** (1878c), p. 186.

747 **Helmholtz** (1878b), p. 648.

748 **Helmholtz** (1878b), p. 642.

749 **Helmholtz** (1878a), p. 229.

750 **Helmholtz** (1878a), p. 392 f. (Germ.)/370 f. (Engl.) (cf. footnote 697). In this appendix titled *Space can be Transcendental, without Axioms being so* [*Der Raum kann transzendental sein, ohne dass es die Axiome sind*] Helmholtz examines **Krause** (1878) critically. The translation of this heading as "Space Can Be Transcendental without There Being Any Axioms" (**Helmholtz** (1995), p. 369) is misleading.

also apply to the formal truths of geometry: they cannot be regarded as unconditional and immutable any more than mechanistic reductions can.

Besides the hypothetization of natural law statements and the theory ladenness of experience, I introduced, as a second criterion of a modern conception of science, the loss of truth that can be traced back to eliminating the distinction between logical-mathematical truths, among which I count truths in geometry and the detection of facts.[751] If the latter is awarded a revisable status, the former lose their a priori validity.

For Helmholtz this kind of modernization does not begin until he finally applies the revisability of scientific statements to mechanics and geometry, thereby undermining their distinguished status. The previously quoted passage is important (I found none similar to it) because some of the formal truths are denied absolute validity, because they are now seen as belonging to empirical science. This, too, can be understood as a result of Helmholtz's expanding the theory of perception to become a general naturalistic epistemology, for the mark of his epistemology consists of denying statements derived by pure thinking, to which traditionally formal truths belong, any kind of autonomy compared to merely relatively valid empirical knowledge.[752] He is not concerned with the loss of truth for analytical statements but with recognizing the approximate nature of synthetic statements that are only wrongly considered analytic and are faced with a residue of irrefutable correct analytic statements.[753] Prior to Quine's modern completed conflation of analytic and synthetic statements,[754] Helmholtz reduces the number of scientifically acknowledged analytical statements.

7.1.3.5 Epistemological Hypotheses

Contemplating the three tendencies of modernization in Helmholtz's conception of science discussed so far – hypothetization of natural laws, the theory ladenness of (spatial) experience and the relativization of logical-mathematical truths – it is conspicuous that Helmholtz himself rarely views this change favorably. What has been said about his conception of geometry should have made it clear that in the first instance his alignment with mechanics had an inhibiting effect on modernization. His critique of the mechanistic theory of matter that stood at the beginning of

---

[751] See Section 4.1.

[752] See Section 7.1.3.2.

[753] Cf. **Helmholtz** (1868b), p. 610, where Helmholtz divides geometric theorems in analytical and synthetic; **Helmholtz** (1870), p. 30, where when examining Kant's conception he implicitly claims that synthetic propositions have empirical content (quoted in Section 6.2, footnote 379); **Helmholtz** (1871b), p. 45, where he distinguishes facts of experience from definitions of terms (quoted here in footnote 635); **Helmholtz** (1885 ff.), p. 590, where he makes a general distinction between analytic and synthetic.

[754] See Section 4.1.

the transitional process did not eliminate mechanics' claim to dominance. Nonetheless, it would be wrong to think that every alignment of a conception of science and nature to a specific discipline is disagreeable to modernity. On the contrary, the relative conditions for validity characteristic of the modern conception of science themselves alone permit continued well-founded adherence to a leading discipline even when there is a lack of empirical confirmation traditionally demanded of the arguments involved. When the limited range of mechanistic reductions becomes obvious, mechanism's only chance for survival as a conception of nature is to be modernized. Only by abandoning its claim to exclusive truth and considering itself one among many possible conceptions of nature can mechanism hope for the right to exist.

A metamorphosis such as this demands not only the modernization of individual aspects of a conception of science but also the modernization of the epistemological fundament that supports it. For Helmholtz the main pillar of that fundament is the realistic assumption of an external world. That this assumption may indeed be questionable has only surfaced once, namely in his popular speech from 1855, only to be immediately eliminated by introducing noumenal causality. Thus it signals both a new quality in reflecting his realism and becomes pivotal for the transition of his conception of science when he explains, at the core of the speech *Facts in Perception:*

> We cannot recognize the realistic opinion as more than a superbly useful and precise hypothesis; we may not ascribe necessary truth to it, since in addition to it still other, irrefutable idealistic hypotheses are possible.[755]

This is an unexpected remark. Up to this point Helmholtz had treated realism as a "necessary truth", now it is allegedly merely a "hypothesis". The radical turnaround for the claim to validity is in keeping with the fact that Helmholtz now uses a concept of hypothesis he had previously employed only on exception, namely to designate entities fictitiously assumed for purposes of explanation.[756] While on the previous understanding of the term the indeterminacy of the truth of hypotheses was considered tentative or capable of minimization, now progress in knowledge can no longer eliminate it at all. To bring the point home, Helmholtz speaks of "metaphysical hypotheses". In contrast to the founding of science metaphysically, which he resolutely dismisses, he considers metaphysical hypotheses indispensable:

> The different shadings of idealistic and realistic opinions are metaphysical hypotheses which, so long as they are recognized as such [...] are completely justified scientifically. [...] Science must discuss all admissible hypotheses in order to retain a full overview of all possible attempts at explanation.[757]

---

[755] **Helmholtz** (1878a), p. 239 (Germ.)/360 (Engl.), **Turner** (1977), p. 57, points out the proximity of this explanation to Fichte's programmatic statement that it is not possible "based on reason, to decide" between realism (Fichte calls it "dogmatism") and idealism (**Fichte** (1797), p. 432).

[756] See Section 7.1.2.

[757] **Helmholtz** (1878a), p. 239 (Germ.)/360 (Engl.). Cf. **Friedman** (1997).

Being a hypothesis, neither should the "realistic opinion" in any way lose its relevance for science. Helmholtz characterizes it by interpreting just that elementary perceptual experience that he had used to support the assumption of an external world: the observation of non-willfully evoked changes.[758] Realism, for which "the material world" – as Helmholtz defines it for the first time in passing – exists "independently of our ideas", makes, in his opinion, generalizations such that "the changes in perception following an action have no mental connection with the previously occurring impulses of the will".[759] In contrast, the idealistic hypothesis interprets the occurrence of changes in perception purely mentally:

> The will's impulse for a definite motion is a mental act and, so too, is the related perceived change in sensation. Now, cannot the first act bring about the second through a purely mental agency? It is not impossible. Something like this happens when we dream. [...]
>
> I do not see how one could refute a system of even the most extremely subjective idealism that wanted to view life as a dream.[760]

The first salient aspect of the "most extreme" subjective-idealistic interpretation of perception that Helmholtz discusses is how little it differs from his own realistic position. The difference between a waking and a dream state may not even be related to the question of an independently existing external world; it can also be understood as the difference between a predominantly outer world-related perception and one that is exclusively inner. Perception, whether it occurs in a waking or dreaming state, presupposes a reality, the existence of which is debatable in terms of dreams just as it is in terms of being awake.[761]

It is hard to say to which extent Helmholtz actually relativizes his realism and considers it a hypothesis. Realism and idealism first become epistemologically comparable by inspecting each one's foundational strategy; in realism that strategy is to assume that an external world exists independent of knowledge of it; in idealism, in contrast, it is to presuppose non-empirical autonomous thinking.[762] Within the framework of the classical conception of science committed to emphatic truth, the two strategies present two mutually exclusive approaches. Classically founded science cannot, to put it roughly, be both idealistic and realistic. Precisely this radically distinguishes the modern concept of science from the classical one. Without suspending the contradistinction between the two strategies, the modern conception of science assumes that both are irrefutable and thus equal in rights. Both the

---

[758] **Helmholtz** (1878a), p. 241 (as footnote 675).

[759] **Helmholtz** (1878a), p. 238 f.

[760] **Helmholtz** (1878a), p. 237 f. (Germ.)/359 (Engl.).

[761] I agree with Kant in **Kant** (1781), A 376. Nonetheless, in the speech **Helmholtz** (1878a) his position on dreamt reality is inconsistent. While here he assumes that the dreamer produces reality himself, on p. 241 he claims that there is no difference between a dreamt reality and one taken to be realistic, since for both it is necessary to assume law-like regularity.

[762] This minimal definition of idealism can rest on the philosophical literature that was relevant at the time and with which Helmholtz was certainly familiar, e.g., **Kant** (1781), A 366 ff., and **Fichte** (1797), p. 425 ff.

idealistic and the realistic approach are, inasmuch as they are used at all in modern times, part of a truth-relativizing diversity of conceptions of science. That is the reason for their hypotheticity.

Inasmuch as Helmholtz abandons the claim to truth for his epistemological assumptions by acknowledging that a different strategy may also be an irrefutable alternative, he gives his own foundation for science overall a modern character. The result of this sort of relativizing is a questioning of the validity of every single statement. The most important indication of such a complete hypothetization may be reflected in the fact that in his speech he mentions Fichte and Hegel as two radical proponents of the idealist position.[763] But if Helmholtz is saying that idealism is the same thing as the conceptual world of a realistically understood dreamer – and there are some good reasons for doing so – then he himself has not understood the real implications of his verbal equalization of idealism and realism.

I would now like to distinguish the modernization effect ensuing from the acknowledgement of alternative strategies for founding science from those circumstances of the transition that result from problems immanent to founding science on realism. While the hypothetization that it generates is easier to prove, it is also of comparatively smaller scope. It has less to do with the status of epistemological preconditions than with that of scientific concepts and theories. By way of explanation I would like to return once more to the argument from the theory of perception that Helmholtz gave as a justification for realism; the problems contained therein lead him to reformulate his classical position.

The argument rests on the claim that it is possible in perception to isolate changes that cannot be actively influenced by the will from those that are produced by an act of will. According to Helmholtz, perceptions of objects originate by the integration of a number of differing observations caused by bodily movements. Thus an observer changes his position, in order to get a different view of an object, or a blind person moves his finger along an object in different directions. What is real appears then to be that which is not changed by bodily movements but which changes itself independent of the observer's movements.[764] As he had done in the first edition of the *Treatise,* in the speech *Facts in Perception* Helmholtz sees a parallel between actions that are constitutive of both perception and reality and scientific experimental procedure:

> Each of our voluntary motions by which we modify the manner of appearance of objects is to be considered as an experiment by which we test whether we have correctly conceived the lawful behavior of the phenomenon in question, that is, its presumed existence in a definite spatial order.
>
> The convincing force of every experiment is, however, in general so much greater than that of the observation of a process occurring without our involvement, because in the experiment the causal chain runs throughout our self-awareness.[765]

---

[763] **Helmholtz** (1878a), p. 238 and 241.

[764] **Helmholtz** (1878a), p. 226ff. and 241. See Sections 6.3.1.5, 6.3.2.2 and 7.1.3.1 here.

[765] **Helmholtz** (1878a), p. 237 (Germ.)/358 (Engl.).

Remarkably, here the experiment can only be a test. It is assumed that something is present in space and merely confirmed by the experiment. This very narrowly limited function does reflect the merely observant attitude of the perceiving subject. The subject does not change the arrangement of perceived objects, he only changes his position in relation to them; he does not create any objects but merely comes across them. Accordingly, in experiments Helmholtz had difficulty demarcating experimental perception from passive observation. Besides the fact that the required changes in perception do not necessarily presuppose a self-awareness reflecting on the intentions of its actions, one can also ask why the "convincing force" of experiments should be traced back to self-awareness at all, without the results being of significance. In terms of their validity the fact is less important that I am the one doing the experiment than – to a much greater extent – the question of whether I or anyone else can produce the specific and perhaps also intended effects.

The one-sided emphasis on awareness only conceals the perceiving subject's powerlessness when faced with the external world. Instead of actively interacting with the world, according to Helmholtz, the subject revolves around allegedly immutable objects. While it is true that by including the constitutive importance of bodily movements Helmholtz incorporates an element of action theory into his theory of perception, but his concept of action remains very reduced. In contrast to Fichte's and Hegel's idealistic systems that he mentions, he places no importance at all on action-guiding principles.[766] Furthermore, it appears as if Helmholtz wants to glean the distinction between voluntarily and non-voluntarily produced changes from the relation between the subject alone and his external world. But within a constellation void of intersubjectivity, it is impossible to make such a demarcation. Because the demarcation is manmade, it rests, as Fichte has shown, on an intersubjective relationship of mutual recognition.[767] However, since Helmholtz gives no criteria for the part of what does change in perception that would have to be considered non-voluntary, his own theory of perception compels him to doubt the realistic assumption of an external world that was supposed to be supported by the theory of perception. His doubt confirms that he is incapable of concretely spelling out the abstract distinction that he claims exists.

If one assumes that it was more likely this immanent difficulty than the – possibly only oral – acknowledgment of other alternative foundations for science that led Helmholtz to hypothesize his realism, the modification of the validity of his epistemological presuppositions looks less like a modern relativization of his own standpoint and more like the expression of difficulties in providing a foundation for a theory of perception and action that is still very much within the framework of a classical conception of science. Thus it is not surprising than in connection with his speech of 1878 he attempts to reformulate the realism that he continues to presuppose is absolutely valid and does so entirely without reference to action. Instead, he

[766] Cf. **Helmholtz** (1878a), p. 239.

[767] **Fichte** (1800), p. 262.

justifies the remaining non-hypothetical assumptions about reality by taking recourse to noumenal causality.

This reformulation can be found in Helmholtz's work *The Origin and Meaning of Geometric Axioms (II)* [*Ueber den Ursprung und den Sinn der geometrischen Sätze; Antwort gegen Herrn Professor Land*], excerpts of which were appended to the printed speech.[768] In this appendix he deals with the refutation of the assumption (which he once more erroneously equates with Kant's conception of geometry) that Euclidean geometry is an innate form of human intuition.[769] In the last part he strives to justify his own empirical foundation by demonstrating an indubitable basis for experience. To do so he "drop[s] the hypothetical part in the realistic viewpoint"[770]:

> The sole assumption which we adhere to is that of the causal law: namely, that the ideas occurring in us with the character of perception occur according to enduring laws, so that, if different perceptions intrude upon us we are justified in drawing the inference to differences in the real conditions under which they have developed. Furthermore, we know nothing about these conditions themselves, about the actual real which underlies the phenomena; all opinions which we may otherwise harbor in this regard are only to be considered as more or less probable hypotheses.[771]

Here Helmholtz understands causality exclusively in the noumenal sense. He is not talking about a regularity of phenomena but about the regularity of how they originate as an effect produced by a reality that exists independent of being known. Relations of ideas are qua causality traced back to relations between "real conditions" that Helmholtz later calls "real conditions that give rise to the idea".[772] These conditions are not transcendental conditions of the possibility of experience; they are "the actual real", whose "difference" is supposed to be unquestionably given.

Helmholtz now subdivides these relations that structure ideas and constitute the non-hypothetical part of the realist's assumption of an external world into "topogenous factors" that are "some sort of relationship [...] which determines at which place in space an object appears to us to be" and in "hylogenous factors" that are the "causes" of temporal change at one place.[773] I interpret his remark that he chose these new designations "so as to separate off any intermixing of additional meanings that could attach themselves to common words,"[774] as an indication of the perhaps non-propositional nature of these factors.[775] They are elementary "causes" or the "determining" of

[768] **Helmholtz** (1878b).

[769] **Helmholtz** (1878b), p. 641 f.

[770] **Helmholtz** (1878b), p. 655.

[771] **Helmholtz** (1878b), p. 655 f. (Germ.)/377 (Engl.).

[772] **Helmholtz** (1878b), p. 656.

[773] **Helmholtz** (1878b), p. 657. For hylogenous and topogenous factors cf. **Hyder** (2002).

[774] **Helmholtz** (1878b), p. 657.

[775] In **Helmholtz** (1878a), p. 244, he remarks, "it was the physiological investigations into sensory perception that led to the final elementary processes of cognition, which cannot be expressed in words". I believe that these elements include the hylogenous and topogenous factors. The importance of components of perception that cannot be expressed linguistically is discussed later in detail in **Helmholtz** (1885 ff.), p. 596 ff. (esp. p. 601). See footnote 779.

perceptions that need be neither accessible for measurement[776] nor capable of being expressed linguistically. They are better ordered than the raw material of sensations but less available than the perception-relevant empirical data and its structure.[777]

Compared to his earlier writing, now the domain of hypothetical statements has been considerably expanded. Even the empirical concepts that are directly connected to elementary sets of data are tied to epistemological preconditions, must be formulated in an idealistic or realistic "language"[778] and are thus, according to Helmholtz, hypothetical.[779] But if even concept production falls under relative conditions of validity, then the scientific theories associated with those concepts and that are quasi even further removed from the empirical data than those concepts, do so a fortiori. It now seems possible that data structures can be represented in various theoretical ways. Such consequences, which are only implicit in Helmholtz's abstract and vague remarks, extend the "theory ladenness" of experience, already expressed in his remarks on geometry, to cover all outer experience.

But in reaction to the difficulties of enunciating a foundation for the theory of perception, Helmholtz conversely drew an untrespassable boundary for hypothetization. Topogenous and hylogenous factors designate empirically given material that in realism exists independent of scientific concept and theory production and is lodged between scientific knowledge and an external world that can never be wholly grasped linguistically. By creating this intermediate position Helmholtz escapes having to hypothesize the empirically given, which would follow, if the reference to experience were made solely on the basis of science, in other words, if a truth-relativizing plurality of kinds of representation were contrivable.

In summary, Helmholtz's statement that his realistic conception has the status of a "metaphysical hypothesis" involves considerable thrust in the hypothetization of scientific concepts and theories. This transition can be traced primarily to difficulties immanent to the justification of his own realism. In the attempt to eliminate the problems of founding a theory of perception, Helmholtz takes recourse to the old strategy of justification based on noumenal causality. This fortifies the classical

---

[776] **Helmholtz** (1878b), p. 658 f.

[777] In this context **Erdmann** (1921), p. 7, speaks of "facts of perceptual awareness".

[778] **Helmholtz** (1878b), pp. 648 and 656.

[779] An objection to this interpretation might be that in **Helmholtz** (1878a), p. 218, he calls natural science, in contrast to philosophy, a speechless undertaking (see footnote 683). But in the aftermath Helmholtz stresses – in my opinion in following the direction pursued in **Helmholtz** (1874b), p. 424 f., and continued in **Helmholtz** (1878a) and **Helmholtz** (1878b) – that the formulation of scientific laws is basically a linguistic endeavor: **Helmholtz** (1892), p. 339, and **Helmholtz** (1897 ff.), Vol. I.1, p. 11 f. But an increased consideration of the linguistic form does not bring scientific knowledge closer to how perception is pervaded by symbols, because for Helmholtz what is law-like continues to be not the signs but the representation of reality (**Helmholtz** (1878a), p. 222, and **Helmholtz** (1885 ff.), p. 586); the sensory signs given in perception need not be expressible linguistically (**Helmholtz** (1885 ff.), p. 598 ff.). Nonetheless, Helmholtz's late work does include passages that place scientific knowledge as a whole quite near symbolicalnes: **Helmholtz** (1885 ff.), p. 593.

assumption of causality compared to approaches to a pragmatic, modernized philosophy of science. How far the modernization process can nonetheless continue depends crucially on the relation between a priori presupposed noumenal causality and the pragmatic factors of his conception of science.

Before I discuss that relation, two aspects should be added to what has been said thus far. First, in an excursus I would like to point out an example of approximately equivalent scientific descriptions, which Helmholtz himself provided. Second, the scope of modernization up to this point cannot be conclusively evaluated without clarifying whether and how his position on the classical goal of science of generating a complete system of knowledge, changed. Answering this question, however, leads directly to the relationship between noumenal causality and pragmatic modernization. It is Helmholtz's endeavor to justify maintaining the classical aim of science as a non-hypothetical basis for experience that reveals the importance and scope of his pragmatic orientation.

Excursus: Mechanic Theories as "Pictures" of Reality

Although Helmholtz's writings in physics bar the option of equally valid theories for specific sets of phenomena,[780] his *Lectures on Theoretical Physics* [*Vorlesungen über theoretische Physik*] do contain one instance of clearly abandoning the belief in an one-to-one presentation of reality taken as independent of knowledge: for the first time he calls scientific theories "pictures", meaning presentations that basically do not correspond to reality.[781]

From the 1860s on he had used the term "picture" [*Bild*] to distinguish the content of truth in scientific statements that do "picture" the laws of reality from the nature of a sign that is typical for the witness of immediate sense perception.[782] As late as 1892, in his second speech on Goethe's work in natural science, he stresses:

> In order to be a sign it is only necessary that the same sign always be given for the same object. No kind of similarity is necessary between a sign and its object [...]
>
> We cannot call sense impressions "images", for an image represents like by means of like.[783]

While Helmholtz here uses the concept of a picture to indicate whether a statement corresponds to a real object or event, in a lecture held in 1894 titled *Dynamics of Equally Dissipated Masses* [*Dynamik continuierlich verbreiteter Massen*] he uses the term to designate statements that only partially correspond to the objects involved:

---

[780] This is conspicuous in Helmholtz's writings on electrodynamics from the 1870s intended to compel a choice between competing theories. Cf. **Woodruff** (1968), p. 308 ff.

[781] I thank **Truesdell** (1984) for pointing this out. He suggests that in the history of science Helmholtz was the first to state that scientific theories are models (loc. cit. p. 490, cf. also p. 31).

[782] Cf. Sections 6.3.1.5 and 6.3.1.6.

[783] **Helmholtz** (1892), p. 357 (Germ.)/496 (Engl.); with almost identical wording in **Helmholtz** (1878a), p. 222.

> Now, by designating something an abstraction or a picture, the claim is not that the idea of equally dissipated masses [...] completely corresponds to the entity found in nature.[784]

Helmholtz contrasts the picture of the continuum mechanics that presupposes that space taken up by mass is continually filled in, with point mechanics' picture of matter thought to be discretely divided.[785] The fact that neither of the two pictures corresponds to reality follows from the limited scope of scientific knowledge. Since nothing "specifically empirical" is known about "the ultimate division of masses", all that can be done is to "devise hypotheses regarding it".[786] In contrast to the former meaning of the concept, now pictures have a hypothetical character.

Besides the negative circumstance that empirical data is (still) lacking for the presupposed "ultimate" structure of an entity, the concept of the picture has a positive aspect, namely: the freedom to present things. In Helmholtz's opinion, different types of mechanical phenomena can be better captured by one kind of picture and worse by another – in other words: one phenomenon can be shown in different ways. For example, elastic phenomena can be described more adequately using continuum mechanics but this does not mean that they cannot be calculated in point mechanics, although that would involve some constraints.[787]

This change in the concept of the picture from being a one-to-one representation of specific properties of an object and the (temporal) pattern of that object's changes to becoming a theory-dependent perspective portrayal of the object under investigation indubitably propels Helmholtz towards a modern position. Note, however, that the scientific theories now called "pictures" are still distinct from the sign character of perception because in contrast to the signs in perception, pictures still correspond to their objects to the extent that they reproduce physically measurable variables. While Helmholtz once again neglects an opportunity to mention the criterion of temporal correspondence, thereby leaving the relation between a picture and reality undefined, his student Heinrich Hertz states an "essential" criterion: the "necessary consequents of the images in thought" must "always [be] the images of the necessary consequents in nature of the things pictured".[788]

The comparison with Hertz shows how far Helmholtz, in spite of modifying his concept of the picture, still was from a modern understanding of the relation between scientific theory and empiricism. While both of his mechanical pictures can describe an identical range of objects, their descriptions are not equivalent. Where mechanics, with its different equivalent axiomatizations by Newton, Lagrange and Hamilton offers a possibility for describing the entire range of objects in varying ways, Helmholtz does not use the concept of the picture.[789] But

[784] **Helmholtz** (1897 ff.), Vol. II, p. 2.

[785] **Helmholtz** (1897 ff.), Vol. II, p. 1 f.

[786] **Helmholtz** (1897 ff.), Vol. II, p. 2.

[787] **Helmholtz** (1897 ff.), Vol. II, p. 1.

[788] **Hertz** (1894), p. 1. See Sections 6.3.1.6 and 7.2.

[789] **Helmholtz** (1897 ff.), Vol. I.2, p. 303 ff.

Hertz, in his way of using the concept of a picture draws upon the plurality of descriptions offered by those axiomatizations. Hertz more adequately expresses the picture and perspective character of scientific theories that in modernity determines knowledge of the world.[790]

7.1.3.6 Regulative Aims for Progress

What remains of the classical demand for a complete system of knowledge, once the difference between laws and hypotheses becomes unclear, mathematic-logical statements can no longer be postulated with unrestricted certainty and scientific concepts like theories lose their truth-ensuring property of picturing reality in an unequivocal way? In the speech *Thought in Medicine* Helmholtz no longer knows whether science can achieve "complete insight into the causal connection of natural processes".[791] What is more, he believes to have discovered a crucial flaw in certain systematizations, here specifically the metaphysical "world-explaining" systems of medicine[792]:

> Yet the essential and fundamental error of these systems was and still continued to be, the false kind of logical conclusion to which it was supposed to lead; the conception that it must be possible to build a complete system which would embrace all forms of disease and their cure, upon any *one* such basis of explanation.[793]

One system, intended to embrace all phenomena, cannot be called complete on the classical understanding of science if it is built upon several independent "bases of explanation".[794] It could no longer even be called a system, because it would resist any inquiry into the conditions and causes of varying co-existing principles. The amount of incompleteness Helmholtz permits is demonstrated by the example he gives here and which has been mentioned above, of the possible effect of differing cosmic forces, for which he requires no necessary explanation for how they are related.[795]

In the 1870s Helmholtz relaxes the demand for a systematic claim, not only for research in special sciences but also for science as a whole. In conclusion he formulates, in the speech *Facts in Perception*, his ideas of the aims of science in a highly ambivalent way. The ambition underlying his mechanism from the start, namely to reduce all phenomena to final causes, is now given the status of a regulative principle:

---

[790] On the concept of picture in Helmholtz and Hertz see **D'Agostino** (1990), **Schiemann** (1997b) and **Leroux** (2001).

[791] **Helmholtz** (1878c), p. 170 and 185, see Section 7.1.3.2.

[792] **Helmholtz** (1878c), p. 171.

[793] **Helmholtz** (1878c), p. 175 (Germ.)/315 (Engl.) (emphasis in German original).

[794] The meaning of the expression "basis of explanation" *(Erklärungsgrund)* follows from his reformulating the concept of explanation as a natural law that enters deductive explanations as a premise (see Section 7.1.3.1).

[795] See Section 7.1.3.3, footnote 721.

> If we assume, however, that the conceptualization [of natural phenomena–G.S.] will be brought to completion, that we will be able to establish a final unchangeable something as the *cause* of the observed changes, then we call the regulative principle of our thinking, that which impels us, the *causal law*. We can say that it expresses the trust in the *complete conceivability* of the world.[796]

Similar to his attempt to reformulate his realistic position, here Helmholtz also takes recourse to a meaning of the concept of causality that he had employed in early writing. In the introduction to *Conservation of Force* he had considered the "final goal of the theoretical natural sciences" to "discover the ultimate invariable causes of natural phenomena".[797] Without upholding his previous adherence to the elementaristic structure of matter and the form of mechanical central forces, reducing phenomena to the final causes remains the strategic main task of science, if it is going to comprehend nature. However, the more difficult and more improbable it becomes for research to find those final causes, the further the goal is postponed to some unknown future. The complete system of explaining nature is now only an ideal and science knows not whether it will ever be achieved. Within the frame of reference I laid out for classical and modern science, this state of affairs is settled between the two conceptions of science. It is classical and thus incompatible with modern science inasmuch as it combines the goal of an explanatory system that captures the entirety of natural phenomena with an exclusive claim to validity. But in terms of truth content, a regulative principle must not lead to the predominance of any one particular explanatory system. Although they may continually and progressively improve their statements, all theories in natural science can be considered equally removed from the ideal goal of objective and comprehensive explanation of nature. In contrast to the optimism that guided the classical conception of science at the beginning of early modern times and that was still at work when Helmholtz wrote in the 1860s,[798] a modern view progress in scientific knowledge is no longer thought of as a finite development but as an open process with undefined content, characterized solely by its inherent dynamics and the methods it uses. Helmholtz adds a modern touch to his conception of science by continuing in the previously quoted passage:

> Conceiving, in the sense that I have described it, is the *method* by means of which our thinking subordinates itself to the world, orders the facts, predetermines the future.[799]

Thus the final causes lose their feature of being real conditions for natural events discoverable in nature that had shaped the Introduction of 1847. Compared to the ideal, every developmental stage in science, every current state of research remains incomplete – and this also clearly reveals the "fragmentary nature of our knowledge" – as expressed in the quote that introduces this section. Paradoxically,

---

[796] **Helmholtz** (1878a), p. 243 (Germ.)/363 (Engl.) (emphasis in German original).

[797] **Helmholtz** (1847a), p. 4, see Section 6.1.1, Part α.

[798] For instance, **Helmholtz** (1862), p. 182, see Section 6.3.1.2.

[799] **Helmholtz** (1878a), p. 243 (Germ.)/363 (Engl.) (my italics).

Helmholtz's conception of science is modernized precisely because he holds fast to the classical idea of a comprehensive system of natural explanation, because now basic incompleteness is essential for each individual phase of knowledge.

7.1.3.7 The Hypothetization of Causality

Just as Helmholtz defines the goal of progress in science using elements of both the classical and the modern conception of science, so also does he use both in his rationale for it. But while for the definition of the goal the classical and modern elements mutually supplement one another, the theoretical underpinnings are characterized by contradictory statements. It appears not possible to establish a consistent position that lies between those two conceptions of science.

At first Helmholtz's intention is thoroughly classical: the law of causality that justifies his goal should be attributed absolute validity. In the reasons he gives for this however, the concept of causality changes its meaning considerably. Now there is no more mention of "final causes" but of how experience comes about:

> The causal law is really an a priori given, a transcendental law. It is not possible to prove it by experience because, as we have seen, not even the first steps of experience are possible without the application of inductive conclusions, i.e., without the causal law; and from the completed experience, when it too taught that everything observed so far has proceeded in a law-like manner – which we are assuredly far from being justified in claiming – would always only be able to follow by an inductive conclusion, i.e., under the assumption of the causal law, that now the causal law would also be valid in the future.[800]

Thus, what is certain a priori is not the law-likeness of phenomena, the existence of which, according to Helmholtz, could only be known in a state of "completed experience". Instead of unrestrictedly presupposing phenomenal causality, he goes on to claim that the scope of it is a matter of empirical investigation. He continues to believe that research may discover in nature phenomena that do not adhere to natural law. By no means does this restriction in scope relativize the validity of phenomenal causality. On the contrary, at this point Helmholtz once again underscores the classical claim to truth for scientific experimental knowledge.

But the "first steps of experience" are an entirely different story, which, even before they have been taken, presuppose the law of causality and thus can only produce causally shaped knowledge. Obviously an everyday type of experience is meant here. Equating "inductive conclusions" with causality also indicates that in this case Helmholtz is talking about noumenal causality. Noumenal causality permits interpreting perceptions as effects of an external world that exists independent of the perceiving subject.[801] Thus this noumenal causality that Helmholtz uses to justify his realism[802] also establishes the regulative principle of research; and the

---

[800] **Helmholtz** (1878a), p. 243 f. (Germ.)/363 (Engl.).

[801] See Section 7.1.3.2.

[802] See Sections 6.3.1.3 and 7.1.3.5.

claim of a priori causal validity is, as in Helmholtz's early writing, related to this special meaning. Giving the precondition of an assumed external world the status of a transcendent cause makes it absolute.

This interpretation of the rationale behind the aim of science reveals how crucially the concept of causality stabilizes Helmholtz's conception of science: in a phase of fundamental transition, when the purposes (in terms of content) of the mechanistic conception of nature have become questionable, "the" law of causality keeps research aligned to a uniform goal, namely, the discovery of the allegedly ultimate causes of all phenomena; the causal law guarantee that an elementary basis of experience ("noumenal causality") exists independently of knowledge, limits the accumulating insight into the hypotheticity that results from epistemological presuppositions; and finally, due to presupposing causality, the acknowledgement that laws are always postulated as hypotheses and cannot be verified with absolute certainty is no threat to the fact that for the empirically researchable part of nature, phenomena follow the laws of nature. Causality thus acts as the last all-embracing stronghold against modernity's tendencies toward hypothetization.

The regulative principle of research only needs an a priori valid reason to the extent that it advocates the classical goal of complete knowledge. However, in the modern conception of science, research regulatives have the status of heuristics and these must be distinguished not by reasons but by successful application. Helmholtz himself appears to have understood the search for final causes in this way, when, in connection with his definition of the goal of science, he remarks:

> For the causal law's applicability we have no additional security than its success.[803]

This pragmatic rationale for causality must, of course, not be related solely to final causes. If one takes "conclusion by induction" to mean not the noumenal-causal interpretation of an external world but the inductive processing of sensations to become ideas and their related actions, then the premise of that conclusion is not a certainty-ensuring law of causality but a set of more or less similar experiences that can do no more than pragmatically suggest action. Seen this way, naturalistic epistemology that subjects thinking to conditions for only relatively valid conclusions by induction[804] backlashes on the meaning of causality.

Helmholtz, however, distinguished the two meanings for conclusions by induction just as little as he ever kept the various meanings for his concept of causality apart. In the speech *Facts in Perception* it is a lack of discrimination – one could agree with Yehuda Elkana and call it progress-instigating conceptual vagueness[805] – that allows him to ignore the contrast between these meanings and abandon his classical standpoint without much ado. Thus he manages to end the passage that began with a claim that causality is valid a priori, with a commitment to a pragmatic concept of truth that is capable of relativizing any epistemological precondition:

---

[803] **Helmholtz** (1878a), p. 243 (Germ.)/363 (Engl.).

[804] See Section 7.1.3.2.

[805] **Elkana** (1974), p. 8 ff.

> Only one piece of advice is valid here: trust and act!
> The insufficiency
> Here grows to Event.
> This would be the response that we would have to offer to the question: what is truth in our ideas?[806]

Subordinated to a pragmatic concept of truth guided by the success of action, the interpretations of causality lose their absolute claim to validity. The assumption of the real existence of an external world prior to all experience becomes a belief to which one is obligated only as long as it is conducive to practical success. The degree of adherence to laws of nature, which was to be determined through experience, becomes a heuristic useful for research: it does not lead to definite knowledge but instead depends on the realization that research results are revisable. What is more: the pragmatic concept of truth that Helmholtz in the 1860s had already developed for his theory of perception but here for the first time uses as a standard for scientific knowledge, fundamentally waives any guarantee for truth established through theory. Questions of truth cannot be decided by the possibility of classifying them in a systematic knowledge context but only through action.

This suddenly confronts the classical definitions with a concept of truth that has not yet eliminated the traditional distinction between theoretical knowledge and practical demands but which does suspend the antagonism of the two that is typical for the classical conception of science. Helmholtz's call to take action does not mean that the conditions under which theoretical statements are deemed valid should depend on the success of their technical implementation. For him, theoretical knowledge continues to be something different than the way it is applied. It suits his conception of science to manifest the reference to action by empirically investigating the statements of natural laws and therefore limiting it to experimental (laboratory) practice. But describing that practice as the execution of an action also means that statements of natural law cannot be definitely confirmed by experience any more.

Whether Helmholtz was aware of the pragmatic concept of truth's consequences for the philosophy of science, we do not know. Applying it to scientific knowledge in the speech given in 1878 might have been motivated by a change in orientation that led him away from the paradigm of mechanics, which he replaces with principles from the theory of perception. However, the resulting modern aspects of his conception of science seem to dangle next to classical definitions that clearly limit modernization. Thus, for instance, neither in this speech, nor in the supplements to it, nor in subsequent publications does he mention that causality may be merely

[806] **Helmholtz** (1878a), p. 244 (Germ.)/363 (Engl.). This is the answer to the question posed at the beginning, regarding the "truth in our intuition and thinking" (**Helmholtz** (1878a), p. 218). In **Helmholtz** (1892), p. 358, he finds that the passage quoted from Goethe expresses the "final result of our physiological epistemology". **Erdmann** (1921), p. 10f., writes that **Helmholtz** (1878a), p. 243f., "obviously" represents "an impact of hypothesis on [Helmholtz's] interpretation of causality" (loc. cit. p. 11). Erdmann, however, does not distinguish the various meanings that Helmholtz uses for the concept of causality.

hypothetically valid. In clear terms he distinguishes both causally understood lawfulness and the topogenous and hylogenous factors understood as causes of perception, from hypotheses.

These distinctions remain effective in §26 of the second edition of the *Treatise,* which had first been published in 1892 and then considerably revised.[807] One almost gets the impression that Helmholtz is shying away from hypothesizing the concept of causality because it has become the last guarantor of his classical conception of science. But once the loss of truth has been set to work it appears unyielding. As reported, in the *Treatise* the postulate of representation is not replaced by a classical idea of causality but by the validity-relativizing, pragmatic concept of truth. In a newly inserted remark for the first time he also says that the origin of causal explanations is hypothetical:

> Naturally, the special kind of causal connection that we must assume for explaining individual cases of lawfulness in the sequence of perceptions can always only be found by way of hypothesis.[808]

At this point emphasis can be placed on the word "found", stressing the contingent circumstances of discovering laws and saving the justificatory reasons that can be distinguished from them from also becoming hypothetical. Otherwise, at this point causal explanation would already surrender its absolute validity and the precondition of lawfulness would lose its meaning.

Not in any of his public speeches or scientific writing but in a note found among his literary remains and which appears to have been written fairly late in life does Helmholtz express what is involved in halfheartedly subjecting his understanding of causality to pragmatic truth.[809] At the end of the process of hypothetization causality – the universal guarantor of certainty – is relinquished and with it the distinction between laws and hypotheses:

> The law of causality (the presupposed lawfulness of natural phenomena) is only a hypothesis and cannot be proven other than as such. None of the lawfulness observed up to this point can prove that it will hold in the future. The only proof of all hypotheses is always: test, whether it is the case and you will find out (preferably experimentally whenever possible). In contrast to the other hypotheses that state specific laws of nature, the law of causality holds only the following exceptional position. 1. It is the precondition for the validity of all others. 2. It provides the only way at all that we can know what is not observable. 3. It is the necessary basis for purposeful action. 4. We are compelled to believe it by the

[807] See the excursus on the postulate of temporal representation in Section 7.1.3.1.

[808] **Helmholtz** (1885 ff.), p. 593.

[809] Koenigsberger, whose biography contains this note (**Koenigsberger** (1902 f.), Vol. 1, p. 247 f.), provides no date for it. Without additional comment he positions it opposite his report on Helmholtz's speech on Kant from 1855. By doing so he appears less to intend to point out contradictions in Helmholtz's early conception of science than to show where it went from there. With respect to **Helmholtz** (1878a), p. 244 (as footnote 806), Erdmann refers to this note from Helmholtz's literary remains (**Erdmann** (1921), p. 11). Riehl thinks that the note was written much later and was a document of the "last phase of Helmholtz's development of epistemological conceptions" (**Riehl** (1904), p. 40).

> natural mechanics that operate among our ideas. Thus the strongest instincts compel us to *desire* it; [...] all we can do then, is *believe* and then *act* and we will find it confirmed by proper investigation [. . .][810]

Helmholtz's attempt to separate causality from "other hypotheses" without claiming that it is a priori is less convincing. As a hypothesis, causality can no longer be the "precondition for the validity" of hypotheses. It also cannot be a "basis for purposeful action" if it is constituted by the actual process of action itself. If it were necessary as a result of human thinking however, there seems to be no reason why there could not be other, equally valuable types of mechanisms that "operate among our ideas". The lack of reflection devoted to this and similar questions demonstrates how far Helmholtz is from the modern philosophy of science, which accommodates a high degree of certainty of validity that is a part of scientific statements to this day by defining various degrees of corroboration for law hypotheses, by accepting various ways of understanding causation and by allowing alternative definitions of the concept of truth. But it would be unreasonable to expect Helmholtz to know that work in modern philosophy of science began with insight into hypotheticity.

The importance of the note from Helmholtz's literary remains lies less in revealing an unsuccessful attempt at maintaining causality's previous special status by redefining it but rather in that it discloses an admission of the loss of absolute validity induced long ago. All of the meanings of causality that Helmholtz had used from the 1840s on are involved: causality as universal "lawfulness" and causality as the link to an otherwise non-perceptible reality ("the only way at all [... to] know what is not observable"), which is not only the cause of all phenomena but also supports the first and final causes of all phenomena.

7.1.4 Summary

At the end of my reconstruction, looking back over the speeches held and forewords written after 1871, one gets the impression that an ideal from the classical conception of science remains that is too weak to prevent the incorporation of modern elements but still sufficiently effective to keep Helmholtz from accepting a modern conception of science. Stated paradoxically, one could say that he modernized his classical conception of science. Like the earlier development of his conception of science this process has less to do with the definition of what constitutes proper science and more to do with its rationale, which in part no longer does justice to preserved classical criteria and in part indicates an altered understanding of science. The new elements can neither be called thoroughly modern, nor are they totally characterized by hypotheticity. At first entirely and then later to a great extent, the upgrading of hypotheses remains within the scope of the classical conception and is compensated, like the modern relativization of the claim to truth, by integrating

[810] **Koenigsberger** (1902f.), Vol. 1, p. 247f. (emphasis in German orginal).

novel definitions of what proper science is, which are usually, as before, absolute. Thus his altered understanding of science includes not only the relativization of the validity of laws but also praise for intuition. This understanding accommodates not only the relativization of what is empirically given and the plurality of options for describing it theoretically but also last, non-hypothetical remnants of reality that is assumed to exist independent of knowledge of it.

By stressing this mutual balance between relativizing liberation and the persistent effort to establish an absolute claim to validity, my reconstruction has strived to present Helmholtz's position as consistently as possible. Instead of placing what is modern and what is classical side-by-side but disconnected, as we sometimes find it in Helmholtz's own writing, the intent was to elaborate the transformation process based on the joint relationship of those extremes. The fact that Helmholtz's late position is not free of contradiction – for instance his various definitions of causality – is only partly due to a certain amateurism in issues of the philosophy of science. It is above all a result of the insecurity with which Helmholtz enters new territory in the philosophy of science and an expression of the historical dimension of the transition he advocated. He was probably one of the first scientists in Germany to make a fundamental issue of the traditional search for truth that went all the way back to Ancient Times and that believed itself so close to its goal. Helmholtz thus prepared the way for the modern process of relativizing truth, which continues until this day.

The structurally formative immanent impetus of the transitional process has been found to almost always be an increased orientation to assumptions from the theory of perception. This increasingly aligned Helmholtz's conception of science to pragmatic points of view. The transition thus has one motive that originated in the late 1860s with the publication of §26 of the *Treatise*. In hindsight this text has a key position. Here for the first time Helmholtz, albeit selectively, suspends the contrast between scientific knowledge and the immediate witness of sensory perception. In a manner typical of the subsequent path of the transition process he also, in preparing the way for this correction, discusses inductive procedures and re-evaluates in turn the importance of the causal definitions of his concept of science. In what follows, causality becomes the guarantor both against losing the validity of induction and against accepting merely relatively valid pragmatic definitions of what is proper science. Finally, §26 also contains his first suggestion for an empirical foundation for geometry.

Nonetheless, in terms of the philosophy of science, the first amendment based on the theory of perception does not lead to modernizing the classical conception but to enunciating its rationale programmatically in the speech delivered to the convention of natural researchers and doctors in 1869. The mark of this commitment to an unbounded claim to validity for scientific knowledge is the link between a mechanistic conception of nature aimed at discovering final causes and a knowledge of laws that aims to capture the temporal relations of changes in observable events.[811] Not until absolute claims to validity have been retracted based on conflicts

[811] See Section 6.3.1.7.

with the theory of perception and until the capacity for thinking is explicitly subordinated to elementary processes of perception, does the classical position begin to unravel. Using the term more frankly than Helmholtz would have, one could say that in the course of the 1870s scientific knowledge took on the character of signs. Basically this erosion can be traced back to the self-referentiality inherent to every argument from perceptual theory that evolves into an argument for the philosophy of science. While perception had formerly been an object of scientific study, whose statements – in terms of their claim to validity – were clearly distinct from the witness of perception itself, now the claim to validity of those scientific statements becomes relativized by pointing out that in principle science cannot transcend the conditions for validity of elementary perceptual processes.

Within the context of my study, the theory of perception became relevant because the basic concepts of Helmholtz's mechanistic conception of nature undergo reinterpretation in terms of perceptual theory. This new definition that he undertakes in his speech *The Facts in Perception* [*Die Tatsachen in der Wahrnehmung*] in 1878 is preceded in 1871 by an almost sudden alienation from the mechanism he had hitherto emphatically advocated. Without referring to his own former position he revises both the contents of core statements (particularly the immutability of atoms) and their claim to validity (particularly that of mechanistic reductions). It is this clear break in the development of Helmholtz's conception of nature that apparently correlates with the change in his conception of science and justifies dividing it into periods. While he subsequently attaches increasing hypothetical character to his mechanistic conception of nature, the concepts of natural law, or phenomenal causality, respectively, remain the classical-scientistic pillars of his conception of science. These two different developments are closely related. In terms of their claim to validity, Helmholtz upgrades empirical laws that only capture observable relations between temporal changes, particularly against merely hypothetically presupposed submicroscopic mechanical processes allegedly at the bottom of all phenomena. By restricting knowledge covered by laws to what is observable – and thus staying within the classical framework – he pursues, as I shall discuss in the closing section, positivism, which in the second half of the nineteenth century spread in general.

In the context of my study the extent to which Helmholtz approximates a modern conception of science can now be determined in two ways. Conclusions can be drawn from the analysis of his conception of science done up to this point, which I shall summarize, supplement and evaluate in concluding this section. But examining the course taken by his mechanistic conception of nature in the 1880s also shows whether Helmholtz attributed a hypothetical character to it only temporarily, or in principle. That is the topic of the next section.

Talk of the modernization of the classical conception of science implies that Helmholtz basically remained a classicist. The most important arguments for this view say that Helmholtz upholds the central definitions of his classical conception of science (particularly rigorous universality, the representational character of natural law statements and the special status of mechanics and geometry) and tries to support their claim to validity with new arguments (the intuitive acquaintance with laws and the reformulation of his realism) and that the detectable elements of a modern conception turn up rather sporadically and unsystematically in speeches, which comprise the main subject matter of my investigation.

As mentioned, even hypothetization has components still taken from the classical framework. In order to classify them it helps to recall the various concepts of hypothesis. Helmholtz first applies the concept only to statements regarding objects (particularly atoms) or properties, whose empirical content is not yet determined, or in some cases cannot be determined at all. He then uses it to characterize an inevitable phase of development in the process of research, during which laws are formulated tentatively. Both of these meanings of the concept of hypothesis are classical, if it means the merely temporary and (in principle) completely eliminable openness of truth. While I have shown that this is true for the greater part of the first meaning, I tried to demonstrate that the hypotheticity of knowledge enunciated as natural laws cannot be eliminated. In an exclusively modern sense Helmholtz uses the concept of hypothesis to designate research-relevant epistemological assumptions whose truth fundamentally cannot be proven.[812]

The concept of hypothesis that I use for evaluating modernization (see Section 4.1) goes beyond Helmholtz's understanding of it. Besides the character of validity for certain theoretical knowledge that Helmholtz discusses in his writing on the theory of science, my concept of hypothesis includes what he only indirectly mentioned, namely the validity of the empirical basis, the plurality of theoretical descriptive options it involves and the validity of mathematical and logical propositions. I listed three criteria that constitute a modern understanding of hypotheses. Inasmuch as each of these criteria can be applied to Helmholtz's conception of science, it can be said to be hypothesized in the modern sense of the term.

- First, natural laws can no longer claim absolute validity, neither with respect to their inductive origin nor with respect to being deductively examined by experiments. Helmholtz's upheld border between laws and hypotheses that lack the empirical confirmation they would need to become laws is pervious.
- Second, regarding axioms in geometry, whose truth he wanted to secure by establishing an empirical foundation, Helmholtz begins to acknowledge that their theorems are just as inexact as those of axioms in natural science. This exemplarily suspends the classical distinction between formal-mathematical and empirical truths in the modern sense.
- Third, the claim that scientific experience is irreducibly linguistic is expressed both by his treatment of the plurality of equivalent systems of axioms in geometry and by his hypothetical understanding of a realistic assumption that an external

[812] **Leiber** (2000) criticized that I missed one meaning of Helmholtz's concept of hypothesis and therefore would not properly describe the hypothetization tendency. However, the places in **Helmholtz** (1867) quoted by Leiber show that Helmholtz's use of the term 'hypothesis' was scarcely consistent at that time and the meanings coming up in this respect were similar to those first two meanings mentioned here. Moreover, Helmholtz did not start using that term just in 1867, as alleged by Leiber but he already used it in **Helmholtz** (1847a) (see footnote 295) and in his public lectures and speeches as of 1852. However, my topic is not the history of terms but the history of concepts.

world exists. Causality, which once had established the one-to-one relation between idea and reality, is now finally subjected to pragmatic truth and called a hypothesis.

Helmholtz's conception of science could be called modern if he had acknowledged inevitable hypotheticity as the primary virtue of proper science. However, for him hypotheses remain an undesirable flaw. If they cannot be eliminated entirely, they must at least be minimized. His closest approximation to a modern conception of science can be found where he openly admits that hypotheticity is inevitable in scientific knowledge, without simultaneously reflecting on how to reduce it. In a few passages he no longer characterizes the process of scientific knowledge as the conversion of hypotheses into laws. Occasionally he even calls the acknowledgement of enduring primal hypotheticity an important factor of the scientific ethos:

> It is, however, unworthy of a thinker wanting to be scientific if he forgets the hypothetical origin of his principles.[813]

Hypotheses denied an openness of truth he calls dogma.[814] They are the mark of "metaphysical systems", which he contrasts with the empirical sciences:

> Characteristic of the schools which built up their system on such hypotheses, which they assumed as dogmas, is [...] intolerance. [...] If, however, the starting-point has been placed upon a hypothesis, which either appears guaranteed by authority, or is only chosen because it agrees with that which it is *wished* to believe true, any crack may then hopelessly destroy the whole fabric of conviction. The convinced disciples must therefore claim for each individual part of such a fabric the same degree of infallibility [. . .].[815]

If "infallibility" is typical of dogmatic systems, then in contrast, refutability should be the distinguishing feature of scientific knowledge. Helmholtz thus arrives at an understanding of science that anticipates the fundamental conviction behind critical rationalism, later to be founded by Karl R. Popper: infallibility becomes a sign of false science.[816]

The appendixes to *Facts in Perception* contain an initial attempt to characterize proper science in this way. Not only are laws of nature, for which "the only approximate proof [... is] by induction,"[817] essentially fallible but, Helmholtz says, axioms of natural science are also essentially fallible, because they are empirically founded:

> But the [... axioms of natural sciences] are either of doubtful validity, or they are mere consequences of the principle of causality, that is to say, of our intellectual impulse to view everything that happens as conforming to law.[818]

[813] **Helmholtz** (1878a), p. 239 (Germ.)/360 (Engl.).

[814] **Helmholtz** (1878c), p. 187.

[815] **Helmholtz** (1878c), p. 175 f. (Germ.)/315 f. (Engl.) (italics in original).

[816] Later Popper will use the same pair that Helmholtz employed to demonstrate the distinction between scientific and unscientific systems, namely astrology and astronomy, to illustrate the criterion for demarcating non-falsifiable statements (**Helmholtz** (1878c), p. 188, **Helmholtz** (1874b), p. 433, and **Popper** (1963), p. 37 f., 188 and in other places). On the relation of Helmholtz to Popper see **Schiemann** (1995a).

[817] See 7.1.3.2, footnote 697.

[818] **Helmholtz** (1878b), p. 642 (Germ.)/361 (Engl.).

If one deletes the second part of axioms, which Helmholtz himself later declared to be hypothetical ("mere consequences of the principle of causality", i.e., presupposing lawfulness, assuming that reality is the cause of perception, reducing all phenomena to final causes and perhaps taking the concepts of matter and force as fundamental concepts of natural research), then the fact that it openly admits that indisputable refutability is immanent to knowledge is what distinguishes science from non-science or false science. Striving to provide its hypotheses with a maximum of logical consistency and empirical confirmation, science must always indicate the remnant of uncertainty it contains.

As near as Helmholtz comes to the modern understanding of science, he remains spellbound by the classical myth of truth. The remote classical ideals seem to be all the more persistent and formative for his basic attitude, the closer he gets to the modern conception.[819] Carl Friedrich von Weizsäcker is mistaken when he says that Helmholtz was "fully aware of the hypothetical character of natural science".[820]

A quasi melancholy note permeates Helmholtz's speeches from the early 1870s on. He does not celebrate the disintegration (that he himself had instigated) of a once axiomatic orientation; he does not see it as an act of enlightenment but experiences it as a loss. Remaining basically a classicist, he nonetheless comforts himself in a very modern way to compensate the loss of certainty for truth. He never tires of stressing the potential and real utility of theoretical knowledge about nature.[821] Compared to the epochal significance of how scientific research revolutionized industrial production and everyday life, Helmholtz finds no great challenge in the inevitable hypotheticity of scientific statements: the consequences of relativizing claims to validity are themselves relative when seen against the backdrop of the historical process that evoked them.

7.2 Helmholtz's Model-Theoretic Mechanism: Mechanistic Analogies and Mathematical Unification

> If we want to get a well rounded, consistent, law-like worldview, we must assume that behind the things we see there are other, invisible things. We must search behind the barriers of our senses for other covert players (Heinrich Hertz, *The Principles of Mechanics*).

While Helmholtz's conception of science is subject to a thorough transitional process, he firmly upholds fundamental principles of a mechanistic conception of nature.

[819] Thus the last quotation is taken from precisely the piece of writing where he attempts to establish the last non-hypothetical assumptions of reality (topogenous and hylogenous factors).

[820] **Weizsäcker** (1974), p. 73. By the "hypothetical character of natural science" Weizsäcker means waiving absolute claims to validity: "Science is not about claiming absolute truth" (ibid.). But the example he provides for this, namely Helmholtz's assumption that matter has a corpuscular structure, illustrates less a radical and more a tentative waiving of truth. See Chapter 8, footnote 949.

[821] Cf. quotations for Section 6.3.1, footnote 430, and Section 6.3.2, footnote 621.

As of the 1870s, dual mechanism's claim to validity, although not its conceptual content, is involved in that transition. While Helmholtz modifies statements on elementary forces and corrects his views on the elementary structure of matter, he basically maintains the notion that phenomena are caused by mechanically moved substance. Yet mechanistic reduction can no longer claim to achieve exclusive and conclusive valid explanation for all natural phenomena. The program has acquired an irrevocable hypothetical character and its completion can now only be considered an ideal, a regulative principle for research.

The historical loss of its claim to truth, as reflected in the transition in Helmholtz's conception of science, also rescued mechanism from its demise. With respect to one aspect of the transitional circumstances, namely growing critique of the contents of mechanism in the latter half of the nineteenth century,[822] Helmholtz appears to have seen the alternative of either giving up dual mechanism or immunizing it against critique by hypothesizing it. The immunization strategy, defending – as science – a conception of nature that is not directly relevant for scientific practice, rescues distressed mechanism by viewing it as heuristic useful for research. It becomes one of a plurality of orientational options and perspective approaches so typical for modern philosophy of nature, promoting the efficacy of the entire system of science.

Of course, Helmholtz did not pursue this strategy straightforwardly. He stands at the onset of a process of relativizing claims to validity that is still incomplete today and for which we do not know whether it will ever end.[823] Helmholtz apparently views the retraction of the claim to validity more as a loss than a gain for interpreting the world through philosophy of nature.[824] Looking over his public speeches one finds that after 1870 all his remarks on the (atomistic) contents of the mechanistic conception of nature are critical. He does mention "substances and forces that underlie" everything but he does not say whether these "final causes" are of a mechanical kind, i.e., whether these forces are instantaneously operative central forces and whether the substances are atoms in motion.[825] He particularly no longer claims, as he had done previously, that the conservation of energy stems from the invisible mechanical structure of reality.[826] On the contrary, he stresses that in the course of proving its "unrestricted universality", the law of the conservation of energy was transformed from a theorem of pure mechanics to a theorem that embraces all of nature.[827] His speeches sometimes seem to suggest that he now considers claims of his former mechanistic position not hypothetical but wrong.

However, a look at his publications in journals of physics shows that the principles of his former conception of nature continue to contribute to theory production.

[822] See Chapter 8.

[823] See the Introduction.

[824] Cf. Section 7.1.4.

[825] **Helmholtz** (1878a), p. 243.

[826] **Helmholtz** (1881a), p. 53 ff.

[827] **Helmholtz** (1887b), p. 283.

Backstage, mechanism with its relativized claim to validity unleashes its research-guiding strength in investigations in the discipline of physics. In the following I would like to select one study from physics to illustrate how Helmholtz develops a heuristic based on his continued orientation in the philosophy of nature and his altered understanding of science. Unlike his former heuristics, these are not rigorously aligned to specific mechanistic-dual ontology, nor are they attached to an explanatory claim. They attempt to illustrate fundamental traits of substance in motion via analogy and strive for mathematical uniformity for empirical laws.

At the start I distinguished two aspects of mechanism: one in the philosophy of nature that is related to the entirety of nature and its basic traits and one in natural science, restricted to certain empirical research work.[828] Up until about the early 1870s Helmholtz's mechanism covers both aspects. His work on the law of the conservation of energy and on the foundations for geometry contains examples of the science-immanent orientation following the paradigm of mechanics.[829] On the other hand, his public speeches written during the same period provide material that permits a reconstruction of a comprehensive mechanistic interpretation of natural phenomena. Most of the speeches were published with some delay and then also reprinted in the 1870s and 1880s.[830] While his old speeches were being recirculated, Helmholtz not only spoke critically of mechanism, he also exercised general restraint in expressing philosophical interpretations of nature.[831] Disregarding the further dispersion of his classical mechanism via his publishing activity, one can say that his interest in the philosophy of nature loses significance and he now concentrates on the scientific aspect. In this situation, the content definitions of mechanistic heuristics reverse. Helmholtz's work on the law of the conservation of energy could be understood as heuristics of mechanism in the sense that empirical investigations of energy-conserving processes are supposed to provide information about the covert mechanical character of nature.[832] Now non-mechanical investigations are substituted by mechanics itself as a heuristic instrument of mechanism. Mechanistic thought reflects its own foundation, its own principles and finds new ways to illustrate movement processes and aspects of system in models. It returns to its – albeit meanwhile relativized – origins.

My choice of a specimen is based on how Helmholtz's later conception of nature was received in the literature.[833] It covers his work on – expressed vaguely – "applying" the principle of least action to mechanical, thermodynamic and

[828] See Section 1.1.

[829] See Sections 6.1.2 and 6.2.

[830] Cf. Chapter 5, footnote 227, and Section 6.3.2, footnotes 535–539.

[831] A typical example is given by the final passages of his speech **Helmholtz** (1878a), where he brings up more questions in the philosophy of nature than he is prepared to answer (loc. cit. p. 246f.).

[832] See Sections 6.1.2, Part γ, and 6.3.1.7.

[833] Helmholtz's later conception of nature has been discussed particularly in connection with the reception of his work in mechanics and thermodynamics: **Koenigsberger** (1895), **Klein** (1972), **Bierhalter** (1981), **Bierhalter** (1983) and **Bierhalter** (1987) (cf. Section 6.3.2, footnote 540).

electrodynamic systems.[834] From the research that Helmholtz pursued during his time in Berlin, this work comes the closest to the issue of mechanism. In this work he develops mechanical models, taking up directly a lively tradition of mechanistic interpretation of nature that continued until the early twentieth century.[835] Nonetheless, limiting this discussion to the work mentioned does not permit a reconstruction of his conception of nature that could easily be generalized and applied to his other projects. It also does not allow a detailed comparison with statements from his classical-mechanistic conception of nature.[836] A comprehensive reconstruction like that would also require considering (only to mention the most important of them) his work on the theory of electrodynamics and the thermodynamics of chemical processes. That would bring up the issue of ether, the question of the structure of elementary reciprocal effects and a critique of the traditional atomistic theory of affinity, all of which are, for the most part, set aside in his work on the principle of least action. I would like to suggest, however, that in terms of content of the fundamental assumptions of mechanism and its altered function within, the theory of science plays a comparable role across all areas of research that Helmholtz worked in, so that this justifies selecting just one of them that suits the purposes of this study.

During the second half of the nineteenth century, electromagnetic phenomena and the irreversibility of the nature posed the greatest difficulty and challenge for the mechanistic explanation of nature.[837] Helmholtz suspects that electrical forces are "by all means identical" to forces of chemical affinity[838] and begins in the early 1880s to occupy himself with the thermodynamics of chemical processes, publishing three essays on it.[839] The approaches to an interpretation of thermal phenomena presented in these studies constitute the crucial starting point for his work on the principle of least action.

Critical of the affinity theory that he himself had formerly advocated and which views "only the expected heat development as the measure for the work value" of chemical forces (affinities), in the first essay Helmholtz introduces the concept of "free energy" (still used today). Instead of basing approaches in energetics solely on heat, Helmholtz suggests splitting the energy of a physical system into two parts: one is "bound energy" – which can only appear as heat; the other is "free energy"

---

[834] **Helmholtz** (1884b), **Helmholtz** (1884c), **Helmholtz** (1886), **Helmholtz** (1887b), **Helmholtz** (1887c) and **Helmholtz** (1892a). **Helmholtz** (1884b) constitutes preparatory work for **Helmholtz** (1884c) as is in part identical to it.

[835] On the history of mechanical models in physics see **Seeliger** (1948) and **Jammer** (1965); on its importance for thermodynamics and atomic physics at the turn of the century and in the early twentieth century see also **Miller** (1984).

[836] See Section 6.3.2.

[837] Cf. **Merz** (1907 ff.). Vol. III, p. 564 ff., **Jaki** (1966), p. 68 ff., **Harman** (1982a), p. 72 ff., and **Jungnickel and McCormmach** (1986), Vol. II, p. 59 ff. and 211 ff.

[838] **Helmholtz** (1881), p. 289.

[839] **Helmholtz** (1882f). On this work see in general **Rosenfeld** (1941), **Ebeling and Hoffmann** (1991) and **Kragh**'s (1994) excellent reconstruction and classification.

that alone can be converted into mechanical work.[840] In the terminology of thermodynamics "free energy" designates that part of internal energy that in reversible (i.e., infinitely slow) and isothermal (i.e., no change in temperature occurs) processes can be transformed into any other forms of energy. Although Helmholtz does not take recourse to hypotheses about the nature of chemical forces or heat in order to arrive at these concepts, his interest in a mechanistic interpretation of thermodynamic processes is clear: he assigns special types of "molecular movements", which are allegedly the basis of all thermodynamic phenomena, to the various forms of energy. He understands free energy as a kind of "ordered motion", the velocities of which can be represented by differentiable functions of space coordinates. Bound energy, on the other hand, is an expression of "disordered motion", where "the movement of any individual particle need not have any similarity with the kind of movement of the next one".[841]

Helmholtz adds to this distinction an interpretation of irreversibility, which he will maintain for the rest of his life. The irreversibility of thermodynamic processes in closed systems is established in physics by the second law of thermodynamics, which postulates that every change of state is accompanied by an increase in a quantity called entropy (symbolized by S such that $\Delta S \geq 0$).[842] Helmholtz acknowledged this law soon after Rudolf Clausius and William Thomson formulated it. Somewhat later, in his lecture *On the Conservation of Force* [*Ueber die Erhaltung der Kraft*] in 1862/63, he expresses the view that one can think of the heat motion of gas as being "analogous to a swarm of gnats".[843] His interpretation from 1882 is also based on the notion of an incalculable process:

> We have every reason to believe that heat motion is of the last [i.e., disordered – G.S.] kind and thus we may call the quantity of entropy the measure of disorderliness. For our tools that are relatively rough compared to molecular structure, only ordered movement can be freely transformed into other forms of work.[844]

Readers familiar with the history of thermodynamics might first think that by using the term "disorderliness" Helmholtz is following statistical thermodynamics established by James Clerk Maxwell and Ludwig Boltzmann. But Helmholtz is far from doing so. None of his writing on thermodynamics and interpreting it mechanistically deals with Maxwell's probability function for the velocities of the components

---

840 **Helmholtz** (1882 f.), p. 959 and 965 ff. Cf. **Kragh** (1994), p. 418.

841 **Helmholtz** (1882 f.), p. 972.

842 "Δ" designates any arbitrary not necessarily infinitesimal difference. A "closed" system is a thermally isolated system that has no material exchange with its surroundings. For a definition of entropy: in case of reversible processes in open (unclosed) systems, the temperature (T) is the integrating factor of the equation, which combines the total differential of entropy (dS) with the non-exact differential of heat (đQ): $đQ = T \cdot dS$. Helmholtz terms this equation not as the defining equation of entropy but as the "second law" (**Helmholtz** (1884b), p. 176 et passim).

843 **Helmholtz** (1862/3), p. 218. Helmholtz uses this metaphor to elucidate disordered motion even in his *Lectures on Theory of Heat* [*Vorlesungen über Theorie der Wärme*], held for the last time in the summer of 1893: **Helmholtz** (1897 ff.), Vol. VI, p. 256.

844 **Helmholtz** (1882 f.), p. 972.

of an (ideal) gas or with Boltzmann's deriving the second law from it, in other words, with approaches whose enormous significance for thermodynamics was already clearly recognizable in Helmholtz's time.[845] The reason for this ignorance is a radical difference in each other's assumptions about the mechanical nature of heat phenomena. While the probability function used by Maxwell and Boltzmann presupposes a statistical distribution of velocities (making very low and very high velocities rather rare), Helmholtz assumes that average values of velocities are adequate for calculation (so that he does not ask how probable the occurrence of specific velocities is).[846] Contrary to Boltzmann, who uses the term "disorder" as synonymous with the calculable probability of the occurrence of states in a given system, Helmholtz uses the term to designate motion that escapes mathematical calculation. In other words, for him irreversibility is due to the (unavoidable?) lack of capacity to recognize and control what are actually reversible configurations in submicroscopic processes. What seems irreversible from a human perspective, he remarks explicitly in a different passage, may be reversible in natural events without any human intervention.[847]

Helmholtz's interpretation of irreversibility is supported by the conviction that no new mathematical expedients are needed for describing the movements of gas particles because well-known mechanics would suffice, if we only knew how to apply them correctly. Their inapplicability implies that the scope of mechanism is limited, which Helmholtz, significantly, asserts like a commitment to a belief. In contrast to the physics contents of the second law, understanding it mechanistically is less a matter of knowledge and, for the time being, more a matter of faith.

Although Helmholtz's work on the thermodynamics of chemical processes was merely a first step towards phenomenological thermodynamics and although it was important and soon appreciated by colleagues in the field,[848] he subsequently no longer elaborated the issue.[849] An overview of the development of all his research activity shows that these investigations were only an intermediate stop along his route from electrodynamics to thermodynamics to his work on the principle of least action. It began by introducing a new way of interpreting phenomena – oriented by mechanics – that substituted explanations that could not, or could not yet, be given: the generation of the analogy between mechanical models and thermodynamic phenomena.

His approach consists in defining conditions for models (models whose internal structure is completely known and whose concrete form has nothing to do with submicroscopic processes), conditions that are allegedly similar to certain fundamental

---

[845] For the history of statistical mechanics in the nineteenth century see **Krajewski** (1974) and **Brush** (1976); on the relationship between Helmholtz and Boltzmann see **Hörz** (1981 ff.) and (1993).

[846] **Helmholtz** (1884b), p. 120, and in other places.

[847] **Helmholtz** (1897 ff.), Vol. VI, p. 259.

[848] On the history of how Helmholtz's work was received see **Kragh** (1994), p. 424 ff.

[849] On plans to continue the work see **Koenigsberger** (1902 f.), Vol. II, p. 298 ff.

features of thermal phenomena. In the model's motion equations he then substitutes mechanical variables with thermodynamic variables in order to obtain the definitional equation for the entropy of reversible changes of state that goes back to Clausius.[850] Helmholtz does not mention the principle of least action when enunciating the motion equations. He starts with Lagrange's formalism that can be derived from the principle of least action just as well as from other principles of mechanics.[851] Not until later, when extending the analogy to cover electrodynamic phenomena, does he explicitly refer to the principle and view the earlier models as part of its broad application.[852]

The conditions that are supposed to correspond to the nature of heat and that determine the specific form of the equations of motion are crucial for the analogies between mechanical models and reversible thermodynamic phenomena. The point of the approach is that the conditions are allegedly independent of the calculation results. The fact that in the end Helmholtz obtains a thermodynamic equation seems to retroactively confirm his choice of conditions.

Helmholtz calls his models "monocyclic systems" and begins his investigations by defining them and explaining his motivation:

> By *monocyclic systems* I mean mechanical systems that contain within them one or more stationary cyclical movements but the velocity of which, if there are several of them, depends on only one parameter. I further assume that only conservative forces operate among the individual bodies that constitute the system [...], while the additional external forces must not necessarily be conservative.
>
> The main reason for investigating this is that heat motion, too, at least in its externally observable effects, shows the essential peculiarities of a monocyclic system.[853]

Thus the familiar vocabulary of the dual conception of mechanics and mechanism[854] provided by the formalism of Newtonian mechanics continues to shape model production: mechanically moved masses (Helmholtz speaks of "atoms" in this context)[855] and the forces that operate among them; although now the forces need only satisfy the law of the conservation of energy ("must [...] be conservative") and need no longer be some kind of central force.

A more special "peculiarity" of heat motion is reflected in the rules for model coordinates that Helmholtz groups in two sorts. One sort (Index a) effects the physical state of the system solely by velocity but not by position in space. For the other sort (Index b) it is the reverse: only position and not velocity, has an effect on the state of the system. By dividing the coordinates thus, Helmholtz wants to do justice

850 **Helmholtz** (1884b), p. 176 et passim.

851 See **Goldstein** (1963).

852 **Helmholtz** (1886), p. 207 f.

853 **Helmholtz** (1884b), p. 119 (italics in original).

854 See Section 3.2 for this conception and line of tradition that go back to Newton and Boscovich.

855 **Helmholtz** (1884b), p. 120.

to the distinction between the quantity of heat in a system and the work done on it. The velocities of elementary particles – of a gas, for instance – determine in his opinion the thermodynamic properties (Index a). In comparison, changes in volume that convert heat into work – for instance piston movements – are negligibly slow and can be essentially recorded in terms of spatial position (Index b).[856]

The prerequisites that Helmholtz postulates for the nature of heat purposely do not extend beyond these few assumptions. Instead of speculating about the alleged mechanical motion of heat, he follows an awareness that emerged along with the crisis in his classical mechanism, the awareness namely, that the nature of heat has not been thoroughly explored and yet that perhaps nothing more will be knowable about it than what is observable in thermodynamic experiments.[857]

In contrast to the unexplored nature of heat – and this made Helmholtz's approach so striking for his contemporaries – it can be said with certainty that some mechanical models entirely meet the demands of monocyclic systems. One of Helmholtz's examples is a top that turns frictionlessly around a fixed axis and whose moment of inertia can be altered by a mass slidable vertically to the axis. Relative to the angular velocity, the change in the distance of the mass to the axis should happen slowly. Then the energy of this system depends solely on the angular velocity (Index a) and the spatial coordinate of the distance (Index b).[858]

Helmholtz himself does not call this kind of mechanical device a model. Calling it a model, however, helps understand the role it played in his investigations. By "model" I mean a representation of properties of a real given object or a sign of these properties. It provides an analogy (and not just a metaphor) for an object, when complete similarity is assumed between the properties of the model and those of the (for the rest possibly entirely dissimilar) object. In an analogy a model thus has a partial real content, a partial claim to truth, if truth is taken – as in the correspondence theory of truth – to mean that a statement corresponds to the facts it designates.[859] Thus Helmholtz's cyclic models represent the movement of the material elements that he thinks constitute warm bodies only to the extent that the motion of heat is cyclic. Since Helmholtz can only assume that the motion of heat is cyclic, the models are (at least temporarily) of a hypothetical nature. All properties of the model that go beyond the similarity relation do not necessarily pertain to the object, so questions of their validity do not apply. Thus the concrete way in which the cyclical movement is manifested in the model need not have anything to do with the concrete way in which the motion of heat is manifested. Expressed more generally, the similarity relation of the analogy need not necessarily hold for the internal

---

[856] **Helmholtz** (1884b), p. 124 and 128 ff.

[857] On this see **Helmholtz** (1884b), p. 176, quoted here in the text related to footnote 870.

[858] **Helmholtz** (1884b), p. 129.

[859] On the historical and systematic significance of mechanical model generation for scientific knowledge see **Hesse** (1963), **Seeliger** (1948), **Jammer** (1965), **Nagel** (1961), p. 106 ff., and **Stegmüller** (1973 ff.), Vol. 1, p. 169 ff.

structure of the mechanical model or for heat. Instead of representing real motion, models only portray aspects of reality; they illustrate an abstract feature of heat.[860]

The hypotheticity of models, however, results not primarily from the fact that they correspond only in part with a reality that is not necessarily mechanical and from the fact that the correspondence relation is revisable. Even in the formalism of mechanics they are not unequivocally related to the equations of motion, which is the only way they can be demonstrated mathematically. It is true, as Helmholtz probably knew, that the number of mechanical models for any given mechanical equation of motion is, in principle, infinite.[861]

I would like to distinguish analogies having real content, for short: real analogies, from formal analogies, where structural similarity exists between the mathematical representation of an object field and the description of the model related to it. This case abstracts from the meaning of physical concepts. For instance, the equations for motion in both mechanical and electrodynamic systems can be written using identical (syntactically isomorphic) structures[862] as differential equations (Lagrange equations). If mathematical propositions – despite the lack of relation or because of being related to reality – are considered correct or true, formal analogies trivially have no hypothetical nature at all.[863]

Helmholtz uses his models not only as real but also as formal analogies. His calculations are precisely supposed to demonstrate the structural identity of the mathematical form of equations of motion taken from mechanical models with that of the definitional equation for entropy.[864] If the mechanical variables are replaced by thermodynamic ones, the result is that the changes of energy in a monocyclic system are of "entirely the same form"[865] as the changes in the quantity of heat in a thermodynamic system. The temperature corresponds to kinetic energy and the entropy corresponds to a function of mechanical momentum.[866]

Helmholtz also calls formal analogies simply "analogies".[867] However, he also extends the term to cover the physical similarities that he assumes exist between heat

---

[860] Earlier analogies hold no comparable position in Helmholtz's investigations. The mechanical vortex motion in **Helmholtz** (1858), for instance, serves not to illustrate electromagnetic phenomena but the latter serves as an analogy for the former.

[861] See **Maxwell**'s (1873), Vol. 2, p. 416, remark and the proof later provided by **Poincaré** (1890), p. IX ff.

[862] Cf. **Stegmüller** (1973 ff.), Vol. 1, p. 171.

[863] With respect to formal analogy production Helmholtz can thus speak of "absolute universality" (**Helmholtz** (1886), p. 209). He had already generated formal analogies in 1847 in *Conservation of Force* by giving expressions of energy for non-mechanical phenomena the form of Lagrange equations (see Section 6.1.2, Part β).

[864] This results from the arrangement of **Helmholtz** (1884b) and (1884c): he first recapitulates the laws of heat (§1), then sets up motion equations for his models (§2) and then immediately continues (§3) to present their formal analogy with the definitional equation of entropy.

[865] **Helmholtz** (1884c), p. 132.

[866] Cf. **Schiemann** (1997), p. 393, and **Klein** (1972), p. 65 f.

[867] Cf. **Helmholtz** (1884b), p. 124, **Helmholtz** (1884c), p. 156 and 186, **Helmholtz** (1886), p. 226, and **Helmholtz** (1892a), p. 480.

motion and the mechanical movements of his models.[868] Precisely this lack of distinction between formal and real analogies encourages the erroneous impression that the indubitably existent formal analogy might be evidence of the assumed truth of the real one. But that even the real analogy would not provide an explanation of anything, even if it were true, Helmholtz does not deny: he stresses explicitly that his analogies do not claim to provide "explanations" for the empirical laws of thermodynamics.[869]

Helmholtz only vaguely alludes to the purpose of analogies, which he categorically distinguishes from the purpose of explaining natural laws:

> The motion of heat first appears to us to be of an unknown kind for which we have only very vague ideas [...] Under such circumstances it seems to me entirely rational to investigate under which most general conditions the already known most general physical features of the motion of heat might occur in other well known classes of movements.
>
> For the purpose proposed to me, the main emphasis in for selecting examples was naturally their complete comprehensibility in terms of mechanics.[870]

The notion that ideas about the nature of heat become clearer by thinking of it as analogous to entirely familiar movements in mechanical models in itself suggests that thermal phenomena are based on mechanical processes. What Helmholtz calls the "most general physical features" are just those mechanical conditions that are supposed to be just as valid for heat as for the models. They are still too general for statements of law (being the prerequisite for explaining individual phenomena) to be derivable from them but they are already concrete enough to be illustrated by hypothetical models. The model-like hypostatization based on the familiar constructions of mechanics is supposed to enhance understanding them mechanically. This is why Helmholtz sees temperature in analogy to the kinetic energy of mechanics and the state variable entropy in analogy to a function of the momentum of mechanically moved parts.

Helmholtz's mechanism remains seminal for his model production, which is why I prefer to speak of "model-theoretic mechanism" in this context. In a similar way, mechanistic analogies influence the works of James Clerk Maxwell and William Thomson. Just as Maxwell arrived at the mathematical formulation for electrodynamics via a mechanical model that he did not claim to be true, Thomson wanted to find new approaches for understanding the structure of matter by using a mechanical model of the atom.[871] Thomson expressed his view in an often quoted pointed phrase:

> It seems to me that the test of "Do we or do we not understand a particular subject in physics?" is "Can we make a mechanical model of it?"[872]

---

[868] Cf. **Helmholtz** (1884b), p. 133 and 176, **Helmholtz** (1884c), p. 155, **Helmholtz** (1892a), p. 500.

[869] **Helmholtz** (1884b), p. 176.

[870] **Helmholtz** (1884b), p. 176.

[871] Cf. **Harman** (1982a), p. 83 f., and **Smith and Wise** (1989), p. 396 ff.

[872] **Thomson** (1884), p. 132. Quoted also in **Duhem** (1904 f.), p. 89.

One could interpret Thomson's statement as saying that what is given by physical laws and boundary conditions is a phenomenon that has been explained but not one that has been understood. Human understanding of nature consists of setting mechanical constructs in analogous relation to the phenomena, in other words, depicting nature using the ideal of mechanics. In contrast to classical mechanism, which presupposed that phenomena are mechanically caused and structured, Thomson views nature as something unknown, possibly something non-mechanical, to which man must first create a relationship by developing models of it. (Notice how the philosophy of nature aspect of mechanism continues to guide science-immanent heuristics). He considers mechanics the one discipline of physics that alone can satisfy the human need to comprehend nature. Without discussing Thomson's opinions in detail, I would like to mention two motives that I believe underlie his standpoint and that Helmholtz shared. For one, I attribute his esteem for mechanics to the fact that intuition was seen as constitutive for understanding nature. In contrast to the other two disciplines of so-called classical physics, namely thermodynamics and electrodynamics,[873] the concepts of mechanics coincide with forms of inner and outer intuition (time and space) and their relation to objective appearance (mass as the quantity of matter). For another, mechanical models are paradigmatic of human products. A phenomenon is understood because the properties that are considered essential to it can be artificially reproduced by a model and modified.

The discrepancy between nature and knowledge of nature is reflected in the discussion of the function of models. Scientistic mechanism's efforts shift from the attempt to prove mechanical causes empirically, to the construction of hypothetical models. Of course, this does not terminate the search for real causes. The fact that a model's actual form says nothing about the cause of thermal phenomena by no means excludes the existence of a cause, whose concrete form might correspond to that of the model such that the physical laws and the thermodynamic boundary conditions that apply to the model could be used to derive thermal phenomena. Explanations of this kind however, have come to be viewed as ideal objectives far removed from prevailing mechanistic heuristics. It is not the rigorous alignment to one uniform goal but the development of an illimitable plurality[874] of analogous relations within the scope of mechanism that is characteristic of model-theoretical mechanism.

By now the altered function that the mechanistic conception of nature has in Helmholtz's research should be evident. Helmholtz does not demand that physics concentrate on producing mechanical analogies. Instead of emphatically demanding that natural science merge into mechanics, as he had still done at the close of the 1860s, he simply defends his own right, which needs no further justification, to preoccupy himself with analogies as an "entirely rational" way of doing science. This is "rational" for someone who wants to comprehend nature guided by the ideal of

[873] See **Schiemann** (1997), p. 38ff.

[874] See footnote 861 and the related passage.

mechanical processes but not for advocates of other conceptions of nature. Mechanism can no longer be taken as the fundamental view of physics. Modifying Fichte's famous words, one could say that the conception of nature one should choose depends primarily on the type of person one is. In spite of the pervasive hypothetization (that this also expresses), the mechanistic conviction should maintain scientific standing and just this is only possible if the conception of science changes to allow making use of mechanism as heuristics in research alongside other conceptions of nature. Borrowing Helmholtz's own terminology, mechanism has become a "metaphysical hypothesis" that is "completely justified scientifically".[875]

Helmholtz also left it up to others to work out the details of his approach for producing mechanical models. In the same year he published the series of articles on monocyclic systems, Ludwig Boltzmann began a more accurate investigation of their properties.[876] Under Boltzmann's guidance model production – as Martin J. Klein once remarked – blossomed opulently.[877] Boltzmann is not content with contriving numerous monocyclic systems for the purpose of better understanding natural phenomena in a mechanical way or with illustrating his many essays with drafts on how to fabricate them. In order to demonstrate the systems in a way that they can be easily remembered, he has models built especially for teaching academic physics. Helmholtz, in contrast, will no longer be detained with such a rough illustration of alleged traits of nature. Instead, he tackles the development of formal analogy generation and how to interpret those analogies for physics.[878] Starting with thermodynamics, he incorporates electrodynamic processes into the formal analogies. He thus returns to one of the problems that had stood at the beginning of his model-theoretic mechanism. Now his goal is to correlate mechanical, electrodynamic and thermodynamic empirical laws combined with the theorem of the conservation of energy to one single, originally mechanistic principle, which will then "provide a complete overview of everything that is essential".[879]

The principle that he selects for this purpose is explicitly the principle of least action. From it he deduces the Lagrange equations and the theorem of the conservation of energy in mechanics; by substituting variables he obtains the entropy equation without any changes; and he arrives at Maxwell's laws of electrodynamics by way of analogy. Thus he again does not reduce empirical laws of thermo- and electrodynamics to laws of mechanics, he simply presents a schema with which one obtains them by way of mathematical manipulation. As dubious as this procedure, which smacks of mnemotechnics, may seem from the perspective of physics, it does contribute to unifying the mathematical presentation of empirical laws. For thinking about experimentally obtained knowledge in terms of philosophy of nature, it has remained significant to this day. For the most part, it has remained difficult

875 **Helmholtz** (1878a), p. 239 (as Section 7.1.3.5, footnote 757).

876 **Boltzmann** (1884), **Boltzmann** (1885) and **Boltzmann** (1886).

877 **Klein** (1972), p. 73.

878 **Helmholtz** (1886), **Helmholtz** (1887b), **Helmholtz** (1887c) and **Helmholtz** (1892a).

879 **Helmholtz** (1886), p. 210.

to explain why empirical laws of physics from the most diverse areas and – as we know today – also from various orders of dimension (from field equations in the theory of gravitation to elementary particle physics) can be shaped to fit Lagrange equations. One can consider it a contingent and unimportant fact, or, as Helmholtz and his student Max Planck did, view it as an expression of a mathematical feature immanent to experimentally researchable nature.[880]

It remains difficult to deliver a philosophy of nature interpretation of this principle that would be useful for so many different purposes. In mechanics it determines, to put it simply, the actual motion of a system by the (unfounded) requirement, that the value of a mathematical quantity, the so-called action, must be extremal.[881] Of all the conceivable mechanical movements, under given initial conditions, only those are realized, whose action is the smallest, or – in a few cases – the largest. In the history of the principle of least action, far-reaching natural philosophical and metaphysical speculation about a teleological force that operates within the world has been linked to this statement taken from mechanics. Pierre Moreau de Maupertuis, for instance, was the first to enunciate the principle (although not correctly in terms of physics) and he saw it "grounded in the properties of a highest intelligence".[882] Leibniz was convinced that the world, as the best of all imaginable worlds, is also constituted extremally in a moral sense.[883]

In a dispute on the conceptions of Maupertuis, Leibniz and other discoverers of the principle, Helmholtz rejected all teleological interpretation of it.[884] However, this leaves a blank where one would hope to find an explanation, in terms of the philosophy of nature, for how the principle is to formally be applied to non-mechanical phenomena, a blank that Helmholtz did not know how to fill satisfactorily. The mechanical interpretation of the principle can still be applied to phenomena that are physically analogous to those of mechanics.[885] But when phenomena, such as those that are minute, are no longer clearly locatable in terms of space and time, not only does (as of yet) the mechanical interpretation fail but attempts to find a uniform meaning for the principle that goes beyond the trivial statement that certain parameters (that have no recognizable relation to the mechanical action) take on extremal values, also fail. Despite being unfamiliar with the peculiarities of submicroscopic and extraterrestrial physics, Helmholtz did not want to forego a physical interpretation of the formal analogy. But his attempt at an interpretation, which picks up the mechanical interpretation, must be considered misleading. He turns the principle into a world formula, whose

[880] **Helmholtz** (1886), p. 209, **Helmholtz** (1894b), p. XXIII, **Planck** (1915), p. 103 f.

[881] **Helmholtz** (1886), p. 205, and in other places; see also the relevant textbooks in mechanics like **Goldstein** (1963).

[882] "Que les Loix selon lequelles le Mouvement se conserve, se distribue et se détruit, sont fondées sur les attributs d'une suprême Intelligence" (*Histoire de l'Académie Royale des Sciences et Belles Lettres* (de Berlin, 2), année 1746. Berlin 1748, p. 277).

[883] **Leibniz** (1705), I, 8 (emphasis in original). For the history of the principle see **Pulte** (1989), **Kneser** (1928) and **Schramm** (1985).

[884] **Helmholtz** (1887b) and **Helmholtz** (1886).

[885] Cf. **Helmholtz** (1887b), p. 291.

properties have salient similarity to the law that he, up to this point, had assumed determined, in combination with the theorem of conservation, the course of all non-mental processes – namely, the second law of thermodynamics[886]:

> Everything that happens is described as the flux, back and forth, of the world's energy supply, which is eternally indestructible and cannot be accreted; and the laws of this flux are entirely summarized by the law of least action.[887]

This inadmissibly generalizes the definition of entropy taken from the equations of motion in monocyclic systems. Flows of energy can only be completely determined by principles that do justice to the temporal asymmetry of natural processes. But in physics (including the physics of the nineteenth century) only the second law of thermodynamics distinguishes a direction in time; no principle in mechanics does.

While Helmholtz used to think that the existence of mechanical substances moved by central forces can be concluded from the conservation of energy, he now stresses that "complete" knowledge of the flux of energy would say little about the processes that underlie it. What is more: the real advantage of applying the principle of least action appears to be the option it provides for abandoning covert causes:

> It is of greater importance for the last fundamental questions of science, for true and legitimate philosophia naturalis [...] that the two most general laws [...] only deal with the two kinds of energy as those that determine the entire course of natural processes and no longer with forces of motion and components of force, of which the latter only appear as derived values for calculation".[888]

By reducing physical quantities to the measurable quantities of energy, Helmholtz does justice to the fact that the recognized empirical laws of mechanics, thermodynamics and electrodynamics have no room for non-observable forces and elements of matter but they can accommodate changes in energetics. By additionally thinking of energy as the essence of the whole of nature, he comes closer – in terms of philosophy of nature – to so-called energetics, which in the second half of the nineteenth century was a widespread alternative to mechanism. Often combined with a positivistic understanding of science, energetics rejects hypothetical assumptions of non-observable entities. It justifies its claim to validity by direct reference to the experimentally detectable quantity of energy, which its advocates often thought of as being a substance.[889] Although in his late years Helmholtz sometimes creates the impression that he is content with formally describing observable "courses of natural processes" in terms of energy,[890] placing him within the energetics trend would be to misconceive his position on the discussion of philosophy of

[886] Cf. Section 6.3.2.3(ii).

[887] **Helmholtz** (1887b), p. 287.

[888] **Helmholtz** (1887b), p. 287.

[889] For energetics, see its most important proponent **Helm** (1898); concerning its competition with mechanism, see **Jungnickel and McCormmach** (1986), Vol. II, p. 211 ff., and **Wise** (1983), p. 19ff., who cites Ernst Mach's positivism and Wilhelm Ostwald's and Georg Helm's energetics as representative positions of the opposition against Helmholtz's (mechanistic – G. S.) program (loc. cit. p. 19).

[890] **Helmholtz** (1887b), p. 287, **Helmholtz** (1878a), p. 246, **Helmholtz** (1892).

nature within physics. As useful as he may have found it to disregard substances and forces in mathematical representations, the search for them remains the true motive even in his final writing on the principle of least action.

His interest in researching invisible mechanical causes is revealed by choosing this principle itself. I shall later elaborate how instability within mechanistic thought in the nineteenth century was in part due to an inflation of principles in mechanics.[891] Since the principle of least action is only one of the principles that can be used to derive mechanical equations of motion (and for formally devising non-mechanical laws of experience), Helmholtz must account for his choice. Remaining entirely within the framework of the formalism in mechanics, he develops an argument that allows him to relate the principle to the theorem of the conservation of energy. In relation to the theorem of the conservation of energy, the principle is the more special law. It restricts mechanical movements satisfying the theorem of conservation to those that exist in nature.[892] Helmholtz may have been thinking of this function, namely, it being a criterion for selecting conceivable (and entirely reversible) trajectories, when he generally reduced the natural order of the flow of energy to the efficacy of the principle of least action.

As if explaining its own restrictive character, the principle covers movements that have no influence on measurable values of energy. Helmholtz calls this motion "hidden motion", without, however, explaining the designation more precisely.[893] The expression only makes sense when applied to non-mechanical phenomena, for which (still) unobservable mechanical causes are postulated. In mechanical models the pertinent motion cannot be defined using energy values; more must be known about the internal structure of the model. However, the internal structure of a model is always known. If, on the other hand, an (as of yet) unknown system that has similarity to the model be at the bottom of a thermodynamic or electrodynamic phenomenon and if it were not possible to completely draw conclusions from the measurable values of energy to the structure of the system, then part of its motion would remain "hidden" in this sense of the word.

Among the structural properties of mechanical systems having hidden motions are conditions that limit the freedom of movement for the parts of the system.[894] For a system made up of points of matter, these could be, for instance, rigid connections between the points that only allow certain movements. Helmholtz's example is a frictionless top spinning at a constant angular velocity, whose axis of symmetry experiences a change of direction caused by external forces (precision). If the structure of the rotating system were hidden, it could not be discovered by observing this change.[895]

---

[891] See Section 8.2.

[892] **Helmholtz** (1886), p. 204, 210f. and 220–222.

[893] **Helmholtz** (1886), p. 211 (first explicit mention), p. 215, and **Helmholtz** (1897ff.), Vol. VI, p. 353–362.

[894] **Helmholtz** (1886), p. 213f., and **Helmholtz** (1897ff.), Vol. VI, p. 353f.

[895] **Helmholtz** (1886), p. 215, **Helmholtz** (1897ff.), Vol. I.2, p. 365f., and Vol. VI, p. 354.

But the fact that hidden motions have no effect on the measurable values of energy does not mean that they have no place in the energy functions in mathematical representations. On the contrary, they modify the difference between kinetic and potential energy, which Helmholtz calls "kinetic potential".[896] Without hidden motions, velocities only exist as a homogeneous quadratic function in the kinetic energy of a mechanical system. This mathematical relation expresses that compared to a velocity reversal, mechanical laws are invariant (the squares of velocities do not change with a change in the direction or algebraic sign of the velocities respectively). In contrast to this, as Helmholtz says, hidden motions can enter into potentials in "a much more complicated way"[897]:

> If one wants to know about the general properties of systems that are governed by the principle of least action, one must drop the older narrower assumption that velocities only occur at the value of a vital force [i.e., of kinetic energy – G.S.], namely as a homogeneous function of the second order; one must investigate what happens when [... the motion equation] is a function of the coordinates and velocities of any form.[898]

Helmholtz now bases the refinement of his mechanistic interpretation of irreversible processes less on the possibility of arbitrary potential functions than on the existence of uneven, namely linear potential functions in the equations of systems having hidden motions. He explained it the clearest in lectures titled *Lectures on the Dynamics of Discrete Mass Points* [*Vorlesungen über die Dynamik discreter Massenpunkte*], held in the winter term before he died. The lectures were published posthumously based on student notes and Helmholtz's own notebooks.[899] He said:

> The occurrence of these linear terms explains an important feature of polycyclic[900] systems. If, namely, one wants to reverse a series of state changes of the system [...] after it is finished, [...] then [...] the algebraic signs of the linear terms change; reversing the series results in a different kinetic potential [...]: The process is not reversible.[901]

Notwithstanding its allowance for the fact that changing algebraic signs for linear terms in the motion equation might be compensated by "simultaneously reversing" other velocities,[902] this interpretation remains untenable. No purely mechanical system has any irreversible processes.[903] Until this day, irreversible laws can only be derived from the reversible laws of mechanics by adding assumptions that are not founded mechanically, in other words, assumptions that are not founded exclusively

[896] **Helmholtz** (1886), p. 205, **Helmholtz** (1897 ff.), Vol. I.2, p. 360, and Vol. VI, p. 357.

[897] **Helmholtz** (1886), p. 207.

[898] Ibid.

[899] **Helmholtz** (1897 ff.), Vol. I.2, p. V.

[900] Polycyclic systems are distinguished from monocyclic systems only by the fact that the cyclical movements depend on more than one parameter. Cf. **Helmholtz** (1884b), p. 130.

[901] **Helmholtz** (1897 ff.), Vol. I.2, p. 365. Earlier traces of this interpretation can be found in **Helmholtz** (1886), p. 215 and 242 f.

[902] **Helmholtz** (1897 ff.), Vol. VI, p. 357. Velocities that can operate in a compensating manner are those disordered velocities that cannot be influenced (by the experimenter). Cf. **Bierhalter** (1994), p. 454.

[903] See the definition of a purely mechanical system in **Schiemann** (1997), p. 45 ff.

on the mechanical concepts of time, space and mass.[904] Faced with the interpretation that says irreversibility is a "disordered" motion, Helmholtz, with reference to hidden motions, paradoxically attempts to more precisely define the motion of heat, which is not discernible, because it is disordered. His method for doing so is not characterized by making certain mechanistic assumptions about the nature of heat plausible by creating a formal analogy between the motion equations of a model and a thermodynamic equation. Instead, Helmholtz modifies his model: he introduces a certain class of motion-restricting conditions from which he derives a physical property that seems to be similar to thermodynamic irreversibility.

As Helmholtz continues to devise models, his interest in explaining nature shifts from experimental phenomena whose conformity to laws of nature is acknowledged, to working on models, i.e., to finding more detailed and sophisticated representations. Entirely in line with Thomson's view mentioned above, he takes less for granted that physical phenomena can always be defined mechanically. The scientist now approaches alienated nature with a model in hand, as if it were a promising light in overcast scenery.

What has been said about hidden motion up to this point does not preclude thinking of it as a hypothesis that does not correspond to any real thing but that simply aids understanding nature in terms of mechanics. However, we find less indication that Helmholtz thought of hidden motions as being fictitious, or as being an imaginary creation of the mind for the purpose of producing a "well-rounded, closed, worldview that conforms to physical laws"[905]; where it can be found is in his student's, Heinrich Hertz's, work. Hertz takes up his teacher's idea but gives hidden motion an entirely different status in terms of the philosophy of science. According to Hertz, the principles, laws and specific boundary conditions of models need only coincide with reality on one point: it must be possible to infer certain predictions from the models and these predictions must be empirically testable.[906] Otherwise, one is free to select the mechanical principles of one's choice (as an alternative to the Newtonian description of force, Hertz lists Hamilton's principle and Gauss's principle of the least constraint). A specific type of structure, a certain "view" of mechanics corresponds to each principle.[907] On this theory of science, natural laws (for example the law of gravity or Maxwell's equations) have no empirical origin; they are not discovered, they are invented.

For Helmholtz though, the principle of least action governs phenomena of nature like a regent rules his subjects. Mechanical, thermal and electrodynamic phenomena are "governed by the principle of least action".[908] Its power is based on the fact that it originates in scientific experience. Although Helmholtz must justify his

---

[904] For a model consisting of freely-moving points of matter this includes neglecting all non-contact correlations among the points ("molecular chaos"). Cf. **Reif** (1976), p. 616.

[905] **Hertz** (1894), p. 30. See the motto introducing this section.

[906] See the quotation taken from **Hertz** (1894), p. 1, in Section 7.1.3.5, footnote 788.

[907] For Hertz's concept of a view see the same section, footnote 790.

[908] See the text to footnote 898.

selection of the principle, for him it is not hypothetical in the sense that it might be merely one starting point among various optional starting points for pursuing mathematical knowledge of nature. The principle's factual efficacy, however, produces another element of mere hypothetical validity. Just as the theorem of the conservation of energy is a result of an inductive and thus only approximately valid generalization, Helmholtz's writings imply that for each individual case it is necessary to examine anew, whether the principle of least effect applies to the empirical laws involved.[909]

It is not known how much Helmholtz assumed that the hidden motion he used in formal analogies for grasping certain kinds of phenomena is real. With respect to the phenomena of thermodynamics, his interpretation of irreversibility misled him to believe that they are established very generally on hidden motions:

> Of much greater importance [than for the movements of ponderable bodies, the principle of least action] became apparent when one began investigating bodies inside which constant hidden motion occurs. The first excellent example [...] occurred in a law in the theory of heat, as enunciated by Sadi Carnot [...] and expressed more accurately by our colleague Mr. Clausius, [and] which was finally traced back to one of the possible expressions of the principle of least action by Mr Boltzmann.[910]

Helmholtz is much less reserved about phenomena in electrodynamics. Although he needs potential functions with non-homogeneous quadratic terms for formally deriving the Maxwell equations, he does not use the term "hidden motion". He merely emphasizes that one has to work with "a system of unknown constitution and unknown inner forces".[911] Of course, the circumstance that two factors are unknown (both the "constitution" and the "forces") indicates that the framework of a dual view of nature is still intact. Helmholtz underscores that "it is not known, which of the inner changes in an [electromagnetic – G.S.] system correspond to changes in position of the individual parts and which correspond to changes in velocity for unknown inner motion".[912] However, if physical parts and their movements are among the basic ideas of the conception of nature in question, then we have mechanism in the narrower sense, if the assumption is that exclusively mechanical principles govern the laws of motion. It is not easily proven that Helmholtz always proceeds from this assumption in the work he did on electrodynamic phenomena after 1870.[913] But inasmuch as he does generally subject this phenomenon to the principle of the least action and equates the efficacy of the principle with the existence of hidden motion,

---

[909] **Helmholtz** (1886), p. 206 ff. For the inductive conception of the conservation of energy see Section 6.1.2. As late as in **Helmholtz** (1892a), Helmholtz thinks it possible that an "investigation might show, whether the theorem of the conservation of energy is [...] satisfied" (loc. cit. p. 479). In **Helmholtz** (1887b) he presents both the theorem of the conservation of energy and the principle of least action as empirical laws gained by induction.

[910] **Helmholtz** (1887b), p. 286.

[911] **Helmholtz** (1892a), p. 478.

[912] **Helmholtz** (1892a), p. 479.

[913] See Chapter 5, footnotes 204 and 216, and Section 7.1, footnote 628, and the passage that corresponds to Chapter 8, footnote 933 ff.

it does seem fair to assume that when working on electrodynamics he also continued to adhere to mechanism in the narrower sense.

As hidden as invisible mechanical movements remain, so also is Helmholtz's conception of nature concealed in his work on electrodynamic phenomena. To the extent that it can be judged within the context of Helmholtz's model theory at all, his work on electrodynamics is limited to devising formal analogies. Addressing relations of physical similarity in real analogies relieves the models of their purpose of enhancing mechanical understanding of nature, which they were to do by providing comprehensible symbolizations of natural phenomena. Marked by growing abstraction, Helmholtz's last work on the principle of least action focuses on developing systems of differential equations, demonstrating that they are isomorphic to Maxwell's equations and otherwise avoiding any further interpretation in terms of physics.[914]

Of course, although it has lost its intuitive power, mechanics is not finished yet for Helmholtz, not even in electrodynamics. In the narrower sense, thinking about nature in terms of mechanics means to start with the independent concepts of time, space and mass and to only employ, for the purpose of explaining nature, those mathematical principles that are equivalent to Newton's formulation. Within the framework of his model-theoretic mechanism, Helmholtz did not have to abandon any of those prerequisites. Whenever he can only connect mechanical principles to empirical laws by using formal analogies, he explicitly foregoes all claims to explanation.[915] However, restricting knowledge in this way is not a matter of principle, as Helmholtz mentions no reasons that speak against the possibility of mechanical explanations for thermodynamic and electrodynamic phenomena.

The acknowledgement of such reasons did not gain ground in physics until after Helmholtz's death: the concepts of mechanics allow no value for mass that is dependent on velocity, as it was surmised in connection with electrodynamics as early as the 1880s and proven in experiments after the turn to the twentieth century.[916] Within the framework of mechanics there is also no reason for the upper limit to the propagation of physical action that Albert Einstein postulated in 1904. Finally, discontinuities, such as those assumed by Helmholtz's student Max Planck for explaining submicroscopic processes, are incompatible with the mathematical structure of mechanical principles.[917]

Helmholtz's model theory reveals why mechanism was so remarkably well armed against its classical critics even before it faced the real challenge. No longer did it have to defend absolute claims to truth and could present itself as a hypothesis.

---

[914] **Helmholtz** (1892a).

[915] See passage related to footnote 869.

[916] Cf. **Kuznecov** (1970), p. 252, and **Hund** (1972), p. 332.

[917] In order to derive the experimentally confirmed formulas for the description of cavity radiation, Planck introduced energy quanta of finite value that were proportional to the frequency of radiation.

Chapter 8
Conditions and Causes for the Change in Helmholtz's Conception of Science and Nature

> *It is not always easy to decide what has caused a specific change in a science. What makes that kind of discovery possible? Why did this new concept turn up? … Questions like this are often very confusing because no absolute methodological principles exist upon which to build such an analysis*
> (Michel Foucault, *The Order of Things*).

The conspicuous change in Helmholtz's conceptions of science and nature raises the question of the conditions and causes responsible for it. Why did Helmholtz not retain the classical position? Why did the tendency towards hypothetization set in between the late 1860s and early 1870s? As much as these questions suggest themselves, clear answers are not to be found. The greatest obstacle for researching it historically was created by Helmholtz himself, by never directly addressing that change and never even hinting at what motivated his change in views.[918] One can thus only continue to attempt a rational reconstruction and will only obtain answers that basically depend on the prerequisites made by that reconstruction.

The complex scientific and cultural contexts surrounding Helmholtz's writing create a web much too complicated than that the change could be traced back to certain identifiable causes. In addition, Helmholtz himself created new sorts of approaches. At the most, one can state under which conditions those approaches emerged but not why they did. It would be entirely inappropriate to narrow the question down to merely grasping the causes within the framework of a historical study, understanding them as conditions that are necessary and sufficient for bringing about the effects. When I speak of causes in the following, I take these merely as conditions that explain – in the sense of making plausible – why something unexpected happened. My intent is to cover a spectrum of factors that is as wide as possible, factors that perhaps determined the transitional process and the direction it took.

[918] The only remarks known to me concern his altered position on Kant's epistemology. These are the passages mentioned in Section 6.1, footnote 261, written in the years 1881 and 1884, where Helmholtz mentions a transition in his understanding of causality, changing from the sole reference to final causes to a phenomenal meaning of causality.

Hypothetization turns up in Helmholtz's writing unexpectedly inasmuch as his early conception of science did not hint at a later transition. Inasmuch as his early conception of science is conceived of in a classical manner, it need not contain clues to a change in favor of a modern conception. This is because the concept of a modern conception of science as I introduced it is independent of its classical counterpart. In particular, I did not claim that the classical conception contains any internal contradictions that would provide reasons for replacing it with a modern one. Forgoing the classical claim to truth must not be preceded by the insight that striving to achieve that truth is in vain.

Looking back over the reconstruction up to this point it can now be shown that Helmholtz's positions on the theory of perception, geometry and the conservations of energy all written prior to 1870 scarcely contain clear hints of the transition in his conception of science that was to follow. In other words, the choice of contexts and subjects in Helmholtz's speeches do not sufficiently explain the hypothetization that took place (1). The search for causes thus leads to contexts he left unmentioned in his stances on the theory of science and nature, or only mentioned sketchily. They concern not only his biography but also touch upon the overall professional, institutional, political and cultural circumstances (2).

1. The two developmental periods are linked by an increased alignment of his conception of science to assumptions made in the theory of perception. I mentioned that this re-orientation began in §26 of the *Treatise* with problems immanent to the reasoning needed for empirically limiting phenomenal causality.[919] As explained above,[920] the effect that the theory of perception had on the transitional process is ambivalent. On the one hand, it was responsible for new definitions for the classical conception of science, particularly the concept of natural law. In the *Treatise*, empirical laws are defined relationally following the postulate of representation. They no longer inform of individual objects but only of the temporal pattern of changes in relations that hold between the properties of objects.[921] On the other hand, pragmatic elements enter into the concept of science. They mitigate the divide between the scientific and a life worldly understanding of truth and also result in a relativization of scientific knowledge, while simultaneously setting limits for that relativization by referring back to life worldly contexts.[922]

Depending on which of the two effects one accentuates, one gets one of two different periodizations. While a change in the definition of the concept of natural law becomes obvious shortly after publishing the *Treatise*, the relevance of the pragmatic elements for the theory of science does not emerge until the late 1870s. The transitional process that I call hypothetization happens during this period. One might be tempted to interpret the prior development of the classical conception of science as the prehistory to the change towards modernity and see the pragmatism

[919] Cf. Section 6.3.1.4.

[920] Cf. Section 7.1.4.

[921] Cf. Sections 6.3.1.6 and 7.1.3.1.

[922] Cf. Section 6.3.1.5.

that, starting in the late 1870s, relativized validity as its preliminary climax. I have already used the latter part of this interpretation in my reconstruction. The first part is supported by the circumstance that the concept of a natural law restricted to observable relations of properties can be seen as an early expression of viewing the classical claim to validity as a problem, in the way that Helmholtz at first viewed it as being linked to the inductive method. In addition, this concept of natural law is contrary to previously unquestioned status of physical statements that referred to individual objects whose non-observable properties are inaccessible for empirical investigation. Within the context of the mechanistic conception of nature it thus acquires a critical potential that questions the unrestricted claim to validity for explanations that make use of a realistic assumption of submicroscopic entities.[923]

But how much reformulating the concept of natural law directly contributes to the hypothetization of scientific knowledge, or whether it counteracts it, depends on what it means for the entirety of knowledge and which kind of validity is claimed for it. For Helmholtz's conception of science until the late 1860s, the explanation of nature is by no means exhausted in formulating empirical laws but wants also to determine the underlying mechanical causes. Indisputably, when writing the first edition of the *Treatise* and his speech to the natural scientists in 1869, Helmholtz still understood the concept of natural law in the classical way. Relations expressed by laws can claim to be valid because they clearly represent real circumstances. Thus Helmholtz's first use of the theory of perception can only conditionally be seen as one of the causes for modernization.

Up to this point I have discussed the theory of perception with respect to the change in his conception of science only and in doing so have assumed that during the period in question his theory of perception itself underwent no substantial change. In fact, in the first edition to the *Treatise* Helmholtz states his physiology of the senses in a conclusive way that he subsequently maintains.[924] The revisions and corrections he makes in §26 for the second edition do not concern his main statements about the theory of perception but deal only with how they apply to scientific knowledge.

Besides the direct effects it had on his conception of science, the theory of perception perhaps also became relevant for his theory of science indirectly via his work on geometry. Helmholtz himself saw the roots of his work in geometry in the investigations he made into the physiology of the senses.[925] Not until the late 1870s

[923] See Section 7.1.4.

[924] The fact that the *Treatise* appeared in a second edition as of 1885 is an expression of this continuity. On the development of Helmholtz's work in the physiology of the senses, in the context of which he formulated his theory of perception, see the literature in Section 6.3.1, footnote 407. Inasmuch as Helmholtz in his theory of perception clearly separates the psychic from the physical, the cause that **Vidoni** (1991) suggests for the transition in Emil Du Bois-Reymond's mechanism probably does not apply here. Vidoni explains, among other things, Du Bois-Reymond's restriction of the scope for mechanistic statements (cf. footnote 985) as a result of the fact that in the course of the 1860s among physiologists the clarification of the relationship between the psychic and the physical had been evaluated with increasing skepticism.

[925] See Section 6.2, footnote 359.

however, did he elaborate concepts in geometry in terms of the theory of perception.[926] The empirical foundations for geometry that he introduced in lectures between 1868 and 1870 relate not to the theory of perception but to the properties of mechanical solid bodies. It takes some reconstructive effort to compare these properties with the conditions that Helmholtz assumes hold for everyday estimations of distances.[927] In contrast, Helmholtz's early writing on geometry indicates that Riemann's *habilitation* paper published in 1867 substantially influenced his foundations for geometry. Aspects of this work, which in contrast to the relational concept of science now become incompatible with his former conception of science and which he will later emphasize (equivalent descriptions of something empirically given and its non-undisturbed manner of existence) can be traced back to Riemann's influence. Since Helmholtz at first was interested in confirming precisely the traditional claim to validity that Riemann had questioned for geometry, I see his work on geometry as still entirely on the classical side. Just like the postulate of representation in the theory of perception, the empirical foundation of geometry can be seen as an answer to the need for justification in the classical conception, which does not end by dismissing it but by corroborating it argumentatively – which in turn lets the hypothetization process that will soon follow emerge all the more abruptly.

When looking for causes within the immediate environment of Helmholtz's own comments on his work, one must also consider his mechanistic explanation of the conservation of energy. Stephen M. Winters has shown that certain elements of the transitional process can be traced back to Helmholtz's own critical assessment (which, however, was externally prompted) of the problem of establishing the conservation of energy (in 1847). Winters's periodization of Helmholtz's development coincides with mine. He too finds that a "turn" occurred in the 1870s.[928] Prior to that time Helmholtz had seen the conservation of energy, understood mechanistically, as a key to the explanation of physical phenomena. Afterwards he considered the theorem of conservation only a necessary boundary condition for physical-chemical processes. Being a restriction for a statement's range of validity, this amendment contains conditionalization and thus belongs among possible early forms of hypothetization. Why did Helmholtz make this correction? Although Winters stresses that Rudolf Clausius brought the shortcomings of the mechanistic concept of energy to Helmholtz's attention in the 1850s, one finds no corresponding reaction in Helmholtz's writing until 1873.

From among Helmholtz's own writings that could have propelled a change in his theory of science and thus also effected the transformation of his mechanism, special attention is due to his studies in electrodynamics.[929] Since 1847 they were

---

[926] See Section 7.1.3.4.

[927] **DiSalle** (1994) also does this sort of reconstruction; see Section 6.2, footnote 363.

[928] Winters erroneously characterizes Helmholtz's early position as aprioristic and the transition that took place in the early 1870s as a "turn towards empiricism" (**Winters** (1985), p. 313 ff.).

[929] On the context of the conservation of energy see Section 6.1.2, Part β.

closely related to his mechanism. Helmholtz's work on questions of electrodynamics spans a broad period from his early research on the physiology of nerves (as of 1850) to his investigations into the fundamental equations of electrodynamics (done between 1870 and 1876), which were highly influential in Germany, to the last essay he published on Maxwell's ether theory (1893). It was quickly identified as being responsible for the change in his conception of nature. As soon as one year after Helmholtz's death, his later biographer Leo Koenigsberger remarked that Michael Faraday's and James Clerk Maxwell's electrodynamics had been "of decisive influence" for Helmholtz's conception of nature as of the early 1880s.[930] Their work had prompted him to ask "whether action at a distance [which up to that point Helmholtz following Newton had assumed to be correct – G.S.] exists at all".[931]

Nonetheless, Helmholtz's work in electrodynamics did not correlate with the transformation of his conception of science and nature for the first time in the 1880s and also not only under the influence of British physicists. Jed Z. Buchwald has said that Helmholtz's work on the fundamental equations of electrodynamics can be used to reconstruct a conception of science that is clearly distinct from his former conception and which also deviates fundamentally from Maxwell's.[932] Designed to directly fit the experimental situation in physics laboratories, according to this conception science deals solely with objects, which, via their measurable physical states, determined the system energy, through which they would interact.[933] Obviously the question here is neither about hypothetisizing what is empirically given, nor about hypothetisizing what constitutes natural law: here the production of scientific theories is being reduced to the processing of observable quantities in a manner appropriate for the laboratory. Along these lines one could say that it is a de-mechanizing of the concept of science still within the framework of the classical conception of science. This seems to be supported by the fact that Helmholtz presented the new conception in a popular science fashion for the first time in a speech that clearly marks the transitional process by distancing itself from the mechanistic program – namely in the speech *Gustav Magnus. In Memoriam*, delivered in 1871.[934] With respect to the concept of science, Buchwald thus covers part of the classical side of the change. By deriving it exclusively from Helmholtz's work on electrodynamics, he provides another (after Winters) independent reason for dividing the entire process of development into periods.

Remarkably, Buchwald also explains the change he noticed as being due to factors that only conditionally are part of the context of Helmholtz's own research.

---

930 **Koenigsberger** (1895), p. 32.

931 **Koenigsberger** (1895), p. 33. Koenigsberger traces a conceptual correction back to this problem that is so typical for mechanism: Helmholtz moves "the concept of force [...] to the background [...] and introduces mass and energy [...] as given, indestructible physical quantities" (loc. cit. p. 38).

932 **Buchwald** (1994a). See Chapter 5, footnotes 204 and 216 with the corresponding text.

933 **Buchwald** (1994a), p. 335 and 339.

934 **Buchwald** (1994a), p. 370. See Section 7.1.1, footnote 628.

Besides influence from British physicists (this time Thomson and Tait),[935] Buchwald particularly investigates Helmholtz's public controversy with Friedrich Zöllner, which goes far beyond an internal dispute in natural science and which I shall therefore discuss later.

Thus neither does a perspective from his work in electrodynamics provide a satisfactory answer to questions of the causes for the change, for why that change accelerated in the early 1870s and why it soon thereafter turned in the direction of a modern understanding of science. On the whole, the reconstruction undertaken based on the topics discussed in his speeches cannot sufficiently explain the transformation process in his theory of science. The fewer preparatory traces can be found in his work, the more sudden does the change in the overall development appear and all the more clearly can one see the inexplicable nature of something new that clearly demarcates Helmholtz's later conception of science from his former one.

2. Unquestionably, the most conspicuous conditions that Helmholtz himself does not discuss concern the switch from a professorship in Heidelberg to taking up a professorship in physics in Berlin in 1870.[936] In closing I shall return to discuss the cultural-political implications of obtaining a university chair in Berlin and how this possibly had repercussions on Helmholtz's ideas of science. At this point I would like to first discuss the development of his scientific work, which without the career change probably would have taken a different course. After taking up the position, which coincides with the beginning of the change in his conception of science, his research activity's emphasis clearly moves from physiology to physics. The studies in electrodynamics just mentioned were among his first projects and remained predominant in his entire work during the 1870s. Although it remains questionable whether occupying himself with topics in electrodynamics contributed to the transitional process, the fact that Helmholtz for the first time concentrated on physical problems of inanimate nature for a longer period of time must have influenced the development of his conception of science. At the time, physics included the closer investigation of those assumptions that were fundamental to his conception of nature: the assumptions concerning phenomena brought forth by mechanical causes. While up to this point Helmholtz had supported his views chiefly with findings of other scientists, now the issue of the existence of atoms and the forces acting between them at a distance became a potential object of study in his own new field of work. It is possible that only then did he realize the full consequences of problems in the mechanistic explanation of nature, which were already known among physicists.[937] Hypothetization would then be the expression of gradual insight into the complexity of physical phenomena that were still largely unresearched and

[935] See Section 7.1.2.

[936] See Chapter 5.

[937] For instance, up to this point Helmholtz had worked predominantly in his studies on electrodynamics on problems related to the validity of the theorem of the conservation of energy (see Section 6.3.2, footnote 593). Not until from the 1870s on he was increasingly confronted with problems of the mechanistic explanation of ether (see Section 6.3.2, footnote 591).

about which one could only speak competently by making reservations about the validity of one's statements.

However, if ruminations of this sort, which trace the transitional process back to individual circumstances in his biography, were to be used to claim that his character was untypical for the entire situation in the theory of science in the second half of the nineteenth century, that view could not remain uncontested. The period in question contained an astounding plurality of intertwined scientific and cultural currents for which both the erosion of the classical and the rejection of modern positions in the theory of science become plausible. Only a few of these drifts can be selected for discussion here but mentioning them briefly cannot do justice to the great amount of historical study devoted to them, nor to the countercurrents of the epoch in question.[938]

First I would like to note circumstances within or close to science that paved the way for questioning the absolute claim to validity held by the dual-mechanistic explanation of nature that had become widespread since the beginning of early modern times. The history of the three mechanistic lines of tradition as I reconstructed them has shown that from the start, within mechanism (in the narrower sense) that claim to validity had not been uncontroversial. Having its origins in the critique of Descartes' materialistic mechanism, the dual explanation was confronted with weighty objections right up to its demise at the end of the nineteenth century. Those objections were formulated both within the context of continued Cartesian attempts to explain nature and within the context of dynamic mechanism based on Leibniz. The latter was particularly influential in natural science during the romantic period. Helmholtz's text from 1847, where he implicitly argues against both the materialistic and the dynamic approach, is itself an example of the obvious (although not overly intense) pressure for justification to which his dual mechanism at the time was subjected.

While it is unmistakable in Helmholtz's writing, during the first half of the nineteenth century additional uncertainty arose for mechanistic thought by the enunciation of principles from which the theory of mechanics could alternatively be deduced. Besides Newtonian laws and d'Alembert's principle there was also Hamilton's principle and Gauss's principle of the least constraint.[939] In terms of the consequences they had for relativizing validity, this plurality of equivalent options was comparable to the development later caused by the possibility of Non-Euclidean representation of physical space. None of the principles could now claim to be exclusively valid and each of them permitted a different and therefore context-dependent portrayal of what is empirically given.[940]

---

[938] Two classical works on the development of natural science in the nineteenth century are **Merz** (1907 ff.) and **Bernal** (1953). Representative of more recent research in the history of science is the anthology compiled by **Treue and Mauel (Eds.)** (1976), **Cahan (Ed.)** 2003, and **Bayertz et al. (Eds.)** (2007). See also: **Stichweh** (1984), **Knight** (1986) and **Schnädelbach** (1983). **Home (Ed.)** (1984) provides a comprehensive bibliography for the history of physics in the nineteenth century.

[939] See pertinent works on the history of mechanics, for example **Dugas** (1957) or **Mach** (1883).

[940] See **Pulte** (1993).

In addition, since the eighteenth century within physics there had been reason to abandon explaining nature entirely in mechanistic terms because successful mathematical description or explanation of certain classes of phenomena did fine without the formalism of mechanics and its ontological baggage. Without questioning the leading role played by mechanism, Joseph Fourier's explanation of heat conduction, Sadi N. L. Carnot's explanation for efficiency in heat engines and Michael Faraday's explanation for the effects of magnetic forces are examples of such cases.[941]

While the conditions discussed so far can in retrospect be seen as general signs of an oncoming crisis for the dual-mechanistic claim to truth, the next two scientific debates to discuss contributed directly to triggering the crisis. These are the development of atomic theories and the controversy over the existence and the structure of elementary natural forces. The course they took is of particular importance for evaluating the change in Helmholtz's conception of science because there hypothetization is accompanied by withdrawal from atomistic ideas.

It was characteristic for the ontology of Helmholtz's dual mechanism that is assumed that all natural phenomena are caused by moved particles of matter among which mechanical forces operated. By equating the particles with chemical elements, Helmholtz interpreted his mechanism atomistically.[942] Taking this equation for granted was in line with the realistic understanding of the chemical hypothesis of atoms that was common in physics and chemistry during the first half of the nineteenth century. The first significant controversial debates over the concept of the atom arose at the international convention of chemists in Karlsruhe in 1860. These debates continued at conventions in London (1869) and Paris (1877) and led in the last two decades to a severe crisis for the traditional idea of the atom. I shall not discuss the experimental findings upon which the proponents of both sides of the debate may have based their views, nor detail the cultural-historical aspects that may have played a role in this dispute.[943] The important point is that the critique was directed largely at the realistic understanding and led to diverging developments on the issue of validity. On the one hand it propelled the modern hypothetization of mechanism. The point was not to deny the existence of atoms but to question the relevance of their existence; instead of questioning whether the idea of the atom is useful, the scientists insisted on recognizing its indubitable hypothetical nature and the possibility that there may be alternative descriptive models.[944] On the other hand, there were also efforts to establish science without making assumptions about

[941] On Fourier see **Herivel** (1975), on Carnot see **Mendoza** (1959) and on Faraday see footnote 968.

[942] See Section 6.1.1, Part γ.

[943] On the development of atomic theory in the nineteenth century see **Farrar** (1965), **Knight** (1967), **Nye** (1976) and a comprehensive documentation of the controversial debates provided in **Nye (Ed.)** (1984).

[944] Poincaré (see **Nye** (1976)) and **Vaihinger** (1920) can be studied as good examples of how the idea of the atom became a hypothesis.

substances. Acknowledging that the idea of the atom is a hypotheses led not only to rejecting the idea of it but also the ideas any substance at all.[945] The extent to which this clean sweep for hypotheses was joined by the classical claim to truth depended on the meaning and function that the concept of the hypothesis had outside of its reference to material causes.

Although Helmholtz did not actively participate in debates on atomic theory, one can assume that he was well informed.[946] One thus has good reasons for interpreting his reluctance to accept the unquestioned realistic idea of the atom as a reaction to the critique of the 1860s. The positions he later took on the issue of atoms reflect the general developments of the times. He continues to maintain classical conceptions, while simultaneously approaching modern ones. Thus he does not abandon his realistic idea but instead makes its claim to validity hypothetical.[947] His concept of "free energy" developed in the 1880s allows a mathematical grasp of certain aspects of chemical reactions without returning to formerly mandatory atomistic presuppositions (connected with the affinity theory).[948] In his work on electrodynamics and mechanics however, Helmholtz does not waive use of the hypothesis of the atom. These contexts do not clearly indicate whether he understood it as tentatively open in terms of truth, or as in principle open in those terms.[949]

Besides the crisis in atomism, among the other factors that were suited for questioning the claim to validity in Helmholtz's conception of nature, one must also count the scientific critique of the mechanical concept of force already mentioned. True to the pattern of Newtonian theory of gravity, Helmholtz equated the forces that operate among the elementary particles of matter with instantly effective mechanical central forces. This assumption can be found even in one of his last essays on electrodynamics.[950]

Uncontested, however, is Koenigsberger's evaluation that under the impression of successful Maxwellian electrodynamics, in the early 1880s Helmholtz begins in

---

[945] **Cassirer** (1910) is representative of this development.

[946] Chemical investigations played an important part in experimental physiology to whose establishment Helmholtz helped to contribute (see Chapter 5). Later his work in phenomenological thermodynamics (see Section 7.2) was groundbreaking for theory generation in chemistry.

[947] Implicitly in **Helmholtz** (1871b), explicitly for the first time in **Helmholtz** (1881), p. 289.

[948] See literature suggestions in Section 7.2, footnote 839.

[949] Helmholtz creates the impression that openness to truth might be tentative by calling the "idea of the atom in modern chemistry" a hypothesis (**Helmholtz** (1881), p. 289). Weizsäcker, who discusses in detail Helmholtz's conceptions of atomic theory in chemistry and the ensuing famous assumption that the structure of electrical charges is corpuscular, probably correctly characterizes Helmholtz's position when he writes "The atom that Helmholtz is talking about [...] is – in a way – a known object in chemistry but for which one is not certain, whether it really exists" (**Weizsäcker** (1974), p. 75, see Section 7.1.4, footnote 820). **Helmholtz** (1897 ff.), Vol. II, p. 2 f., provides an example of an unclear formulation and arguments for hypotheticity in principle can be found in his works on monocyclic systems written in the 1880s. For these, see Section 7.2.

[950] See **Helmholtz** (1892b) and also **Buchwald** (1985), p. 237 ff., who shows that this writing contains a peculiar mixture of Helmholtz's old and new views.

part to question the structure of the elementary interaction that he had previously assumed.[951] In Maxwell's theory, which Helmholtz first without reservation acknowledges as the correct empirical description of electrical and magnetic phenomena, physical actions have finite velocity and are not structured like mechanical forces. If their structures determine appearances, then it can no longer be ruled out – as Helmholtz suggests in various passages – that they also apply to the elementary level of "final causes".[952]

If one assumes that from the beginning of early modern times onward the dual-mechanistic explanation of nature had been problematic and that this difficulty climaxed in the second half of the nineteenth century, then it looks as if the increasing pressure of critique on the tenets of mechanism led for Helmholtz to a change of his conception of science. On this interpretation, once he found it questionable whether matter is comprised of immutable atoms and whether elementary interactions consist only of instantaneous operative central mechanical forces, he abandoned the claim to truth that he had once advocated for his mechanistic conception of nature, in order to hold fast to its fundamental assumptions as hypotheses. If Helmholtz had changed his conception of science in this way as an answer to the crisis in mechanism, he would have been following a pattern that has often been discussed in portrayals of the history of modern philosophy of science. Attempts have also been made to explain the rise of conventionalism, logical empiricism and critical rationalism – only to mention the most important developments – as the processing of fundamental crisis in physics and mathematics by doing philosophy of science.[953]

But the change need be seen neither as exclusively a result (or even an effect) of his struggle with his conception of nature, nor as related to it at all. The fact that his philosophy of nature distances itself from his former ideas of atoms and forces may just as well be an expression of re-orientation in the theory of science. This could be supported by the way that his conception of science remains independent of his conception of nature from the start,[954] as well as by circumstances that had nothing

---

[951] See footnote 931 and the related passage.

[952] As early as 1871 (see footnote 970) Helmholtz points out that Faraday's theory (taken up subsequently by Maxwell's electrodynamics) is of a new type. Initial doubts as to the structure of fundamental forces can be found in the amendments written in 1881 to **Helmholtz** (1847a), p. 53 ff. (see Section 6.1.1, Part γ, footnote 298), and in **Helmholtz** (1882 ff.), Vol. 2, p. 1003 ff.

[953] On the rise of conventionalism and critical rationalism see, for instance **Schäfer** (1986 ff.) and **Schäfer** (1992), on logical empiricism see **Stegmüller** (1978 ff), Vol. I, p. 351 ff.

[954] According to Helmholtz's classical conception, in order to do science one need not be convinced of mechanism (in the narrower sense) but must be convinced of the principles that the basis for the validity of scientific knowledge lies solely in experience and that non-mental natural phenomena are causally constituted, etc. Solely the fact that scientific knowledge is established independently guarantees mechanism's claim to validity (in the narrower sense). Consequently, awareness of the difficulties in establishing it and of the problem that it is relative compared to other rationales is an independent factor in the crisis of the scientistic conception of nature.

to do with certain explanations of nature but may have influenced the development of his conception of science.

Before I next discuss the conditions and causes that may be most useful for explaining the change in Helmholtz's conception of science; I would like to mention a factor that may have had a significant part in the creation of Helmholtz's original position in the theory of science and may also have affected its further development: the influence of his teacher Johannes Müller. At first Helmholtz strictly separates the objectivity of scientific knowledge from the illusory witness of sensory perception. In order to establish why sensory perception is only relatively valid, he draws upon (and is not entirely wrong in doing so) Müller's law of "specific nerve energies".[955] The later turn towards modernity could be interpreted as applying this positive aspect taken from Müller to scientific knowledge. In this respect one could say that Müller's influence was persistent and increased in significance for the philosophy of science and epistemology as time progressed.

The merely relative validity of sensory perception can be applied to scientific statements to the extent that Helmholtz questions the initial division of the two and realizes that perception precedes all experimentally generated knowledge. Thus it is a change within the empiricist conception of science that gradually retracts the special status awarded to experimental experience against that of elementary perception. Helmholtz thus adopts a basic insight of all epistemology. It is so fundamental that one almost has the impression that Helmholtz did not begin to think seriously about epistemology until the 1860s.[956]

In this period we also find the first mention of John Stuart Mill, the outstanding theorist in empiricism at the time. In the first edition to the *Treatise* Helmholtz refers to Mill in a central passage.[957] Besides Kant (who gets little mention in this phase of his work) and Fichte, Mill is the only non-scientific epistemologist that Helmholtz explicitly and repeatedly discusses in his writing. Mill's main work in the philosophy of science, the *System of Logic, Ratiocinative and Inductive,* published in 1843 and translated into German as early as 1849, gained considerable attention from German scientists from the moment it appeared. Particularly, Justus Liebig was soon committed to spreading Mill's philosophy of science.[958] For Mill the only acceptable method for obtaining knowledge in natural science is the inductive procedure that can be traced back to Bacon. However, in contrast to early modern expectations of certainty, for Mill neither the precondition (the assumption

---

[955] Cf. Sections 6.3.1.1 and 6.3.2.2.

[956] Naturally, Helmholtz's interest in epistemology can be traced back much further. Early proof is his writing on fundamental concepts of natural science (see Section 6.2, footnote 359), which is undated but was probably written before 1847. See also Section 7.1, footnote 700.

[957] **Helmholtz** (1862), p. 178, and **Helmholtz** (1856 ff.), p. 447 (as Section 6.3.1.4, footnote 475) and 453. For Helmholtz's other references to Mill see Chapter 5, footnote 247.

[958] Cf. **Poggi and Röd** (1989), p. 127 ff. On Liebig's importance for young Helmholtz see Chapter 5, footnote 186.

of a uniform course of nature) nor the results (among which Mill counts the law of causality) have unrestricted validity.[959]

In the *Treatise* Helmholtz refers to Mill within the discussion of the inductive method.[960] Inasmuch as he there factually revises the notion of "logical induction" that he had advocated up to that point, one could say that Mill had a crucial part in inducing the transitional process. In addition, in the *Treatise* Helmholtz also adopts Mill's theory of induction and bases his own conclusions by induction (of the second kind) in his psychology of perception on it. Since Helmholtz subsequently continues to maintain these positions,[961] one can assume that Mill's influence was enduring. The definition of the inductive method affects the core of both his theory of perception and his conception of science. While in the first edition of the *Treatise* Helmholtz still tries to deny Mill's empiricist conception of causality, in order to compensate the loss of validity for inductively obtained knowledge, the fragment found in his literary remains and that was mentioned in the last section indicates that in later years he (probably) no longer deviated much from Mill's standpoint on this issue.[962]

Other points that Mill and Helmholtz had in common could be discussed; among them the critique of metaphysical, particularly idealistic systems, understanding logic in terms of psychology, a concept of science guided by practical success, the hypothetical conception of material causes, the concept of explanation related to empirical laws and finally the effort to achieve public acknowledgment of the educational value of scientific knowledge.[963] Since Helmholtz does not explicitly mention Mill all too frequently however, the effort required to reconstruct such a comparison would exceed the framework of this book. Hence, I shall make only two comments on the relationship between Helmholtz and Mill. For one, Mill represents the influence in the philosophy of science that most clearly leans towards modernity. His radical empiricism follows the tradition of Hume's skepticism, is directed fundamentally against absolute claims to validity (including claims of realism) and

---

[959] Once Mill had introduced a uniform course of nature as a precondition for the inductive method, he continued: "In reality the course of nature is not only uniform but also infinitely diverse" (**Mill** (1869ff.), Vol. 2, p. 363). According to Mill the law of causality that says that every consequence has an irrevocable antecedent (Vol. 3, p. 13f.) is a result of "induction by means of simple enumeration" [*Induktion durch einfache Abzählung*] (loc. cit. p. 299f.) and must therefore not be valid without reservation: first, "it would be foolish to want to claim with certainty" that the law also governs "distant parts of fixed star regions" (loc. cit. p. 309) and second, there exist "uniformities of coexistence, that do not rest on causal circumstances" (loc. cit. p. IX, cf. p. 310ff.).

[960] Cf. Section 6.3.1.4.

[961] Only the positive (and not the negative; see the next footnote) reference to Mill is used in the second edition of the *Treatise*: **Helmholtz** (1885ff.), p. 581.

[962] On the fragment found in Helmholtz's literary remains and the problem of dating it see Section 7.1.3.7, footnote 809. In the second edition of the *Treatise* Helmholtz omitted his critique of Mill's idea of causality (**Helmholtz** (1856ff.), p. 453).

[963] Mill's anti-idealistic attitude is particularly conspicuous in his early works, for example in **Mill** (1963ff.), Vol. XIII, p. 489ff.; on the psychological conception of logic see **Mill** (1865), p. 359, on the concepts of explanation, cause and hypothesis see **Mill** (1869ff.), Book 3, Chapters 12 and 13, and on his concept of science in general see **Ryan** (1970).

prepares the way for a pragmatic conception of science.[964] Yet Mill simultaneously continues to uphold classical absoluteness. Both Mill and Helmholtz only conditionally acknowledge alternative ways of representing what is empirically given and also uphold – albeit with certain limitations – the special status of scientific method compared to other ways of obtaining knowledge.

Thus, corresponding to his influence on Helmholtz, Mill will have also contributed to reshaping classical elements in Helmholtz's conception of science. However, I do not see any significant components of this possible secondary effect as being responsible for Helmholtz basically remaining a classicist. Above all, his sustained orientation to the claim of exclusiveness for scientific knowledge might rest on a different current in the philosophy of science that was in many ways related to the empiricism advocated by Mill but which nonetheless was unmistakably determined by other principles. This would be the positivistic drift that had been very influential in natural research from the end of the eighteenth century on and that was expressed programmatically by Auguste Comte's theory of science related work.[965]

The classical conception entered Helmholtz's writing (who to my knowledge never mentions Comte) in the shape of positivism via his discussions with other natural scientists, who are associated with that current less because of their own commitment than out of hindsight in the history of science. One thinks first of Faraday, whom Helmholtz knew personally and no doubt highly esteemed.[966] Faraday shared Fourier's[967] (whom he mentioned several times) opinion that physical explanations must do without recourse to non-observable substances and forces. Instead, he founded the truth of his geometrical concept on magnetic lines of force by devising and successfully executing experiments. Experimental practice was directive for the basic assumptions of the concept as well as crucial for the proof that it corresponded to the phenomena being examined.[968] I see this justification based exclusively on what is observable as the characteristic of claims to validity in positivism. Helmholtz unmistakably follows the positivistic strategy of justification by opposing the classical concept of natural law with the merely hypothetically valid reduction of phenomena to non-observable substances and forces.[969]

While from the start Helmholtz's definition of the concept of natural law is based on the experimental investigation of phenomena, he does not pit its claim to validity against that of mechanistic reduction until the early 1870s. This is precisely the time when and the context in which in public speeches he for the first time (and once

---

[964] According to Mill laws of reason can only claim to be valid to the extent that they are acknowledged in concrete usage (**Mill** (1869 ff.), Vol. 4, p. 231). On Hume see Chapter 2, footnote 59.

[965] **Comte** (1830 ff.). On positivism in the nineteenth century see **Blühdorn and Ritter (Eds.)** (1971).

[966] Helmholtz visited Faraday on his first trip to England in 1853 (**Koenigsberger** (1902 f.), Vol. 1, p. 196, and here Chapter 5, footnote 192). **Heidelberger** (1994a) and (1994b) particularly stresses Faraday's influence on Helmholtz.

[967] **Comte** (1830 ff.), Vol. 1, p. 78 and 406, and Vol. 2, p. 287 ff., made frequent reference to Fourier.

[968] **Faraday** (1855), p. 328 ff., 366 ff. and 402 ff. Friedrich Steinle contributed this information. On the importance of experimental practice for Faraday see **Steinle** (1994) and (1995).

[969] Cf. Section 7.1.3.1.

again, it is the speech *Gustav Magnus. In Memoriam*) mentions Faraday. He praises him for "clearly addressing the contrast to the physical theories hitherto held, which treated of atoms and forces acting at a distance".[970] He later sees Faraday's "essential progress in the principles of scientific method" in his effort to only involve "observable and observed facts in theoretical ideas [...] and to carefully avoid adding any hypothetical elements".[971] By phrasing the concept of hypotheses very generally,[972] in this speech Helmholtz emphasizes the classical claim to validity for scientific knowledge by drawing upon Faraday's positivistic idea of physics. Aside from how Faraday's lines of force concept influenced Helmholtz's work in electrodynamics (which has not yet been settled in the history of science),[973] Helmholtz stressed Faraday's understanding of physics for compensating the increase in hypothetization by restricting knowledge of laws to the observable.

Besides Faraday, Gustav Kirchhoff needs mention. He helped shape Helmholtz's conviction that natural research must focus increasingly on simply describing phenomena. From the early 1870s on Helmholtz frequently quotes Kirchhoff's famous saying that the task of physics is to "*describe* motion in nature, completely and in the simplest way".[974] According to this, a physical theory can be said to be complete without involving causes based on substances. Kirchhoff thinks of completeness as of a thoroughly attainable goal that can be fulfilled in a clear-cut way. Additionally, one should notice that this idea so typical for positivism is hooked to an easily overlooked restriction. His use of the superlative in the second criterion shows that Kirchoff thinks it is possible to have different simple and comparable (complete) descriptions of a field. Presumably unknowingly, he thus provides an example of how restricting the task of a science to pure description can involve relativizing its claims to validity. In spite of the basically classical tendency of positivism, which is also apparent in Kirchhoff, it prepared the way for modern relativizing of validity.[975]

---

[970] **Helmholtz** (1871b), p. 47. A year earlier Helmholtz and Gustav Wiedemann published a German rendition of John Tyndall's book on Faraday that their wives had translated (**Tyndall** (1870)). In terms of topics, the speeches Helmholtz had written before then would have provided ample opportunity to mention Faraday.

[971] **Helmholtz** (1881), p. 252.

[972] In this speech the concept of hypothesis covers all three meanings found in Helmholtz's writings: first non-observable or not yet observed entities (in this case atoms, **Helmholtz** (1881), p. 289); second, law-like statements (in this case the alternative mathematical formulations of the law of induction that go back to Neumann and Faraday, loc. cit. p. 258); third, the "last remainders of metaphysics" (loc. cit. p. 252), which can also mean epistemological prerequisites ("metaphysical hypotheses", cf. Section 7.1.3.5) in research.

[973] See the passage related to footnote 932 and Chapter 5.

[974] **Kirchhoff** (1876), p. V (emphasis in original), cf. Section 7.1.1, footnote 633.

[975] Another example for positivistic relativization of validity (which, however, goes unmentioned in Helmholtz) is Ernst Mach's early anti-mechanistic essay on the conservation of energy. There Mach develops a phenomenological definition of the task of physics, which is to portray "every phenomena as a function of other phenomena" (**Mach** (1872), p. 35). The type of link that holds for each case is historically entirely contingent: "History made everything and history can change everything" (loc. cit. p. 3). But overall Helmholtz did not refer to Mach, whose philosophy of science only weakly reflected tendencies toward hypothetization (see **König** (1968)).

However, this is hardly the reason why Helmholtz distances himself from positivism just when he is struggling with Kirchhoff's thought.[976] In explicitly distinguishing his own position from Kirchhoff's, Helmholtz does not demand that science unconditionally restrict itself to describing nature but simply increase focus on it. Instead of adopting positions from positivism, he draws upon its proponents for the purpose of redefining his own standpoint. For Helmholtz, higher esteem for the descriptive tasks of physics does not lead – as Kirchhoff demands – to abandoning the concept of cause. In Helmholtz's later conception of science causes, as (tentative or unquestionable) hypotheses, are among the necessary prerequisites for explaining nature.[977]

Just as Helmholtz does not wholly adopt Mill's empiricist ideas, he also does not wholeheartedly take up Faraday and Kirchhoff's positivistic views. His late conception of science tries to mediate between the extremes of empiricism, which already preludes modernity and a kind of positivism that still desires to fulfill classical demands. Rising insight into the uncertain nature of empirical foundations for science is not supposed to relativize the traditional claim to truth in general; in spite of upholding that claim, it should be possible to acknowledge in principle hypotheticity in scientific statements. Difficult intermediate positions of this kind were not uncommon among scientists representative for the second half of the nineteenth century. Above all, work done by the English physicists James Clerk Maxwell and William Thomson, with whom Helmholtz maintained lively scientific exchange, may have provided him with an example of how to balance the positivistic restriction, which science determine empirical laws only, with the ontological search for hypothetical mechanical causes. Like Maxwell and Thomson, Helmholtz also assumes that the ultimate aim of explaining the entirety of nature cannot be attained solely by formulating rigorously valid empirical laws. Their conviction that an explanation of nature is not complete until one knows the mechanical causes agrees with his dictum that comprehension of natural phenomena is not complete until science has postulated "something final and unchangeable".[978] The development of Maxwell and Thomson's own conception of science was probably considerably shaped by the crisis afflicting their mechanistic conception of nature. Their influence on Helmholtz thus also documents the close interplay of factors from the philosophy of nature and the theory of science.

Within the philosophy of science context surrounding Helmholtz, the importance of English philosophers of science and physicists is striking. He valued them perhaps not only for reasons related to the theory of science: in 1872, shortly after

[976] Cf. particularly **Helmholtz** (1897 ff.), Vol. I.1, p. 13 f.

[977] **Helmholtz** (1878a), p. 239.

[978] Cf. Section 7.1.3.6, footnote 796. For Maxwell's conception of science and nature see **Harman** (1982b), p. 127 ff., and for Thomson's see **Smith and Wise** (1989) and corresponding passages in Section 7.2 of this book. More on the relationship between Thomson and Helmholtz might be found in their voluminous correspondence, which has not yet been edited.

mentioning the name Faraday for the first time, his person was publicly heavily attacked. German-national resentment accused him of Anglophilia. Not until – as one would say today – Friedrich Zöllner "outed" him, did Helmholtz see reason to abandon the reservation he had exercised up to that point and acknowledge, in vein with the cosmopolitan character of science that he held the ideas of the English physicists to be correct.[979]

Zöllner's fanaticism is not atypical of a large part of the discourse in the philosophies of nature and science of the latter half of the nineteenth century. The pathos for truth with which Zöllner spoke on a conglomeration of topics like metaphysics, apriorism, mysticism, romantic philosophy of nature and nationalism well suited the contemporary atmosphere of worldviews in which scientific findings and success were discussed. Against that background of cultural-political debate, the change in Helmholtz's conception of science can also be understood as a position that strives for balance. In closing I would like to discuss the cultural-historical situation in which his conception of nature and science grew and explain how his own efforts at mediation and those of other cultural institutions in the newly created German Reich at least prepared the way for modernizing the conception of science.

The second half of the nineteenth century was not only a time when relativizing claims to scientific validity began. It was also still a time of classical conceptions of science and nature, even a time of their last renaissance. Helmholtz's early mechanism is an excellent example. The circumstances are also well illustrated by the "materialism debate" [*Materialismusstreit*] between spiritually minded physiologist Rudolph Wagner and the materialist Carl Vogt at the convention of natural researchers and doctors in 1854. The vulgar materialism advocated by Carl Vogt, Ludwig Büchner and Jocob Moleschott represented both a tendency that was influential in the natural sciences, denying any assumptions of independent spiritual principles for the explanation of nature, as well as a public mood in favor of science, which found increasing support among a labor force critical of capitalism.[980] The epochal rise of an extreme form of materialism that seemed to be at no loss for reasons for truth, was set off by a revival of classical metaphysical approaches, including the Schopenhauer cult that began in the late 1850s, the spiritualistic opposition to the scientific worldview led by Gustav Theodor Fechner and Rudolf Hermann Lotze, an animosity to progress expressed by

---

[979] On the dispute between Helmholtz and Zöllner see **Stromberg** (1989), **Cahan** (1994c), **Buchwald** (1994a) and Section 7.1.2.

[980] On the materialism debate, on vulgar materialism and the social influence it had, additional literature recommendations in Chapter 5, footnote 234, and also **Lübbe** (1963), p. 127 ff. Regarding the 1850s, **Nipperdey** (1983), p. 446, states that the "labor movement [...] was typically penetrated by scientific ('vulgar') materialism", In the final quarter of the century the worldview accommodation of the Darwinian theory of evolution "was like a cosmic following wind" for social democrats (**Nipperdey** (1992 f.), Vol. 1, p. 626). Examples of how natural science was welcomed by the labor movement in the last third of the nineteenth century can be found in **Bayertz** (1983).

radical clerical circles and Zöllner's conglomerate of nature and science that was reprinted several times.[981]

After a phase of relatively unhampered spread of extreme world schemes, in the 1870s a change in attitude seemed to occur that cannot be understood without considering the political circumstances of the times. An effort to objectify the discussion on materialism arose, aimed particularly at warding off the linking of an allegedly scientifically proven materialistic conception of nature to radical critique on capitalism. Eminent natural scientists publicly stressed that the scope of the materialistic (later monistic) conception of nature is limited, pointed out the epistemological assumptions connected to it and which alternatives were available. Besides Emil Du Bois-Reymond and Rudolf Virchow, Helmholtz participated in this critical debate, which helped counteract the growing influence of materialism on the working populace that particularly middle class circles found threatening.[982] Within the context of the conflict between the Catholic Church and the Prussian government, namely the so-called "cultural struggle" [*Kulturkampf*] that Bismarck initiated in 1871, the claim to exclusiveness formerly upheld by religious, spiritual and metaphysical conceptions of the world lost ground to a degree hitherto unknown. Overall interpretations of the world, which were meant to provide meaning for it all, increasingly became the private matter of each individual and just as contingent as individual existence.[983]

Neo-Kantianism provides a good example of how worldviews that claimed to be absolute were questioned. The Neo-Kantian trend, which in the 1870s advanced to become the dominant school in the philosophy of science, was, as Christian Köhnke has convincingly shown, understood by contemporaries foremost as a "critique of naturalism or materialism, of clericalism and the pessimistic movement". "While the Neo-Kantians themselves were as little disinclined to a mechanistic, scientific

---

[981] On the Schopenhauer cult that was soon to also influence Helmholtz (cf. **Conrat** (1904), p. 229 ff.), see **Safranski** (1990), who lets Schopenhauer's fame begin in 1853 (loc. cit. p. 548). On Fechner see **Heidelberger** (1993), on Lotze see **Poggi and Röd** (1989), p. 295 ff., and on the clerical animosity to progress see the chapter *Der Aufmarsch gegen den Zeitgeist* in **Buchheim** (1963), p. 131–154.

[982] Neither Helmholtz (when distancing himself from materialism in **Helmholtz** (1878c), p. 186 f.) nor Du Bois-Reymond (when discussing the limited scope of materialistic explanations of nature in **Du Bois-Reymond** (1872)) directly addresses the social effects of scientific worldviews. Helmholtz perhaps saw no reason to do so, since he basically did not want to make theoretical knowledge depend on its practical consequences and also assumed that in Germany there was "much less fear of the consequences of the complete and full truth" than "elsewhere" (**Helmholtz** (1869), p. 397). Du Bois-Reymond's reservation could be explained by the fact that he believed "culture generally [...] has nothing to fear from the red Internationals" (**Du Bois-Reymond** (1877), p. 141). Virchow on the other hand viewed Darwin's "theory of descent" as a real political threat. In his speech to the convention of natural researchers and doctors in 1877 he remarked: "After all, in all its consequences this theory has a very alarming side and the fact that socialism sought contact to it has, I hope, not escaped your attention" (as quoted in **Sudhoff** (1922), p. 191).

[983] On the "cultural struggle" [*Kulturkampf*] and the change of religious behavior that ensued see **Nipperdey** (1992 f.), Vol. 1, p. 428 ff.

worldview [...] as they were to Christian belief or doubt about qualitative progress, they contended everywhere the tendency to understand these opinions, findings and ideas as absolute and in doing so in a way took on the part of the advocate of the ideals of civil freedom".[984] Despite maintaining an a priori foundation for scientific knowledge, Neo-Kantianism and its global criticism of worldviews increased the momentum for a thorough relativization of claims to validity.

Not only do these examples of attempts to objectify the discussion in philosophy and the philosophy of science illustrate the common tendency to reject emphatic claims to validity, they also shed light on the plurality of motivations that promoted the change in attitude and the type of hypothetization that was "modern" only to a certain extent. Virchow and Du Bois-Reymond were miles away from assuming that truth in the natural sciences is open to the extent suggested by Helmholtz in his public speeches. While Virchow continued to represent an early modern kind of optimism and only objected to inadmissible generalizations of specialist knowledge while not doubting its certainty, Du Bois-Reymond was basically content to point out that materialistic explanation of nature has its limits, without otherwise fundamentally questioning it.[985] Although neither of them rejected the classical claim to validity for scientific knowledge, they both discussed its limited scope and objected to exaggerated worldviews.

Typically, positions like these did not in principle forego scientific worldviews. If one does not take the concept of a worldview[986] too narrowly but extends it to also cover cultural interest in broad scientific education, then particularly when seen against the background of the political situation in Germany, moderate attachment to standardized habits of world interpretation borrowed from science acquires an important ideological function that is more encouraged than impeded by limiting claims to validity, if not even first made possible by doing so: scientific worldviews were extremely well suited for contributing to the desired ideological unification of the German Reich founded in 1870.[987] The founding of the Reich, which first cre-

---

[984] **Köhnke** (1986), p. 321.

[985] According to Virchow, the realism in materialism is a result of empirical research: "These are perceptible things [...] that we can prove with all the tests of the experiment, so that there can be no doubt of their existence" (**Virchow** (1863), p. 38). According to **Du Bois-Reymond** (1872) natural research can advance to neither the final causes of material events nor to the phenomena of mental experience. He also points out the hypothetical nature of the idea of the atom, which is only "under certain circumstances" a "very useful fiction" (loc. cit. p. 60). This rather casual formulation does not exclude the possibility of other, perhaps contradictory fictions, which may also be useful heuristically in a similar way. Thus one cannot agree with **Malter** (1981), p. 53, who says that Du Bois-Reymond "relativized the kind of mechanistic procedure aimed at the totality of knowledge without touching its validity". The comprehensive public debate that **Du Bois-Reymond** (1872) triggered is characteristic of the overall intellectual situation. See **Engelhardt** (1972) and **Mann (Ed.)** (1981).

[986] See Section 4.2.

[987] **Cahan** (1994b), p. 598, pointed out this cultural function of Helmholtz's public speeches: "By teaching established scientific laws and by establishing a universal set of physical units and standards–more generally, by constructing a rational world order–science created a common ground for all Germans". Cahan thinks this is the reason why Heinrich von Treitschke places so much importance on his friend Helmholtz's speeches.

ated the formal, legal conditions for unification, could not itself overcome the social and cultural diversity of the population. This diversity prevented political and economical integration. In contrast, methodically supported and technically tested scientific contents, such as those discussed in scientific worldviews, provided a common point of reference. Although they were neutral with respect to complex national issues, they could nonetheless be utilized to support national interests. Restricting the validity of scientific worldviews promoted subordinating them to national demands and also let them oppose materialism's exclusive claim.

This context sheds new light on Helmholtz's appointment to a chair in Berlin. In his person, the Prussian government not only gains an excellent scientist to succeed Gustav Magnus. By appointing Helmholtz, the government demonstrates its interest in the advancement of science and concern that the new center of power should also be the national center of scientific thought.[988] In Berlin, Helmholtz is equipped accordingly. He is paid a sensational salary of 4,000 talers annually with an additional 1,000 talers for housing costs. Once the war is over, his demand for a new and representative institute for physics under his sole administration is fulfilled – one that finds no equal in the rest of Germany. Becoming the president of the *Physikalisch-Technische Reichsanstalt*, founded in 1887, brings him another considerable raise. As of 1889 he resides in a grand villa built especially for him on the grounds of the *Reichsanstalt*.[989] Issued from Berlin, Helmholtz's speeches take on an air of being half-official bulletins on matters of scientific research. Franz von Lenbach quipped that Helmholtz had become "the chancellor of the sciences".[990]

Naturally, the transition in his conception of science that first began in Berlin cannot be traced to the science-political and cultural-political importance of his appointment there. However, Helmholtz does fulfill the position's profile ideally. He very effectively propagates a scientific worldview that with a gesture of authority negates extreme positions. Whatever may have motivated the change – its contemporaneousness is striking. Helmholtz combines distancing himself from atomism with explicitly distancing himself from the anti-establishment figure called materialism, although it had "proven to be very fertile in science".[991] By claiming that both realism and idealism are hypothetical, Helmholtz helps – just as Du Bois Reymond had done some years before – mitigate the struggle between

[988] A letter from the secretary of education, Von Mühler, dated July 14, 1870, written for the purpose of obtaining finances for Helmholtz's professorship, reads: "Considering the fame that Helmholtz generally and indisputably enjoys in the scientific community, gaining him as a professor would also be a political move of great importance" (quoted from **Hoffmann and Laitko** (1995), p. 256).

[989] Cf. **Koenigsberger** (1902 f.), Vol. 2, p. 178 ff. and 345 ff., and Vol. 3, p. 12, the letter from Du Bois-Reymond to Helmholtz dated June 23, 1870, in **Helmholtz** (1986), p. 241 f., stating that his predecessor for the Berlin chair for physics, Gustav Magnus, drew a salary of just 1,500 talers. See also **Cahan** (1992), p. 153 ff.

[990] **Koenigsberger** (1902 f.), Vol. 3, p. 97.

[991] **Helmholtz** (1878c), p. 186 f.

worldviews. His second speech on Goethe also reaches out to neo-romantic currents. Relativizing claims to validity seems to follow an imperative for unification, an imperative less interested in doctrinarian enforcement of selected (in terms of content) principles and more concerned with public esteem for the scientific method within the limits of its heteronomously determined applicability.

Taking these ideological functions into consideration makes the historical emergence of the problem of validity a phenomenon that fits into the entire context of the cultural-historical situation. But a historical analysis can do no more than record such contingent correlations. The motivations of the persons involved in the historical process are various and have no common denominator. Also, they need not be related to the conditions that could make the emergence of socially effective hypothetization historically plausible and which I discussed under the headings of "autonomization" and "functionalization".[992] In Helmholtz, who remained tenaciously silent on his motives, modern-inclined criticism of the emphatic claim to truth appears simply for what it is: genuine new insight that does not necessarily follow from his own research or from previous early modern traditions in the predominant understanding of science but which nonetheless is able to express itself and make itself understood through reference to issues immanent to science.

Against the backdrop of the historical cultural context, Helmholtz's altered conception of science is clearly distinct from other tendencies to relativize scientific knowledge. Barely visible in his model production in later years, the main feature of the transition is that Helmholtz evaluates scientific claims to truth in terms of the theory of perception. The conditions of validity for human sensory perception that he investigates experimentally enter into the discussion of the conditions of validity for scientific knowledge. This places Helmholtz at the head of a development in epistemology, during the course of which the natural sciences claim to make the conditions for the possibility of their knowledge the object of their own research. The subject of knowledge becomes one of the experimentally determinable objects of knowledge. By thus losing its autonomy, it loses any absolute point of reference for questions of truth. From the circular structure of the empirically mediated investigations into itself arises a relativization of the scientific claim to validity just as it undergoes potentialization. However, human perception, from which Helmholtz derives the crucial arguments for his modernized conception of science, is life worldly experience and thus the basis that sets an indisputable boundary for the hypothetization of scientific statements. Just as observation, which is not free of illusion, has proven itself excellent for everyday purposes, so also does the scientific claim to truth that is based on the performance of the senses, have, for the purposes of experimental practice, a secure – albeit also not entirely protected from error – foundation. Although the relativization of validity, still ongoing today, has just begun in Helmholtz, his position already includes aspects of a pragmatic standpoint that lie beyond the dichotomy of absolute and relative claims to truth.

---

992 See Section 4.1.

Bibliography

I Writings of Hermann von Helmholtz

1 Books

Helmholtz, H. von (1847a): Über die Erhaltung der Kraft (Ostwald's Klassiker der exacten Wissenschaften No. 1). Leipzig 1889. Engl. trans.: The Conservation of Force: A Physical Memoir, in: Helmholtz (1971), p. 3 ff.

— (1856 ff.): Handbuch der Physiologischen Optik. Leipzig 1856 (1. installment: p. 1–192), 1860 (2. installment: p. 193–432) and 1867 (3. installment: p. 433–874). Engl. trans.: Helmholtz's Treatise on Physiological Optics, in: Helmholtz (1962). Trans. J.P.C. Southall. 2 Vols. New York 1962.

— (1863): Die Lehre von den Tonempfindungen als physiologische Grundlage für die Theorie der Musik. Braunschweig [6]1913.

— (1865): Populäre wissenschaftliche Vorträge. Booklet 1. Braunschweig.

— (1871): Populäre wissenschaftliche Vorträge. Booklet 2. Braunschweig.

— (1876): Populäre wissenschaftliche Vorträge. Booklet 3. Braunschweig.

— (1882 ff.): Wissenschaftliche Abhandlungen. 3 Vols. Leipzig 1882 (Vol. 1), 1883 (Vol. 2) and 1895 (Vol. 3).

— (1884a): Vorträge und Reden. 2 Vols. Braunschweig [5]1903.

— (1885 ff.): Handbuch der Physiologischen Optik. Hamburg/Leipzig [2]1896.

— (1895): Popular Lectures on Scientific Subjects. Trans. E. Atkinson. 2 Vols. London 1996.

— (1897 ff.): Vorlesungen über theoretische Physik. Ed. by A. König et al. 6 Vols. Leipzig 1903 (Vol. I.1: Einleitung zu den Vorlesungen), 1898 (Vol. I.2: Die Dynamik discreter Massenpunkte), 1902 (Vol. II: Dynamik continuirlich verbreiteter Massen), 1898 (Vol. III: Die mathematischen Principien der Akustik) and 1907 (Vol. IV: Elektrodynamik und Theorie des Magnetismus). Hamburg/Leipzig 1897 (Vol. V: Elektromagnetische Theorie des Lichts). Leipzig 1903 (Vol. VI: Theorie der Wärme).

— (1921): Schriften zur Erkenntnistheorie. Ed. and explained by P. Hertz and M. Schlick. Berlin. Engl. trans.: Epistemological Writings, ed. with an introduction and bibliography by R. Cohen. Dordrecht 1977.

— (1971): Selected Writings of H.v. Helmholtz. Trans. R. Kahl. Middletown.

— (1983): Über die Erhaltung der Kraft. Transkription des handschriftlichen Textes. Revised by C. Kirsten and with an introduction by H.J. Treder. Weinheim.

— (1986): Dokumente einer Freundschaft: Briefwechsel zwischen Hermann von Helmholtz und Emil du Bois-Reymond; 1846–1894. Ed. by C. Kirsten et al. Classified in terms of the history

of science as one of the scientific and philosophical movements of their period by H. Hörz and S. Wollgast. Berlin.
— (1990): Letters of Hermann von Helmholtz to his Wife. 1847–1859. Ed. by R.L. Kremer. Stuttgart.
— (1993): Letters of Hermann von Helmholtz to his Parents. The Medical Education of a German Scientist. 1837–1846. Ed. by D. Cahan. Stuttgart.
— (1995): Science and Culture. Popular and Philosophical Essays. Trans. D. Cahan. Chicago (IL).

2 *Lectures and Articles*

Helmholtz, H. von (1842): De Fabrica Systematis nervosi Evertebratorum, in: Helmholtz (1882 ff.), Vol. 2, p. 663 ff.
— (1843): Über das Wesen der Fäulniss und Gährung, in: Helmholtz (1882 ff.), Vol. 2, p. 726 ff.
— (1845): Über den Stoffverbrauch bei der Muskelaction, in: Helmholtz (1882 ff.), Vol. 2, p. 735 ff.
— (1846): Wärme, physiologisch, in: Helmholtz (1882 ff.), Vol. 2, p. 680 ff.
— (1847b): Bericht über die Theorie der physiologischen Wärmeerscheinungen für 1845, in: Helmholtz (1882 ff.), Vol. 1, p. 3 ff.
— (1847c): Über die Wärmeentwickelung bei der Muskelaction, in: Helmholtz (1882 ff.), Vol. 2, p. 745 ff.
— (1850a): Über die Fortpflanzungsgeschwindigkeit der Nervenreizung, in: Helmholtz (1882 ff.), Vol. 3, p. 1 ff.
— (1850b): Messungen über den zeitlichen Verlauf der Zuckung animalischer Muskeln und die Fortpflanzungsgeschwindigkeit der Reizung in den Nerven, in: Helmholtz (1882 ff.), Vol. 2, p. 764 ff.
— (1850c): Über die Methoden, kleinste Zeittheile zu messen, und ihre Anwendung für physiologische Zwecke, in: Helmholtz (1882 ff.), Vol. 2, p. 862 ff.
— (1851): Beschreibung eines Augenspiegels zur Untersuchung der Netzhaut im lebenden Auge, in: Helmholtz (1882 ff.), Vol. 2, p. 229 ff.
— (1852): Über die Natur der menschlichen Sinnesempfindungen, in: Helmholtz (1882 ff.), Vol. 2, p. 591 ff.
— (1852a): Messungen über Fortpflanzungsgeschwindigkeit der Reizung in den Nerven. Second column, in: Helmholtz (1882 ff.), Vol. 2, p. 844 ff.
— (1853): Über Goethe's naturwissenschaftliche Arbeiten, in: Helmholtz (1884a), Vol. 1, p. 23 ff. Engl. trans.: On Goethe's Scientific Researches, in: Helmholtz (1995), p. 1 ff.
— (1854a): Über die Wechselwirkung der Naturkräfte und die darauf bezüglichen neuesten Ermittelungen der Physik, in: Helmholtz (1884a), Vol. 1, p. 48 ff. Engl. trans.: On the Interaction of the Natural Forces, in: Helmholtz (1995), p.18 ff.
— (1854b): Erwiderung auf die Bemerkungen von Hrn. Clausius, in: Helmholtz (1882 ff.), Vol. 1, p. 76 ff.
— (1855): Über das Sehen des Menschen, in: Helmholtz (1884a), Vol. 1, p. 85 ff.
— (1856): Theorie der Wärme, in: Fortschritte der Physik im Jahre 1853, IX. Vol., Berlin.
— (1857): Über die physiologischen Ursachen der musikalischen Harmonie, in: Helmholtz (1884a), Vol. 1, p. 119 ff. Engl. trans.: On the Physiological Causes of Harmony in Music, in: Helmholtz (1995), p. 46 ff.
— (1858): Über Integrale der hydrodynamischen Gleichungen, welche den Wirbelbewegungen entsprechen, in: Helmholtz (1882 ff.), Vol. 1, p. 101 ff.
— (1861): On the Application of the Law of the Conservation of Force to Organic Nature, in: Helmholtz (1882 ff.), Vol. 3, p. 565 ff.

— (1862): Über das Verhältniss der Naturwissenschaften zur Gesammtheit der Wissenschaften, in: Helmholtz (1884a), Vol. 1, p. 157 ff. Engl. trans.: On the Relation of Natural Science to Science in General, in: Helmholtz (1995), p. 76 ff.

— (1862/3): Über die Erhaltung der Kraft, in: Helmholtz (1884a), Vol. 1, p. 187 ff. Engl. trans.: On the Conservation of Force, in: Helmholtz (1995), p. 96 ff.

— (1865a): Eis und Gletscher, in: Helmholtz (1884a), Vol. 1, p. 231 ff. Engl. trans.: Ice and Glaciers, in: Helmholtz (1895), Vol. I, p. 95 ff.

— (1867): Mittheilung, betreffend Versuche über die Fortpflanzungsgeschwindigkeit der Reizung in den motorischen Nerven des Menschen, welche Herr N. Baxt aus Petersburg im Physiologischen Laboratorium zu Heidelberg ausgeführt hat, in: Helmholtz (1882 ff.), Vol. 2, p. 932 ff.

— (1868a): Die neueren Fortschritte in der Theorie des Sehens, in: Helmholtz (1884a), Vol. 1, p. 265 ff. Engl. trans.: The Recent Progress in the Theory of Vision, in: Helmholtz (1995), p.127 ff.

— (1868b): Über die thatsächlichen Grundlagen der Geometrie, in: Helmholtz (1882 ff.), Vol. 2, p. 610 ff.

— (1868c): Über die Thatsachen, die der Geometrie zum Grunde liegen, in: Helmholtz (1882 ff.), Vol. 2, p. 618 ff.

— (1869): Über das Ziel und die Fortschritte der Naturwissenschaft, in: Helmholtz (1884a), Vol. 1, p. 367 ff. Engl. trans.: On the Aim and Progress of Physical Science, in: Helmholtz (1995), p. 204 ff.

— (1869a): Zusatz (1868), in: Helmholtz (1882 ff.), Vol. 2, p. 617.

— (1870): Über den Ursprung und die Bedeutung der geometrischen Axiome, in: Helmholtz (1884a), Vol. 2, p. 1 ff. Engl. trans.: On the Origin and Significance of Geometrical Axioms, in: Helmholtz (1995), p. 226 ff.

— (1870a): Neue Versuche über die Fortpflanzungsgeschwindigkeit der Reizung in den motorischen Nerven der Menschen, ausgeführt von N. Baxt aus Petersburg, in: Helmholtz (1882 ff.), Vol. 2, p. 939 ff.

— (1870b): The Axioms of Geometry, in: The Academy. Vol. 1, p. 128 ff.

— (1870 ff.): Über die Theorie der Elektrodynamik, in: Helmholtz (1882 ff.), Vol. 1, p. 545 ff. (first treatise 1870), p. 647 ff. (second treatise 1873) and p. 702 ff. (third treatise 1874).

— (1871a): Über die Entstehung des Planetensystems, in: Helmholtz (1884a), Vol. 2, p. 53 ff. Engl. trans.: On the Origin of the Planetary System, in: Helmholtz (1995), p. 249 ff.

— (1871b): Zum Gedächtniss an Gustav Magnus, in: Helmholtz (1884a), Vol. 2, p. 33 ff. Engl. trans.: Gustav Magnus. In Memoriam, in: Helmholtz (1895), Vol. II, p. 1 ff.

— (1871c): Vorwort, in: Thomson und Tait (1871 ff.), Vol. 1.1.

— (1871 ff.): Optisches über Malerei, in: Helmholtz (1884a), Vol. 2, p. 93 ff. Engl. trans.: On the Relation of Optics to Painting, in: Helmholtz (1995), p. 279 ff.

— (1873b): Über ein Theorem, geometrisch ähnliche Bewegungen flüssiger Körper betreffend, nebst Anwendung auf das Problem, Luftballons zu lenken, in: Helmholtz (1882 ff.), Vol. 1, p. 158 ff.

— (1874a): Vorwort, in: Thomson und Tait (1871 ff.), Vol. 1.2. Under the title *Induction und Deduction* also found in: Helmholtz (1884a), Vol. 2, p. 413 ff. Quotations are taken from this edition.

— (1874b): Vorrede, in: Tyndall (1874). Under the title *Ueber das Streben nach Popularisirung der Wissenschaft* also found in: Helmholtz (1884a), Vol. 2, p. 422 ff. Quotations are taken from this edition. Engl. trans.: The Endeavour to Popularize Science, in: Helmholtz (1971), p. 330 ff.

— (1875): Nachschrift zu Helmholtz (1853), in: Helmholtz (1884a), Vol. 1, p. 46 f.

— (1875a): Wirbelstürme und Gewitter, in: Helmholtz (1884a), Vol. 2, p. 137 ff.

— (1875b): Zur Theorie der anomalen Dispersion, in: Helmholtz (1882 ff.), Vol. 2, p. 213 ff.

— (1877a): Das Denken in der Medicin. Berlin. Engl. trans.: On Thought in Medicine, in: Helmholtz (1995), p. 309 ff.

— (1877b): Über die akademische Freiheit der deutschen Universitäten, in: Helmholtz (1884a), Vol. 2, p. 191 ff. Engl. trans.: On Academic Freedom in German Universities, in: Helmholtz (1995), p. 328 ff.
— (1878a): Die Thatsachen in der Wahrnehmung. Mit drei Beilagen in: Helmholtz (1884a), Vol. 2, p. 213 ff. and 387 ff.; the 3rd supplement is an excerpt from: Helmholtz (1878b) and is used here for quotational purposes. Engl. trans.: The Facts in Perception, in: Helmholtz (1995), p. 342 ff.
— (1878b): Über den Ursprung und Sinn der geometrischen Sätze; Antwort gegen Herrn Professor Land, in: Helmholtz (1882 ff.), Vol. 2, p. 640 ff. Engl. trans.: The Origin and Meaning of Geometrical Axioms, in: Helmholtz (1971), p. 360 ff.
— (1878c): Das Denken in der Medicin. 2. New revised edition. Berlin. Also found in: Helmholtz (1884a), Vol. 2, p. 165 ff. Quotations after this edition. Engl. trans.: On Thought in Medicine, in: Helmholtz (1995), p. 309 ff.
— (1881): Die neuere Entwickelung von Faraday's Ideen über Elektricität, in: Helmholtz (1884a), Vol. 2, p. 249 ff. Engl. trans.: The Modern Development of Faraday's Conception of Electricity, in: Helmholtz (1971), p. 409 ff.
— (1881a): Zusätze zu Helmholtz (1847a), in: Helmholtz (1847a), p. 53 ff.
— (1882 f.): Die Thermodynamik chemischer Vorgänge (1882), in: Helmholtz (1882 ff.), Vol. 2, p. 958 ff. and 979 ff.; und: Zur Thermodynamik chemischer Vorgänge (1883), in: Helmholtz (1882 ff.), Vol. 3, p. 92 ff.
— (1883): Anhang zu Helmholtz (1854a): Robert Mayer's Priorität, in: Helmholtz (1884a), Vol. 1, p. 401 ff.
— (1884b): Studien zur Statik monocyklischer Systeme, in: Helmholtz (1882 ff.), Vol. 3, p. 119 ff., 163 ff. and 173 ff.
— (1884c): Principien der Statik monocyklischer Systeme, in: Helmholtz (1882 ff.), Vol. 3, p. 142 ff. and 179 ff.
— (1886): Über die physikalische Bedeutung des Princips der kleinsten Wirkung, in: Helmholtz (1882 ff.), Vol. 3, p. 203 ff.
— (1886a): Antwortrede, gehalten beim Empfang der Graefe-Medaille, in: Helmholtz (1884a), Vol. 2, p. 311 ff.
— (1887a): Zählen und Messen, erkenntnisstheoretisch betrachtet, in: Helmholtz (1882 ff.), Vol. 3, p. 356 ff. Engl. trans.: An Epistemological Analysis of Counting and Measurement, in: Helmholtz (1971), p. 437 ff.
— (1887b): Rede über die Entdeckungsgeschichte des Princips der kleinsten Action, in: Adolf Harnack (Ed.): Geschichte der königl. preuss. Akademie der Wiss. zu Berlin. Vol. 2. Berlin 1900, p. 282 ff.
— (1887c): Zur Geschichte des Princips der kleinsten Action, in: Helmholtz (1882 ff.), Vol. 3, p. 249 ff.
— (1891): Erinnerungen, in: Helmholtz (1884a), Vol. 1, p. 1 ff.
— (1892): Goethe's Vorahnungen kommender naturwissenschaftlicher Ideen, in: Helmholtz (1884a), Vol. 2, p. 335 ff. Engl. trans.: Goethe's Anticipation of Subsequent Scientific Ideas, in: Helmholtz (1971), p. 479 ff.
— (1892a): Das Princip der kleinsten Wirkung in der Elektrodynamik, in: Helmholtz (1882 ff.), Vol. 3, p. 476 ff.
— (1892b): Elektromagnetische Theorie der Farbenzerstreuung, in: Helmholtz (1882 ff.), Vol. 3, p. 505 ff.
— (1893): Adresse an Hrn. E. du Bois-Reymond bei Gelegenheit seines 50jährigen Doctorjubiläums verfasst im Auftrage der Königlichen Akademie der Wissenschaften zu Berlin, in: Berliner Sitzungsberichte der Königlichen Akademie der Wissenschaften vom 16.2.1893, p. 93 ff.
— (1894a): Über den Ursprung der richtigen Deutung unserer Sinneseindrücke, in: Helmholtz (1882 ff.), Vol. 3, p. 536 ff. Engl. trans.: The Origin and Correct Interpretation of Our Sense Impressions, in: Helmholtz (1971), p. 501 ff.
— (1894b): Vorwort zu Hertz (1894), in: Hertz (1894), p. IX ff.

II Other Literature

Allen, D. (1983): Mechanical Explanations and the Ultimate Origin of the Universe According to Leibniz, in: Studia Leibnitiana. Special edition 11. Wiesbaden.

Andersson, G. (1981): Sind Falsifikationismus und Fallibilismus vereinbar?, in: G. Radnitzky and G. Andersson (Eds.), Voraussetzungen und Grenzen der Wissenschaft. Tübingen.

Apel, K.-O. (1970): Szientismus oder transzendentale Hermeneutik?, in: R. Bubner et al. (Eds.), Hermeneutik und Dialektik. Vol. 1. Tübingen.

Archibald, T. (1989): Energy and the Mathematization of Electrodynamics in Germany. 1845–1875, in: Archives Internationales d'Histoire des Sciences 39, p. 276 ff.

Aristoteles (1984a): Metaphysik. Trans. and ed. by F.F. Schwarz. Stuttgart.

Bachelard, G. (1938): Die Bildung des wissenschaftlichen Geistes. Beitrag zu einer Psychoanalyse der objektiven Erkenntnis. Frankfurt am Main 1978.

Bachelard, G. (1940): The Philosophy of No. A Philosophy of the New Scientific Mind. New York 1968.

Bacon, F. (1620): The New Organon. Ed. by L. Jardine. Cambridge etc. 2000.

Bacon, F. (1858 ff.): De augmentis scientiarum, in: F. Bacon, Works. Ed. by J. Spedding, R.L. Ellis and D.D. Heath. London, Vols. 3–5.

Banasiewicz, W., et al. (1973): Hermann von Helmholtz' philosophische und naturwissenschaftliche Leistungen, in: Wissenschaftliche Zeitung der Humboldt-Universität zu Berlin, Mathematisch-naturwissenschaftliche Reihe XXII, p. 277 ff.

Bayertz, K. (1983): Naturwissenschaft und Sozialismus. Tendenzen der Naturwissenschafts-Rezeption in der deutschen Arbeiterbewegung des 19. Jahrhunderts, in: Social Studies of Science 13, p. 355 ff.

Bayertz, K., M. Gerhard and W. Jaeschke (Eds.) (2007): Weltanschauung, Philosophie und Wissenschaft im 19. Jahrhundert, 3 Vols.: 1. Der Materialismus-Streit, 2. Der Darwinismus-Streit, 3. Ignorabimus-Streit. Hamburg.

Bechtel, P.W. (1980): Indeterminacy and Underdetermination: Are Quine's Two Thesis Consistent?, in: Philosophical Studies 38, p. 309 ff.

Beer, J.J. and W.D. Lewis (1963): Aspects of the Professionalization of Science, in: Daedalus 92, p. 764 ff.

Bernal, J.D. (1953): Science, Industry and Society in the Nineteenth Century, in: Centaurus 3, p. 138 ff.

Bernal, J.D. (1961): Die Wissenschaft in der Geschichte. Darmstadt.

Bevilacqua, F. (1994): Helmholtz's *Über die Erhaltung der Kraft*: The Emergence of a Theoretical Physicist, in: Cahan (Ed.) (1994).

Bierhalter, G. (1981): Zu Hermann von Helmholtzens mechanischer Grundlegung der Wärmelehre aus dem Jahre 1884, in: Archive for History of Exact Sciences 25, p. 71 ff.

Bierhalter, G. (1983): Die v. Helmholtzschen Monozykel-Analogien zur Thermodynamik und das Clausiussche Disgregationskonzept, in: Archive for History of Exact Sciences 29, p. 95 ff.

Bierhalter, G. (1987): Wie erfolgreich waren die im 19. Jahrhundert betriebenen Versuche einer mechanischen Grundlegung des zweiten Hauptsatzes der Thermodynamik?, in: Archive for History of Exact Sciences 37, p. 77 ff.

Bierhalter, G. (1994): Helmholtz's Mechanical Foundation of Thermodynamics, in: Cahan (Ed.) (1994).

Blaschke, W. (1942): Nicht-Euklidische Geometrie und Mechanik (Hamburger Mathematische Einzelschriften. Booklet 34). Leipzig/Berlin.

Blühdorn, J. and J. Ritter (Ed.) (1971): Positivismus im 19. Jahrhundert. Frankfurt am Main.

Blumenberg, H. (1973): Der Prozess der theoretischen Neugierde. Enhanced and revised edition by *Die Legitimität der Neuzeit*, 3rd part. Frankfurt am Main.

Blumenberg, H. (1981): Die Genesis der kopernikanischen Welt. Frankfurt am Main.

Blumenberg, H. (1986): Lebenszeit und Weltzeit. Frankfurt am Main.

Böhm, K. (1904): Rev. of Helmholtz (1897 ff.), Vol. I.1, in: Physikalische Zeitschrift 5, p. 140 ff.
Böhme, G. (1986): Philosophieren mit Kant. Frankfurt am Main.
Böhme, G. (1992): Aporien unserer Beziehung zur Natur, in: G. Böhme, Natürlich Natur. Frankfurt am Main.
Böhme, G. (1993): Ist das Perpetuum mobile unmöglich?, in: G. Böhme, Am Ende des Baconschen Zeitalters. Frankfurt am Main.
Böhme, G. (Ed.) (1989): Klassiker der Naturphilosophie. München.
Bohring, E.G. (1942): Sensation and Perception in the History of Experimental Psychology. New York.
Boltzmann, L. (1884): Über die Eigenschaften monozyklischer und anderer damit verwandter Systeme, in: Boltzmann (1909), Vol. III.
Boltzmann, L. (1885): Über einige Fälle, wo die lebendige Kraft nicht integrierender Nenner des Differentials der zugeführten Energie ist, in: Boltzmann (1909), Vol. III.
Boltzmann, L. (1886): Neuer Beweis eines von Helmholtz aufgestellten Theorems betreffend die Eigenschaften monozyklischer Systeme, in: Boltzmann (1909), Vol. III.
Boltzmann, L. (1899): Über die Entwicklung der Methoden der theoretischen Physik in neuerer Zeit, in: Boltzmann, Populäre Schriften. Leipzig 1905.
Bos, H.J.M. et al. (Eds.) (1980): Studies on Christiaan Huygens. Invited Papers from the Symposium on the Life and Work of Christiaan Huygens. Lisse.
Boyer, C.B. (1956): A History of Analytic Geometry. New York.
Brachner, A. (1985): German Nineteenth-Century Scientific Instrument Makers, in: P.R. de Clercq (Ed.), Nineteenth-Century Scientific Instruments and their Makers. Leiden/Amsterdam.
Brain, R.M. and M.N. Wise (1994): Muscles and Engines: Indicator Diagrams and Helmholtz's Graphical Methods, in: Krüger (Ed.) (1994).
Breger, H. (1982): Die Natur als arbeitende Maschine. Zur Entstehung des Energiebegriffs in der Physik 1840–1850. Frankfurt am Main/New York.
Brocke, vom B. (1996): Hermann von Helmholtz und die Politik, in: Eckart and Volkert (Eds.) (1996).
Brugger, W. (Ed.) (1950): Philosophisches Wörterbuch. Freiburg.
Brunner, O. et al. (Eds.) (1977): Geschichtliche Grundbegriffe. Historisches Lexikon zur politisch-sozialen Sprache in Deutschland. Stuttgart.
Brush, St.G. (1976): The Kind of Motion We Call Heat. 2.Vols. Amsterdam etc.
Buchdahl, G. (1969): Metaphysics and the Philosophy of Science. The Classical Origins. Descartes to Kant. Oxford.
Buchheim, K. (1963): Ultramontanismus und Demokratie. Der Weg der deutschen Katholiken im 19. Jahrhundert. München.
Buchwald, J.Z. (1985), From Maxwell to Microphysics: Aspects of Electromagnetic Theory in the Last Quarter of the Nineteenth Century. Chicago (IL)/London.
Buchwald, J.Z. (1994a): Electrodynamics in Context: Object States, Laboratory Practice, and Anti-Romanticism, in: Cahan (Ed.) (1994).
Buchwald, J.Z. (1994b): The Creation of Scientific Effects. Heinrich Hertz and Electric Waves. Chicago (IL)/London.
Bunge, M. (1987): Kausalität. Geschichte und Probleme. Tübingen. Engl.: Causality. The Place of the Causal Principle in Modern Science. Cambridge 1959.
Burrichter, C. et al. (Eds.) (1987): Zum Wandel des Naturverständnisses. Paderborn etc.
Butts, R.E. (Ed.) (1986): Kant's Philosophy of Physical Science. *Metaphysische Anfangsgründe der Naturwissenschaft*. 1786–1986. Dordrecht.
Cahan, D. (1985): The Institutional Revolution in German Physics, 1865–1914, in: Historical Studies in the Physical Sciences 15, p. 1 ff.
Cahan, D. (1992): Meister der Messung. Die Physikalisch-Technische Reichsanstalt im Deutschen Kaiserreich. Weinheim.
Cahan, D. (1993): Introduction, in: Helmholtz (1993).
Cahan, D. (1994a): Introduction, in: Cahan (Ed.) (1994).

Cahan, D. (1994b): Helmholtz and the Civilizing Power of Science, in: Cahan (Ed.) (1994).
Cahan, D. (1994c): Anti-Helmholtz, Anti-Zöllner, Anti-Dühring: The Freedom of Science in Germany during the 1870s, in: Krüger (Ed.) (1994).
Cahan, D. (1995): Helmholtz als Kulturträger, in: PTB-Mitteilungen 105, p. 249 ff.
Cahan, D. (2005): Helmholtz and the Shaping of the American Physics Elite in the Gilded Age, in: Historical Studies in the Physical and Biological Sciences 35, p. 1 ff.
Cahan, D. (2006): The Imperial Chancellor of the Sciences. Helmholtz between Science and Politics, in: Social Research 73, p. 1093 ff.
Cahan, D. (Ed.) (1994): Hermann von Helmholtz and the Foundations of Nineteenth-Century Science. Berkeley etc.
Cahan, D. (Ed.) (2003): From Natural Philosophy to the Sciences: Writing the History of Nineteeth-Century Science. Chicago (IL)/London.
Cardwell, D.S.L. (1971): From Watt to Clausius. The Rise of Thermodynamics in the Early Industrial Age. London.
Carnap, R. (1936 f.): Testability and Meaning, in: Philosophy of Science 3 (1936), p. 419 ff., and 4 (1937), p. 1 ff.
Carnap, R. (1966): Einführung in die Philosophie der Naturwissenschaft. Frankfurt am Main/ Berlin 1986.
Carrier, M. (1984): Atome und Kräfte. Die Entwicklung des Atomismus und der Affinitätstheorie im 18. Jahrhundert und die Methodologie Irme Lakatos'. Münster.
Carrier, M. (1994): Geometric Facts and Geometric Theory: Helmholtz and 20th-Century Philosophy of Physical Geometry, in: Krüger (Ed.) (1994).
Carrier, M. and J. Mittelstraß (1989): Johannes Kepler, in: Böhme (Ed.) (1989).
Cassirer, E. (1902): Leibniz' System in seinen wissenschaftlichen Grundlagen. Marburg.
Cassirer, E. (1910): Substanzbegriff und Funktionsbegriff. Untersuchungen über die Grundlagen der Erkenntniskritik. Darmstadt 1994.
Cassirer, E. (1923 ff.): Philosophie der symbolischen Formen. Darmstadt 1977.
Cassirer, E. (1944): The Concept of Group and the Theory of Perception, in: Philosophy and Phenomenological Research V, p. 1 ff.
Cassirer, E. (1957): Das Erkenntnisproblem in der Philosophie und Wissenschaft der neueren Zeit. 4. Vol.: Von Hegels Tod bis zur Gegenwart (1832–1932). Darmstadt 1973.
Chalcidius (1962): Timaeus a Clidio translatus commentarioque instructus, ed. by J.A. Waszink. London/Leiden.
Cohen, I.B. (1985): Revolution in Science. Cambridge etc.
Cohen, R.S. and Y. Elkana (Eds.) (1977): Introduction, in: Hermann von Helmholtz. Epistemological Writings. The Paul Hertz/Moritz Schlick Centenary Edition of 1921. Edited, with an Introduction and Bibliography, by R.S. Cohen and Y. Elkana. Dordrecht/Boston (MA).
Cohen, R.S., et al. (Eds.) (1976): PSA 1974. Boston Studies in the Philosophy of Science 32. Dordrecht/Boston (MA).
Comte, I.-A.-M.-X. (1830 ff.): Cours de philosophie positive. Paris 1864.
Conrat, F. (1904): Hermann von Helmholtz' psychologische Anschauungen. Halle.
Cranefield, P.F. (1957): The Organic Physics of 1847 and the Biophysics of Today, in: Journal of the History of Medicine and Allied Sciences XII, p. 407 ff.
Cranefield, P.F. (1966): The Philosophical and Cultural Interests of the Biophysics Movement of 1847, in: Journal of the History of Medicine and Allied Sciences XXI, p. 1 ff.
Croce, P.J. (1995): Science and Religion in the Era of William James, in: P.J. Croce, Eclipse of Certainty. 1820–1880, Vol. 1. London.
Curd, M.V. (1978): Ludwig Boltzmann's Philosophy of Science: Theories, Pictures and Analogies. Phil. Diss. University of Pittsburgh.
D'Agostino, S. (1975): Hertz's Researches on Electromagnetic Waves, in: Historical Studies in the Physical Sciences 6, p. 261 ff.
D'Agostino, S. (1990): Boltzmann and Hertz on the *Bild*-Conception of Physical Theory, in: History of Science 28, p. 380 ff.

Darrigol, O. (1994): Helmholtz's Electrodynamics and the Comprehensibility of Nature, in: Krüger (Ed.) (1994).

Debru, C. (2001): Helmholtz and the Psychophysiology of Time, in: Science in Context 14, p. 471 ff.

Dellian, E. (1988): Einleitung, in: Isaac Newton, Mathematische Grundlagen der Naturphilosophie. Hamburg.

Descartes, R. (1644): Principia philosophiae, in: Descartes (1969 ff.), Vol. VIII-1.

Descartes, R. (1969 ff.): Œuvres. Pub. par Ch. Adam et P. Tannery. Paris.

Descartes, R. (1984a): Abhandlung über die Methode des richtigen Vernunftgebrauchs. German by K. Fischer, ed. by H. Glockner. Stuttgart.

Diemer, A. (1964): Was heisst Wissenschaft? Meisenheim am Glan.

Diemer, A. (1967 ff.): Art. "Wahrheit", in: A. Diemer and I. Frenzel: Philosophie (Das Fischer-Lexikon. Vol. 11). Reprint. Frankfurt am Main.

Diemer, A. (1968): Die Begründung des Wissenschaftscharakters der Wissenschaft im 19. Jahrhundert – Die Wissenschaftstheorie zwischen klassischer und moderner Wissenschaftskonzeption, in: A. Diemer (Ed.), Beiträge zur Entwicklung der Wissenschaftstheorie im 19. Jahrhundert. Meisenheim am Glan.

Diemer, A. (1970): Der Wissenschaftsbegriff in historischem und systematischem Zusammenhang, in: A. Diemer (Ed.), Der Wissenschaftsbegriff. Historische und systematische Untersuchungen. Meisenheim am Glan.

Diemer, A. and G. König (1991): Was ist Wissenschaft?, in: A. Hermann and C. Schönbeck (Eds.), Technik und Wissenschaft. Düsseldorf.

Diemer, A. and H. Seiffert (1989): Art. "Wissenschaft", in: H. Seiffert and G. Radnitzky (Eds.), Handlexikon zur Wissenschaftstheorie. München.

Dijksterhuis, E.J. (1956): Die Mechanisierung des Weltbildes. Trans. by H. Habicht. Berlin etc.

Dingler, H. (1934): H. Helmholtz und die Grundlagen der Geometrie, in: Zeitschrift für Physik 90, pp. 348 ff. and 674 ff.

DiSalle, R. (1994): Helmholtz's Empiricist Philosophy of Mathematics: Between Laws of Perception and Laws of Nature, in: Cahan (Ed.) (1994).

DiSalle, R. (2006): Kant, Helmholtz, and the Meaning of Empiricism, in: Friedman and Nordmann (Eds.) (2006).

Dobbs, B.J.T. (1975): The Foundations of Newton's Alchemy, or "The Hunting of the Greene Lyon". Cambridge (Mass.) etc.

Du Bois-Reymond, E. (1848): Über die Lebenskraft, in: Du Bois-Reymond (1974).

Du Bois-Reymond, E. (1872): Über die Grenzen des Naturerkennens, in: Du Bois-Reymond (1974).

Du Bois-Reymond, E. (1877): Kulturgeschichte und Naturwissenschaft, in: Du Bois-Reymond (1974).

Du Bois-Reymond, E. (1895): Gedächtnisrede auf Hermann von Helmholtz. Berlin 1896.

Du Bois-Reymond, E. (1974): Vorträge über Philosophie und Gesellschaft. Ed. by S. Wollgast. Hamburg.

Du Bois-Reymond, E. (Ed.) (1918): Jugendbriefe von Emil Du Bois-Reymond an Eduard Hallmann. Berlin.

Dugas, R. (1957): A History of Mechanics. London.

Duhem, P. (1904 f.): Ziel und Struktur der physikalischen Theorien. Trans. by F. Adler. With a foreword by E. Mach. Reproduction of the edition from 1908. Hamburg 1978.

Duhem, P. (1906 ff.): Études sur Léonard de Vinci. 3 Vols. Paris.

Duhem, P. (1912): Die Wandlungen der Mechanik und der mechanischen Naturerklärung. Trans. and with a foreword by Ph. Frank. Leipzig.

Dühring, E. (1873): Kritische Geschichte der allgemeinen Principien der Mechanik. Berlin.

Düring, I. (1966): Aristoteles. Heidelberg.

Ebeling, W. and D. Hoffmann (1991): The Berlin School of Thermodynamics Founded by Helmholtz and Clausius, in: European Journal of Physics 12, p. 1 ff.

Eckart, W.U. and Ch. Gradmann (1994): Hermann Helmholtz und die Wissenschaft im 19. Jahrhundert, in: Spektrum der Wissenschaft (December 1994), p. 100ff.

Eckart, W.U. and K. Volkert (Eds.) (1996): Hermann von Helmholtz. Vorträge eines Heidelberger Symposiums anläßlich des einhundertsten Todestages. Pfaffenweiler.

Eisler, R. (1899ff.): Wörterbuch der Philosophischen Begriffe. Berlin [2]1904 and [4]1929.

Elkana, Y. (1971): Newtonianism in the Eighteenth Century, in: British Journal for the Philosophy of Science 22, p. 297ff.

Elkana, Y. (1974): The Discovery of the Conservation of Energy. London.

Engelhardt, D. von (1972): Naturphilosophie und Wissenschaftstheorie auf den Versammlungen Deutscher Naturforscher und Ärzte in der zweiten Hälfte des 19. Jahrhunderts, in: Querner and Schipperges (Eds.) (1972).

Engelhardt, D. von (1979): Historisches Bewusstsein in der Naturwissenschaft von der Aufklärung bis zum Positivismus. Freiburg/München.

Enskat, R. (1978): Kants Theorie des geometrischen Gegenstandes. Untersuchungen über die Voraussetzungen der Entdeckbarkeit geometrischer Gegenstände bei Kant. Berlin/New York.

Erdmann, B. (1877): Die Axiome der Geometrie. Eine philosophische Untersuchung der Riemann-Helmholtz'schen Raumtheorie. Leipzig.

Erdmann, B. (1921): Die philosophischen Grundlagen von Helmholtz' Wahrnehmungstheorie. Abh. Preuss. Akad. phil.-hist. Klasse. Berlin.

Falkenburg, B. (1987): Die Form der Materie. Zur Metaphysik der Natur bei Kant und Hegel. Frankfurt am Main.

Faraday, M. (1855): Experimental Researches in Electricity. Vol. 3. London.

Farrar, W.V. (1965): Nineteenth-Century Speculations on the Complexity of the Chemical Elements, in: British Journal of the History of Science 2, p. 297ff.

Farwell, R. and Ch. Knee (1990): The End of the Absolute: A Nineteenth-Century Contribution to General Relativity, in: Studies in History and Philosophy of Science 21, p. 91ff.

Feyerabend, P.K. (1967): Bemerkungen zur Geschichte und Systematik des Empirismus, in: P.K. Feyerabend, Der wissenschaftstheoretische Realismus und die Autorität der Wissenschaften (Ausgewählte Schriften. Vol. 1). Braunschweig/Wiesbaden 1978.

Feyerabend, P.K. (1976): Wider den Methodenzwang. Skizze einer anarchistischen Erkenntnistheorie. Frankfurt am Main.

Feyerabend, P.K. (1978): Kuhns *Struktur wissenschaftlicher Revolutionen*, in: P.K. Feyerabend, Der wissenschaftstheoretische Realismus und die Autorität der Wissenschaften (Selected Works, Vol. 1), Braunschweig/Wiesbaden 1978.

Fichte, J.G. (1797): Erste Einleitung in die Wissenschaftslehre, in: J.G. Fichte, Werke. Ed. by I.H. Fichte. Reprint. Berlin 1971, Vol. I.

Fichte, J.G. (1800): Die Bestimmung des Menschen, in: J.G. Fichte, Werke. Ed. by I.H. Fichte. Reprint. Berlin 1971, Vol. II.

Fichte, J.G. (1812): Ueber das Verhältniß der Logik zur Philosophie oder transcendentale Logik, in: J.G. Fichte, Werke. Ed. by I.H. Fichte. Reprint. Berlin 1971, Vol. IX.

Foucault, M. (1966): The Order of Things. An Archology of the Human Sciences. London/New York 2004.

Frank, M. (1987): Zwei Jahrhunderte Rationalitäts-Kritik und ihre "postmoderne" Überbietung, in: D. Kamper and W.v. Reijen (Eds.), Die unvollendete Vernunft: Moderne versus Postmoderne. Frankfurt am Main.

Frank, Ph. (1932): Das Kausalgesetz und seine Grenzen. Ed. by A.J. Kox. Frankfurt am Main 1988.

Freudenthal, G. (1982): Atom und Individuum im Zeitalter Newtons. Frankfurt am Main.

Friedman, M. (1997): Helmholtz's Zeichentheorie and Schlicks Allgemeine Erkenntnislehre. Early Logical Empiricism and its Nineteenth Century Background, in: Philosophical Topics 25, p.19ff.

Friedman, M. and A. Nordmann (Eds.) (2006): The Kantian Legacy in Nineteenth-Century Science. Cambridge (MA)/London etc.

Fullinwider, S.P. (1990): Hermann von Helmholtz: The Problem of Kantian Influence, in: Studies in History and Philosophy of Science 21, p. 41 ff.

Gabbey, A. (1980): Huygens and Mechanics, in: Bos et al. (Eds.) (1980).

Galilei, G. (1890a ff.): Le Opere. Edizione Nazionale. 22 Vols., Florenz.

Gassendi, P. (1727): Opera onmia in sex tomos divisa cur. Florenz.

Gawlick, G. (1962): Art. "Wahrheit", in: K. Galling (Ed.), Die Religion in Geschichte und Gegenwart. Handwörterbuch für Theologie und Religionswissenschaft. 6 Vols. Tübingen.

Gehlhaar, S.S. (1991): Die frühpositivistische (Helmholtz) und phänomenologische (Husserl) Revision der Kantischen Erkenntnislehre. Cuxhaven.

Geldsetzer, L. (1975): Einführung, in: K. Vorländer, Geschichte der Philosophie. Vol. III/1. Hamburg.

Gloy, K. (1976): Die Kantische Theorie der Naturwissenschaft. Berlin/New York.

Goldschmidt, L. (1898): Kant und Helmholtz – Populärwissenschaftliche Studie. Hamburg/Leipzig.

Goldstein, H. (1963): Klassische Mechanik. Wiesbaden 1981.

Goodman, N. (1975): Tatsache, Fiktion, Voraussage. Frankfurt am Main.

Grattan-Guiness, I. (1987): From Laplacian Physics to Mathematical Physics. 1815–1826, in: Burrichter (Ed.) (1987).

Gregory, F. (1977): Scientific Materialism in Nineteenth Century Germany. Dordrecht/Boston (MA).

Grimm, J. and W. Grimm (1854 ff.): Deutsches Wörterbuch. Ed. by the "Deutsche Akademie der Wissenschaften". Leipzig.

Grossmann, H. (1935): Die gesellschaftlichen Grundlagen der mechanistischen Philosophie und die Manufaktur, in: Zeitschrift für Sozialforschung IV, p. 161 ff.

Haas, A.E. (1909): Die Entwicklungsgeschichte des Satzes von der Erhaltung der Kraft. Wien.

Haas, A.E. (1914): Die Grundgleichungen der Mechanik, dargestellt auf Grund der geschichtlichen Entwicklung. Leipzig.

Habermas, J. (1974): Können komplexe Gesellschaften eine vernünftige Identität ausbilden?, in: J. Habermas and D. Henrich. Zwei Reden. Frankfurt am Main.

Hacker, P.M.S. (1978): Einsicht und Täuschung. Frankfurt am Main.

Hacking, I. (1984): Die Bedeutung der Sprache für die Philosophie. Königstein/Ts.

Hagner, M. and B. Wahrig-Schmidt (Eds.) (1992): Johannes Müller und die Philosophie. Berlin.

Hall, A.R. (1990): Was Galileo a Metaphysicist?, in: Levere and Shea (Eds.) (1990).

Hall, M.B (1965): Robert Boyle on Natural Philosophy. An Essay with Selection from his Writings. Bloomington.

Hamm, J. (1937): Das philosophische Weltbild von Helmholtz (Diss. Bonn). Bielefeld.

Harman, P.M. (1982a): Energy, Force, and Matter: The Conceptual Development of Nineteenth-Century Physics. Cambridge.

Harman, P.M. (1982b): Metaphysics and Natural Philosophy: The Problem of Substance in Classical Physics. Brighton.

Hatfield, G. (1979): Mind and Space from Kant to Helmholtz: The Development of the Empiristic Theory of Spatial Vision. Diss. University of Wisconsin-Madison.

Hatfield, G. (1990): The Natural and the Normative. Theories of Spatial Perception from Kant to Helmholtz. Cambridge/London.

Hatfield, G. (1994): Helmholtz and Classicism: The Science of Aesthetics and the Aesthetics of Science, in: Cahan (Ed.) (1994).

Heckmann, H.-D. (1981): Was ist Wahrheit? Heidelberg.

Heidegger, M. (1955): Die Frage nach der Technik, in: M. Heidegger, Die Technik und die Kehre. Pfullingen 1962.

Heidelberger, M. (1993): Die innere Seite der Natur. Gustav Theodor Fechners wissenschaftlich-philosophische Weltauffassung. Frankfurt am Main.

Heidelberger, M. (1994a): Force, Law, and Experiment: The Evolution of Helmholtz's Philosophy of Science, in: Cahan (Ed.) (1994).

Heidelberger, M. (1994b): Helmholtz' Erkenntnis- und Wissenschaftstheorie im Kontext der Philosophie und Naturwissenschaft des 19. Jahrhunderts, in: Krüger (Ed.) (1994).

Heidelberger, M. (1995): Helmholtz als Philosoph, in: Deutsche Zeitschrift für Philosophie. 43, p. 835 ff.
Heidelberger, M. (1996): Helmholtz, in: K.v. Meyenn (Ed.), Klassiker der Physik. München.
Heimann, P.M. (1974): Helmholtz and Kant: The Metaphysical Foundations of *Über die Erhaltung der Kraft*; in: Studies in History and Philosophy of Science 5, p. 205 ff.
Helm, G. (1898): Die Energetik nach ihrer geschichtlichen Entwickelung. Leipzig.
Hempel, C.G., and P. Oppenheim (1948): Studies in the Logic of Explanation, in: Philosophy of Science 15, p. 135 ff.
Herivel, J. (1975): Joseph Fourier. The Man and the Physicist. Oxford.
Hermann, A. (1976): Physik im 19. Jahrhundert, in: Treue and Mauel (Eds.) (1976).
Herneck, F. (1973): Die Stellung von Hermann von Helmholtz in der Wissenschaftsgeschichte, in: Banasiewicz et al. (1973).
Hertz, H. (1894): Die Prinzipien der Mechanik in neuem Zusammenhange dargestellt. Leipzig. Reprint. Darmstadt 1963. Engl. trans.: The Principles of Mechanics Presented in a New Form, trans. by D.E. Jones and J.T. Walley. New York 1956.
Hertz, P. and M. Schlick (1921): Vorrede, in: Helmholtz (1921).
Hesse, M.B. (1961): Forces and Fields. London.
Hesse, M.B. (1963): Models and Analogies in Science. London.
Heyfelder, V. (1897): Über den Begriff der Erfahrung bei Helmholtz. Diss. Berlin.
Hiebert, E. (1962): Historical Roots of the Principle of Conservation of Energy. Madison.
Hoffmann, D. (2000): Art. "Hermann von Helmholtz", in: A. Hessenbruch (Ed.), Reader's Guide to the History of Science. London/Chicago (IL).
Hoffmann, D. and H. Laitko (1995): Kompetenz, Autorität und Verantwortung: Helmholtz und die Wissenschaftspolitik im Wilhelminischen Deutschland, in: PTB-Mitteilungen 105, p. 255 ff.
Hoffmann, D. and H. Lübbig (Ed.) (1996): Hermann von Helmholtz (1821 bis 1894). Berliner Kolloquium zum 100. Todestag. Bremerhaven.
Höller, H. (1925): Der Gegenstand der Sinneswahrnehmung bei Aristoteles und Helmholtz. Zur Geschichte und Erläuterung des Wahrnehmungsproblems. Diss. Gießen.
Holmes, F.L. (1994): The Role of Johannes Müller in the Formation of Helmholtz's Physiological Career, in: Krüger (Ed.) (1994).
Home, R.W. (Ed.) (1984): The History of Classical Physics. New York/London.
Hoppe, H. (1969): Kants Theorie der Physik. Eine Untersuchung über das Opus postumum. Frankfurt am Main.
Horkheimer, M. (1947): Zur Kritik der instrumentellen Vernunft. Ed. by A. Schmidt. Frankfurt am Main 1967.
Hörz, H. (1981 ff.): Helmholtz und Boltzmann, in: Ludwig Boltzmann, Gesamtausgabe. Ed. by R. Sexl. Vol. 8.
Hörz, H. (1993): Hermann von Helmholtz und Österreich, in: Zentralbibliothek für Physik in Wien (Ed.), Engelbert Broda (1910–1983). Wissenschaft und Gesellschaft. Wien.
Hörz, H. (1994): Physiologie und Kultur in der zweiten Hälfte des 19. Jahrhunderts. Briefe an Hermann von Helmholtz. Marburg.
Hörz, H. (2000): Naturphilosophie als Heuristik? Korrespondenz zwischen Hermann von Helmholtz und Lord Kelvin (William Thomson). Marburg.
Hörz, H. and S. Wollgast (1971): Einleitung, in: Hermann von Helmholtz, Philosophische Vorträge und Aufsätze. Berlin.
Hoven, A. (1989): Wege zur Wahrheit. Frankfurt am Main etc.
Hughes, Th.P. (1983): Networks of Power. Baltimore (MD)/London.
Hume, D. (1739 f.): Ein Traktat über die menschliche Natur. Trans. by. T. Lipps and ed. by R. Brandt. Hamburg 1973.
Hume, D. (1748): Eine Untersuchung über den menschlichen Verstand. Trans. and ed. by H. Herring. Stuttgart 1982.
Hund, F. (1972): Geschichte der physikalischen Begriffe. Mannheim/Wien/Zürich.
Huygens, Ch. (1690a): Discourse de la cause de la pesanteur, in: Huygens (1888 ff,), Vol. XXI.

Huygens, Ch. (1690b): Traité de la lumière, in: Huygens (1888 ff,), Vol. XIX.
Huygens, Ch. (1888 ff.): Œuvres Complètes de Christiaan Huygens. Publiées par la Société hollandaise des Sciences. 22 Vols. La Haye.
Huygens, Ch. (1896): Abhandlung über die Ursache der Schwere. Trans. and ed. by R. Mewes. Berlin.
Huygens, Ch. (1964): Abhandlung über das Licht. Ed. by E. Lommel (Ostwalds Klassiker No. 20). Darmstadt.
Hyder, D.J. (2002): The Mechanics of Meaning. Propositional Content and the Logical Space of Wittgenstein's Tractatus. Berlin/New York.
Jaki, St.L. (1966): The Relevance of Physics. Chicago (IL)/London.
Jammer, M. (1957): Concepts of Force. Cambridge (MA).
Jammer, M. (1964): Der Begriff der Masse in der Physik. Darmstadt.
Jammer, M. (1965): Die Entwicklung des Modellbegriffes in den physikalischen Wissenschaften, in: Studium Generale 18, p. 166 ff.
Janik, A. and St. Toulmin (1973): Wittgensteins Wien. München/Wien 1985.
Juhos, B. and H. Schleichert (1963): Die erkenntnislogischen Grundlagen der klassischen Physik. Berlin.
Jungnickel, Ch. and R. McCormmach (1986): Intellectual Mastery of Nature. Theoretical Physics from Ohm to Einstein. 2 Vols. Chicago (IL)/London.
Jurkowitz, E. (2002): Helmholtz and the Liberal Unification of Science, in: Historical Studies in the Physical and Biological Sciences 32, p. 291 ff.
Kahl, H.R. (1951): The Philosophical Work of Hermann von Helmholtz. Diss. Columbia.
Kaiser, W. (1981): Theorien der Elektrodynamik im 19. Jahrhundert. Hildesheim.
Kaiser, W. (1984): Zur Struktur wissenschaftlicher Kontroversen. Habilitationsschrift. Mainz.
Kaiser, W. (1994): Helmholtz's Instrumental Role in the Formation of Classical Electrodynamics, in: Cahan (Ed.) (1994).
Kanitscheider, B. (1971): Geometrie und Wirklichkeit. Berlin.
Kanitscheider, B. (1993): Von der mechanistischen Welt zum kreativen Universum. Darmstadt.
Kanitscheider, B. (Ed.) (1984): Moderne Naturphilosophie. Würzburg.
Kant, I. (1781): Kritik der reinen Vernunft, in: Kant (1900 ff.), Vol. IV (1. edition, "A") and Vol. III (2. edition, "B"). Engl. trans. of the quotation in footnote 742: Critique of Pure Reason, trans. by N.K. Smith, New York 2003.
Kant, I. (1786): Metaphysische Anfangsgründe der Naturwissenschaft, in: Kant (1900 ff.), Vol. IV.
Kant, I. (1788): Über den Gebrauch teleologischer Principien in der Philosophie, in: Kant (1900 ff.), Vol. VIII.
Kant, I. (1788a): Kritik der praktischen Vernunft, in: Kant (1900 ff.), Vol. V.
Kant, I. (1790): Kritik der Urtheilskraft (2. edition, "B"), in: Kant (1900 ff.), Vol. V.
Kant, I. (1900 ff.): Kant's gesammelte Schriften. Ed. by der Königlich Preußischen (later: Deutschen) Akademie der Wissenschaften (zu Berlin). Berlin.
Kern, U. (1996): Hermann von Helmholtz und die Physikalisch-Technische Reichsanstalt, in: Eckart and Volkert (Eds.) (1996).
Kirchhoff, G. (1865): Über das Ziel der Naturwissenschaften. Heidelberg.
Kirchhoff, G. (1876): Vorlesungen über Mechanik. Vol. 1 of Vorlesungen über mathematische Physik. Ed. by W. Wien. Leipzig ^{4}1897.
Kirchner, F. (1833): Wörterbuch der Philosophischen Grundbegriffe. Heidelberg.
Kirk, R. (1973): Underdetermination of Theory and Indeterminacy of Translation, in: Analysis 33, p. 195 ff.
Klages, H. (1997): Hermann von Helmholtz: Klassiker an der Epochenwende. Vorträge zur Ausstellung. Braunschweigisches Landesmuseum. Bremerhaven.
Klein, F. (1926 f.): Vorlesungen über die Entwicklung der Mathematik im 19. Jahrhundert. 2 parts. Berlin.
Klein, M.J. (1972): Mechanical Explanation at the End of the Nineteenth Century, in: Centaurus 17, p. 58 ff.

Kneser, A. (1928): Das Prinzip der kleinsten Wirkung von Leibniz bis zur Gegenwart. Leipzig.
Knight, D.M. (1967): Atoms and Elements. London.
Knight, D.M. (1986): The Age of Science: The Scientific World View in the 19th Century. Oxford.
Knobloch, E., et al. (1995): "... das Wesen der reinen Mathematik verherrlichen": Reine Mathematik und mathematische Naturphilosophie bei C.G.J. Jacobi. Mit seiner Rede zum Eintritt in die philosophische Fakultät der Universität Königsberg aus dem Jahre 1832, in: Mathematische Semesterberichte 45, p. 99ff.
Knorr Cetina, K. (1984): Die Fabrikation des Wissens. Frankfurt am Main.
Koenigsberger, L. (1895): Hermann von Helmholtz's Untersuchungen über die Grundlagen der Mathematik und Mechanik. Heidelberg.
Koenigsberger, L. (1902f.): Hermann von Helmholtz. 3 Vols. Braunschweig.
Koenigsberger, L. (1921): Die Erweiterung des Helmholtzschen Prinzips von der verborgenen Bewegung und den unvollständigen Problemen auf kinetische Potentiale beliebiger Ordnung, in: Sitzungsberichte der Heidelberger Akademie der Wissenschaften. Mathe.-nat. Klasse. Heidelberg.
Köhnke, K.Ch. (1986): Entstehung und Aufstieg des Neukantianismus. Die deutsche Universitätsphilosophie zwischen Idealismus und Positivismus. Frankfurt am Main.
König, G. (1968): Der Wissenschaftsbegriff bei Helmholtz und Mach, in: A. Diemer (Ed.), Beiträge zur Entwicklung der Wissenschaftstheorie im 19. Jahrhundert. Meisenheim am Glan.
Koslowski, P. (1986): Die Baustellen der Postmoderne – Wider den Vollendungszwang der Moderne, in: P. Koslowski et al. (Eds.), Moderne oder Postmoderne? Weinheim.
Koyré, A. (1957): From the Closed World to the Infinite Universe. Baltimore (MD).
Krafft, F. (1982): Das Selbstverständnis der Physik im Wandel der Zeit. Weinheim.
Kragh, H. (1994): Between Physics and Chemistry: Helmholtz's Route to a Theory of Chemical Thermodynamics, in: Cahan (Ed.) (1994).
Krajewski, W. (1974): The Idea of Statistical Law in Nineteenth Century Science, in: R.S. Cohen and M.W. Wartofsky (Eds.), Boston Studies in the Philosophy of Science. Vol. XIV. Dordrecht, p. 397ff.
Krause, A. (1878): Kant und Helmholtz über den Ursprung und die Bedeutung der Raumanschauung und der geometrischen Axiome. Lahr.
Kremer, R.L. (1990a): The Thermodynamics of Life and Experimental Physiology, 1770–1880. New York/London.
Kremer, R.L. (1990b): Introduction, in: Helmholtz (1990).
Kremer, R.L. (1994): Innovation Through Synthesis: Helmholtz and Color Research, in: Cahan (Ed.) (1994).
Krohn, W. (1990): Einleitung, in: Francis Bacon, Neues Organon. Ed. by W. Krohn. Hamburg.
Krug, W.T. (1827ff.): Allgemeines Handwörterbuch der philosophischen Wissenschaften, nebst ihrer Literatur und Geschichte, nach dem heutigen Standpuncte der Wissenschaft. 4 Vols. Leipzig.
Krüger, L. (1994): Helmholtz über die Begreiflichkeit der Natur, in: Krüger (Ed.) (1994).
Krüger, L. (Ed.) (1994): Universalgenie Helmholtz. Rückblick nach 100 Jahren. Berlin.
Kuhn, Th.S. (1959): Energy Conservation as an Example of Simultaneous Discovery, in: M. Clagett (Ed.), Critical Problems in the History of Science. Madison.
Kuhn, Th.S. (1962): Die Struktur wissenschaftlicher Revolutionen. Frankfurt am Main 1976.
Küppers, B.O. (Ed.) (1987): Ordnung aus dem Chaos. Prinzipien der Selbstorganisation des Lebendigen. München.
Kuznecov, B.G. (1970): Von Galilei bis Einstein. Entwicklung der physikalischen Ideen. Basel/Braunschweig.
Lagrange, J.-L. (1887): Analytische Mechanik. Ed. and trans. by H. Servus. Berlin.
Lakatos, I. (1970): Falsification and the Methodology of Scientific Research Programmes, in: Lakatos (1978).
Lakatos, I. (1971): History of Science and its Rational Reconstructions, in: Lakatos (1978).
Lakatos, I. (1973): Introduction: Science and Pseudoscience, in: Lakatos (1978).
Lakatos, I. (1974): The Role of Crucial Experiments in Science, in: Studies in History and Philosophy of Science 4, p. 309ff.

Lakatos, I. (1978): The Methodology of Scientific Research Programmes. Philosophical Papers. Vol. I. Cambridge.

Lange, F.A. (1866): Geschichte des Materialismus. In the second, enhanced edition from 1873–1875. Ed. and with an introduction by A. Schmidt. Frankfurt am Main 1974.

Lasswitz, K. (1890): Geschichte der Atomistik vom Mittelalter bis Newton. 2 Vols. Hamburg/ Leipzig. New edition. Darmstadt 1963.

Laudan, L. (1977): Progress and its Problems. Berkeley.

Laudan, L. (1981): Science and Hypothesis. Historical Essays on Scientific Methodology. Dordrecht etc.

Laue, M. (1944): Zum 50. Todestage von Hermann von Helmholtz, in: Die Naturwissenschaften 32, p. 206 f.

Laue, M. (1947): Geschichte der Physik. Bonn 1950.

Lavoisier, A.L. (1790): Elements of Chemistry. Trans. by R. Kerr. New York 1965.

Le Grand, H.E. (1978): Galileo's Matter Theory, in: Butts and Pitt (Eds.) (1978).

Lehmann, G. (1953): Die Philosophie des 19. Jahrhunderts. 2 Vols. Berlin.

Leiber, T. (2000): Vom mechanistischen Weltbild zur Selbstorganisation des Lebens. Helmholtz' und Boltzmanns Forschungsprogramme und ihre Bedeutung für Physik, Chemie, Biologie und Philosophie. Freiburg.

Leibniz, G.W. (1692): Bemerkungen zum allgemeinen Teil der Kartesischen Prinzipien, in: Leibniz (1904 ff.), Vol. I.

Leibniz, G.W. (1695a): Specimen dynamicum, in: Leibniz (1904 ff.), Vol. I. Engl. trans. of the quotation in footnote 118: Philosophical Texts, trans. by R. Francks and R.S. Woolhouse, Oxford/New York 1998, p. 154.

Leibniz, G.W. (1695b): Neues System der Natur und der Gemeinschaft der Substanzen, wie der Vereinigung zwischen Körper und Seele, in: Leibniz (1904 ff.), Vol. II.

Leibniz, G.W. (1702): (Gegen Descartes), in: Leibniz (1904 ff.), Vol. I.

Leibniz, G.W. (1705): Die Theodizee. Trans. by A. Buchenau. Hamburg 1968.

Leibniz, G.W. (1714): Monadologie. Trans. and ed. by H. Glockner. Stuttgart 1979. Engl. trans.: The Monadology and Other Philosophical Writings, trans. with an introduction and notes by R. Latta, New York/London 1985.

Leibniz, G.W. (1849 ff.): Mathematische Schriften. Ed. by C.J. Gerhardt. 7 Vols. Berlin.

Leibniz, G.W. (1875 ff.): Die philosophischen Schriften von G.W. Leibniz. Ed. by C.J. Gerhardt. 7 Vols. Berlin. Engl. trans. of the quotation in footnote 120: Philosophical Papers and Letters, A Selection, trans. and ed., with an introduction by L.E. Loemker, Dordrecht/Boston (MA) 1969, p. 173.

Leibniz, G.W. (1904 ff.): Hauptschriften zur Grundlegung der Philosophie. Trans. by A. Buchenau. Ed. by E. Cassirer. 2 Vols. Berlin.

Leinfellner, W. (1967): Einführung in die Erkenntnis- und Wissenschaftstheorie. Mannheim etc.

Lenk, H. (1979): Zur Wissenschaftstheorie der Sozialwissenschaften, in: H. Lenk, Pragmatische Vernunft. Stuttgart.

Lenk, H. (1989): Art. "Szientismus", in: H. Seiffert and G. Radnitzky, Handlexikon zur Wissenschaftstheorie. München.

Lenoble, R. (1943): Mersenne ou la naissance du mécanisme. Paris.

Lenoir, T. (1982): The Strategy of Life. Teleology and Mechanics in Nineteenth-Century German Biology. Dordrecht/Boston (MA).

Lenoir, T. (1988): Soziale Interessen und die organische Physik von 1847, in: T. Lenoir, Politik im Tempel der Wissenschaft. Forschung und Machtausübung im deutschen Kaiserreich. Frankfurt am Main 1992.

Lenoir, T. (1992): Helmholtz, Müller und die Erziehung der Sinne, in: Hagner and Wahrig-Schmidt (Eds.) (1992).

Lenoir, T. (1994): The Eye as Mathematician: Clinical Practice, Instrumentation, and Helmholtz's Construction of an Empirist Theory of Vision, in: Cahan (Ed.) (1994).

Lenoir, T. (2006): Operationalizing Kant. Manifolds, Models, and Mathematics in Helmholtz's Theory of Perception, in: Friedman and Nordmann (Eds.) (2006).

Lenzen, V.F. (1944): Helmholtz's Theory of Knowledge, in: M.F. Ashley Montague (Ed.), Studies and Essays in the History of Science and Learning. Offered in Homage to George Sarton on the Occasion of His Sixtieth Birthday 31 August 1944. New York 1946.

Lepenies, W. (1976): Das Ende der Naturgeschichte. Wandel kultureller Selbstverständlichkeiten in den Wissenschaften des 18. und 19. Jahrhunderts. Frankfurt am Main 1978.

Leroux, J. (1995): Helmholtz and Modern Empirism, in: M. Marion and R.S. Cohen (Eds.), Québec Studies in the Philosophy of Science I, p.287 ff.

Leroux, J. (2001): "Picture Theories" as Forerunners of the Semantic Approach to Scientific Theorising, in: International Studies in the Philosophy of Science 15, p. 189 ff.

Levere, T.H. and W.R. Shea (Eds.) (1990): Nature, Experiment, and the Sciences. Dordrecht etc.

Lie, S. (1888 ff.): Theorie der Transformationsgruppen. Leipzig.

Lipman, T.O. (1966): The Response to Liebig's Vitalism, in: Bulletin of the History of Medicine XI, p. 511 ff.

Lipschitz, R. (1870): Untersuchungen in Betreff der ganzen homogenen Funktionen von n Differentialen, in: Borchardt's Journal für reine und angewandte Mathematik, Vol. LXX and Vol. LXXII.

Lipschitz, R. (1986): Briefwechsel mit Cantor, Dedekind, Helmholtz, Kronecker, Weierstrass et al. Revised by W. Scharlau. Dokumente zur Geschichte der Mathematik. Vol. 2. Braunschweig/Wiesbaden.

Loeck, G. (1986): Der cartesische Materialismus, Maschine, Gesetz und Simulation. Frankfurt am Main etc.

Loh, W. (1990): War Kant naiver Realist?, in: Prima Philosophia 3, p. 365 ff.

Lorenzen, P. (1960): Die Entstehung der exakten Wissenschaften. Berlin etc.

Lübbe, H. (1963): Politische Philosophie in Deutschland. Studien zu ihrer Geschichte. Basel/Stuttgart.

Lübbe, H. (1986): Religion nach der Aufklärung. Graz etc.

Lübbig, H. (Ed.) (1995): The Inverse Problem. Symposium ad memoriam Hermann von Helmholtz. Weinheim.

Lyotard, J.-F. (1979): Das postmoderne Wissen. Wien etc. 1986.

Mach, E. (1872): Die Geschichte und die Wurzel des Satzes von der Erhaltung der Arbeit. Prag.

Mach, E. (1883): Die Mechanik, historisch-kritisch dargestellt. Leipzig [9]1933. Reprint. Darmstadt 1963.

Maier, A. (1938): Die Mechanisierung des Weltbilds im 17. Jahrhundert. Leipzig.

Maier, A. (1949): Die Vorläufer Galileis im 14. Jahrhundert. Rom.

Maier, A. (1951): Zwei Grundprobleme der scholastischen Naturphilosophie. Rom.

Mainzer, K. (1980): Geschichte der Geometrie. Mannheim etc.

Malter, R. (1981): "Kausalitätstrieb" und Erkenntnisschranke. Zur philosophischen Grundposition Emil Du Bois-Reymonds, in: Mann (Ed.) (1981).

Manegold, K.H. (1976): Das Verhältnis von Naturwissenschaft und Technik im 19. Jahrhundert im Spiegel der Wissenschaftsorganisation, in: Treue and Mauel (Eds.) (1976).

Mann, G. (Ed.) (1981): Naturwissen und Erkenntnis im 19. Jahrhundert. Hildesheim.

Mathieu, V. (1989): Kants Opus postumum. Frankfurt am Main.

Mauel, K. (1976): Die Aufnahme naturwissenschaftlicher Erkenntnisse und Methoden durch die Ingenieure im 19. Jahrhundert, in: Treue and Mauel (Eds.) (1976).

Maxwell, J.C. (1873): A Treatise on Electricity and Magnetism. 2. Vols. Oxford.

McClelland, E.Ch. (1985): Zur Professionalisierung der akademischen Berufe in Deutschland, in: W. Conze and J. Kocka (Eds.), Bildungsbürgertum im 19. Jahrhundert. Part 1: Bildungssystem und Professionalisierung in internationalen Vergleichen. Stuttgart.

McCormmach, R. (1982): Night Thoughts of a Classical Physicist. Cambridge (MA)/London.

McDonald, P.J. (2001): Epistemology of Experiment and the Physiological Acoustics of Hermann von Helmholtz. Diss. Notre Dame.

McDonald, P.J. (2003): Demonstration by Simulation. The Philosophical Significance of Experiment in Helmholtz's Theory of Perception, in: Perspectives on Science 11, p. 170 ff.

Mehrtens, H. (1990): Moderne – Sprache – Mathematik. Eine Geschichte des Streits um die Grundlagen der Disziplin und des Subjekts formaler Systeme. Frankfurt am Main.

Melson, A.G.M. van (1957): Atom – Gestern und Heute. Die Geschichte des Atombegriffs von der Antike bis zur Gegenwart. Freiburg/München.
Mendelsohn, E. (1964): The Biological Sciences in the Nineteenth Century: Some Problems and Sources, in: History of Science 3, p. 39 ff.
Mendelsohn, E. (1965): Physical Models and Physiological Concepts: Explanation in the Nineteenth Century Biology, in: R.S. Cohen and M.W. Wartofsky (Eds.): Boston Studies in the Philosophy of Science 2. New York.
Mendoza, E. (1959): Contributions to the Study of Sadi Carnot and his Work, in: Archives Internationales d'Histoire des Sciences 12, p. 377 ff.
Merz, J.Th. (1907 ff.): A History of European Thought in the Nineteenth Century. 4 Vols. Edinburgh/London.
Meyering, T.C. (1989): Historical Roots of Cognitive Science. Dordrecht etc.
Mill, J.St. (1865): An Examination of Sir William Hamilton's Philosophy, in: Mill (1963 ff.), Vol. IX.
Mill, J.St. (1869 ff.): Gesammelte Werke. Ed. by Th. Gomperz. Leipzig. Reprint. Aalen 1968.
Mill, J.St. (1963 ff.): The Collected Works. Ed. by J.M. Robson et al. 25 Vols. Toronto/London.
Miller, A.I. (1984): Imagery in Scientific Thought Creating 20th Century Physics. Boston (MA)/Basel/Stuttgart.
Mittelstaedt, P. (1970): Klassische Mechanik. Mannheim etc.
Mittelstaedt, P. (1976): Philosophische Probleme der modernen Physik. Mannheim etc.
Mittelstraß, J. (1970): Neuzeit und Aufklärung. Berlin/New York.
Mittelstraß, J. (1974): Die Möglichkeit von Wissenschaft. Frankfurt am Main.
Mittelstraß, J. (1981): Das Wirken der Natur. Materialien zur Geschichte des Naturbegriffs, in: F. Rapp (Ed.), Naturverständnis und Naturbeherrschung. München.
Mittelstraß, J. (1992): Leonardo-Welt. Über Wissenschaft, Forschung und Verantwortung. Frankfurt am Main.
Moll, K. (1978 ff.): Der junge Leibniz. 2 Vols. Stuttgart/Bad Cannstatt.
Moulines, C.U. (1981): Hermann von Helmholtz: A Physiological Approach to the Theory of Knowledge, in: H.N. Jahnke and M. Otte (Eds.), Epistemological and Social Problems of the Sciences in the Early Nineteenth Century. Dordrecht/Boston (MA)/London.
Müller, J. (1824): Antrittsvorlesung, in: Müller (1826).
Müller, J. (1826): Zur vergleichenden Physiologie des Gesichtssinnes des Menschen und der Thiere nebst einem Versuch über die Bewegungen der Augen und über den menschlichen Blick. Leipzig.
Müller, J. (1826a): Über die phantastischen Gesichtserscheinungen. Coblenz.
Müller, J. (1833 ff.): Handbuch der Physiologie des Menschen für Vorlesungen. Coblenz 1833 (Vol. 1) and 1840 (Vol. 2).
Nagel, E. (1939): The Formation of Modern Conceptions of Formal Logic in the Development of Geometry, in: Osiris 7, p. 142 ff.
Nagel, E. (1961): The Structure of Science. Problems in the Logic of Scientific Explanation. London.
Newton, I. (1704): Opticks, in: Newton (1779 ff.), Vol. 4.
Newton, I. (1726): Philosophiae Naturalis Principia Mathematica. The third Ed. with variant readings. Ass. and ed. by A. Koyré and I.B. Cohen. 2 Vols. Cambridge 1972.
Newton, I. (1779 ff.): Opera omnia. Ed. by S. Horsley. London.
Newton, I. (1872): Mathematische Prinzipien der Naturlehre. With notes and explanations. Ed. by J. Ph. Wolfers. Darmstadt 1963. Engl.: The Principia. Mathematical Principles of Natural Philosophy. Berkley etc. 1999.
Newton, I. (1898): Optik. Trans. and ed. by W. Abendroth. Nachdruck. Braunschweig/Wiesbaden 1983.
Nietzsche, F. (1980): Werke. 6 Vols. Ed. by K. Schlechta. München/Wien.
Nipperdey, Th. (1983): Deutsche Geschichte. 1800–1866. Bürgerwelt und starker Staat. München.
Nipperdey, Th. (1992 f.): Deutsche Geschichte. 1866–1918. 2 Vols. München.

Nobis, H.M. (1967): Frühneuzeitliche Verständnisweisen der Natur und ihr Wandel bis zum 18. Jahrhundert, in: Archiv für Begriffsgeschichte 11, p. 37 ff.

Nobis, H.M. (1969): Die Umwandlung der mittelalterlichen Naturvorstellung, in: Archiv für Begriffsgeschichte 13, p. 34 ff.

Nye, M.J. (1976): The Nineteenth-Century Atomic Debates and the Dilemma of an 'Indifferent Hypothesis', in: Studies in History and Philosophy of Science 7, p. 245 ff.

Nye, M.J. (1984): Introduction, in: Nye (Ed.) (1984).

Nye, M.J. (Ed.) (1984): The Question of the Atom. From the Karlsruhe Congress to the First Solvay Conference, 1860–1911. Los Angeles/San Francisco.

Okruhlik, K. and J.R. Brown (Eds.) (1985): The Natural Philosophy of Leibniz. Dordrecht.

Olesko, K.M. (1980): The Emergence of Theoretical Physics in Germany: Franz Neumann and the Königsberg School of Physics, 1830–1890. Diss. Cornell University.

Olesko, K.M. and F.L. Holmes (1994): Experiment, Quantification, and Discovery: Helmholtz's Early Physiological Researches, 1843–50, in: Cahan (Ed.) (1994).

Paulsen, F. (1902): Die deutschen Universitäten und das Universitätsstudium. Berlin.

Pitcher, G. (1967): Wittgenstein. Freiburg/München.

Plaaß, P. (1965): Kants Theorie der Naturwissenschaft. Göttingen.

Planck, M. (1887): Das Princip der Erhaltung der Energie. Leipzig.

Planck, M. (1915): Das Prinzip der kleinsten Wirkung, in: M. Planck, Vorträge und Erinnerungen. Darmstadt 1970.

Platon (1959): Sämtliche Werke. Trans. by F. Schleiermacher and H. Müller. Hamburg.

Plessner, H. (1924): Zur Soziologie der modernen Forschung und ihrer Organisation in der Deutschen Universität – Tradition und Ideologie, in: H. Plessner, Ges. Schriften. Frankfurt am Main, p. 1980 ff., Vol. X.

Poggi, S. and W. Röd (1989): Die Philosophie der Neuzeit 4. Positivismus, Sozialismus und Spiritualismus im 19. Jahrhundert (Geschichte der Philosophie. Volume X). München.

Poincaré, H. (1890): Electricité et Optique. Paris.

Poincaré, H. (1902): Wissenschaft und Hypothese. Reproduction of the 3rd, improved edition from 1914. Darmstadt 1974.

Popper, K.R. (1935): Logik der Forschung. Tübingen ^{9}1989.

Popper, K.R. (1963): Conjectures and Refutation. The Growth of Scientific Knowledge. London.

Post, K. (1905): Johannes Müller's Philosophische Anschauungen. Halle.

Pot, J.H.J. van der (1985): Die Bewertung des technischen Fortschritts. Eine systematische Übersicht der Theorien. 2 Vols. Maastricht.

Pulte, H. (1989): Das Prinzip der kleinsten Wirkung und die Kraftkonzeptionen der rationalen Mechanik. Eine Untersuchung zur Grundlegungsproblematik bei Leonhard Euler, Pierre Louis Moreau de Maupertuis und Joseph Louis Lagrange (Studia Leibnitiana. Special edition 19). Stuttgart.

Pulte, H. (1993): Zum Niedergang des Euklidianismus in der Mechanik des 19. Jahrhunderts, in: Allgemeine Gesellschaft für Philosophie (Ed.), Neue Realitäten. XVI. Deutscher Kongreß für Philosophie. Berlin.

Pulte, H. (1994): C.G.J. Jacobis Vermächtnis einer 'konventionellen' analytischen Mechanik. Vorgeschichte, Nachschriften und Inhalt seiner letzten Mechanik-Vorlesung, in: Annals of Science 51, p. 487 ff.

Pulte, H. (1996): Einleitung zu: Carl G.J. Jacobi, Vorlesungen über analytische Mechanik. Wiesbaden.

Pulte, H. (2000): Beyond the Edge of Certainty. Reflections on the Rise of 'Physical' Conventionalism in the 19th Century, in: Philosophiae Scientiae. Travaux d'histoire et de philosophie des sciences 4, p. 87 ff.

Pulte, H. (2005): Axiomatik und Empirie. Eine wissenschaftstheoriegeschichtliche Untersuchung zur Mathematischen Naturphilosophie von Newton bis Neumann. Darmstadt.

Pulte, H. and G. König (1998): Art. "Theorie II: 20. Jahrhundert (Wissenschaftstheorie)", in: Ritter and Gründer (Eds.) (1971 ff.).

Puntel, L.B. (1978): Wahrheitstheorien in der neueren Philosophie. Darmstadt.

Querner, H. and H. Schipperges (Ed.) (1972): Wege der Naturforschung 1822–1872 im Spiegel der Versammlungen Deutscher Naturforscher und Ärzte. Berlin etc.

Quine, W.V. (1953): Von einem logischen Standpunkt. Frankfurt am Main etc. 1979.

Quine, W.V. (1960): Wort und Gegenstand. Stuttgart 1980.

Quine, W.V. (1969): Ontologische Relativität und andere Schriften. Stuttgart 1975.

Quine, W.V. (1970): On the Reasons for Indeterminacy of Translation, in: Journal of Philosophy 67, p. 178 ff.

Quine, W.V. (1975): On Empirically Equivalent Systems of the World, in: Erkenntnis 9, p. 313 ff.

Rabinbach, A. (1990): The Human Motor: Energy, Fatigue, and the Origins of Modernity. New York.

Rechenberg, H. (1994): Hermann von Helmholtz. Bilder seines Lebens und Wirkens. Weinheim etc.

Redner, H. (1987): The Ends of Science. An Essay in Scientific Authority. Boulder/London.

Reichenbach, H. (1928): Philosophie der Raum-Zeit-Lehre, in: Reichenbach (1977 ff.), Vol. 2.

Reichenbach, H. (1977 ff.): Gesammelte Werke. Braunschweig/Wiesbaden.

Reif, F. (1976): Grundlagen der Physikalischen Statistik und der Physik der Wärme. Berlin/New York.

Rey, A. (1909): Die Theorie der Physik bei den modernen Physikern. German by R. Eisler. Leipzig.

Richards, J.L. (1977): The Evolution of Empiricism: Hermann von Helmholtz and the Foundations of Geometry, in: British Journal for the Philosophy of Science 28, p. 235 ff.

Riehl, A. (1904): Hermann von Helmholtz in seinem Verhältnis zu Kant. Berlin.

Riemann, B. (1867): Über die Hypothesen, welche der Geometrie zugrunde liegen. Darmstadt 1959. Engl. trans.: On the Hypotheses which lie at the Foundation of Geometry, in: W.B. Ewald (Ed.), From Kant to Hilbert: A Source Book in the Foundations of Mathematics, 2 Vols. Oxford 1996.

Ritter, J. (1971 ff.): Art. "Fortschritt", in: Ritter and Gründer (Eds.) (1971 ff.).

Ritter, J. and K. Gründer (Eds.) (1971 ff.): Historisches Wörterbuch der Philosophie. Darmstadt.

Rorty, R. (1989): Kontingenz, Ironie und Solidarität. Frankfurt am Main.

Rorty, R. (1991): Ist Naturwissenschaft eine natürliche Art? in: R. Rorty, Eine Kultur ohne Zentrum. Stuttgart 1993.

Röseberg, U. (1994): Ontologische und erkenntnistheoretische Dimensionen des Gesetzesproblems in den Helmholtzschen Reflexionen über Naturgesetze, in Krüger (Ed.) (1994).

Rosenfeld, L. (1941): La genèse des principes de la thermodynamique, in: Bulletin de la Société Royale des Sciences de Liège 10, p. 199 ff.

Ryan, A. (1970): The Philosophy of John Stuart Mill. London.

Safranski, R. (1990): Schopenhauer und die wilden Jahre der Philosophie. Reinbek.

Schäfer, L. (1966): Kants Metaphysik der Natur. Berlin.

Schäfer, L. (1986 ff.): Der Konventionalismus des beginnenden 20. Jahrhunderts, in: H. Stachowiak (Ed.), Pragmatik: Handbuch des pragmatischen Denkens. 5 Vols. Hamburg, 2. Vol.

Schäfer, L. (1992): Karl R. Popper. München.

Schäfer, L. (1993): Das Bacon-Projekt. Von der Erkenntnis, Nutzung und Schonung der Natur. Frankfurt am Main.

Schaffner, K.F. (1972): Nineteenth-Century Aether Theories. Oxford etc.

Scheibe, E. (1971 ff.): Art. "Kausalgesetz", in: Ritter and Gründer (Eds.) (1971 ff.).

Scheibe, E. (1982): Zum Theorienvergleich in der Physik, in: K.M. Meyer-Abich (Ed.), Physik, Philosophie, Politik. Festschrift für C.F. von Weizsäcker zum 70. Geburtstag. München.

Schelling, F.W.J. (1799): Einleitung zu seinem Entwurf eines Systems der Naturphilosophie. Ed. and with an introduction by W.G. Jacobs. Stuttgart 1988.

Schiemann, G. (1992): Totalität oder Zweckmässigkeit? Kants Ringen mit dem Mannigfaltigen der Erfahrung im Ausgang der Vernunftkritik, in: Kant-Studien 83, p. 294 ff.

Schiemann, G. (1994): Die Hypothetisierung des Mechanismus bei Hermann von Helmholtz. Ein Beitrag zum Wandel der Wissenschafts- und Naturauffassung im 19. Jahrhundert, in: Krüger (Ed.) (1994).

Schiemann, G. (1995a): Zwischen klassischer und moderner Wissenschaftstheorie: Hermann von Helmholtz und Karl R. Popper, erkenntnistheoretisch verglichen, in: Deutsche Zeitschrift für Philosophie 43, p. 845 ff. Engl. in: Lübbig (Ed.) (1995).

Schiemann, G. (1996a): Was heißt moderne Physik?, in: F. Bevilacqua et al. (Ed.), Proceedings of the Conference "Emergence of Modern Physics", Berlin 23.3.–24.3.1995.

Schiemann, G. (1996b): The Loss of World in the Image. Origin and Development of the Concept of Image in the Thought of Hermann von Helmholtz and Heinrich Hertz, in: D. Baird et al. (Ed.), Heinrich Hertz. Classical Physicist, Modern Philosopher. Boston Studies in the Philosophy of Science. Dordrecht.

Schiemann, G. (1997): Wahrheitsgewißheitsverlust. Hermann von Helmholtz' Mechanismus im Anbruch der Moderne. Eine Studie zum Übergang von klassischer zu moderner Naturphilosophie. Darmstadt.

Schiemann, G. (1998): Physik und Natur. Zu Hermann von Helmholtz' Begründung des Energieprinzips in der Einleitung zu seiner Schrift "Über die Erhaltung der Kraft", in: H. Klages and H. Lübbig (Eds.): Hermann von Helmholtz. Klassiker der Epochenschwelle. Braunschweig.

Schiemann, G. (1999a): 100 Jahre danach: Zum Stand der Dinge in Sachen Hermann von Helmholtz. Rev. of: D. Cahan (Ed.), Hermann von Helmholtz and the Foundations of Nineteenth-Century Science. Berkeley/Los Angeles/London 1994, and L. Krüger (Ed.), Universalgenie Helmholtz. Rückblick nach 100 Jahren. Berlin 1994, in: Journal for General Philosophy of Science 30, p. 179 ff.

Schiemann, G. (1999b): Implikationen des Energieprinzips bei Hermann von Helmholtz. Erkenntnistheoretische und naturphilosophische Voraussetzungen, in: Max-Planck-Institut für Wissenschaftsgeschichte (Ed.), Symposium 150. Jahrestag des Vortrages "Über die Erhaltung der Kraft" von Hermann von Helmholtz. Berlin.

Schiemann, G. (2004): Art. "Hermann von Helmholtz", in: D. Hoffmann, H. Laitko and S. Müller-Wille (Eds.), Lexikon bedeutender Naturwissenschaftler. Heidelberg.

Schiemann, G. (2007): Experimental Knowledge and the Theory of Producing it: Hermann von Helmholtz, in: U. Feest, G. Hon, H.-J. Rheinberger, J. Schickore, F. Steinle (Eds.), Generating Experimental Knowledge. Preprints of the Max Planck Institute for the History of Science Berlin. Berlin.

Schimank, H. (1949): Der Aspekt der Naturgesetzlichkeit im Wandel der Zeiten, in: Joachim Jungius-Gesellschaft der Wissenschaften zu Hamburg e.V. (Ed.), Das Problem der Gesetzlichkeit. Hamburg. Vol. 2.

Schimank, H. (1976): Die experimentelle Physik im 19. Jahrhundert und ihre handwerklich-technischen Hilfsmittel, in: Treue and Mauel (Eds.) (1976).

Schirra, N. (1991): Die Entwicklung des Energiebegriffs und seines Erhaltungskonzepts. Diss. Frankfurt am Main/Thun.

Schlick, M. (1922): Helmholtz als Erkenntnistheoretiker, in: Warburg et al. (1922).Schmidt, H. (1912 ff.): Philosophisches Wörterbuch. Revised by G. Schischkoff. Stuttgart 1982.

Schmidt-Biggemann, W. (1971 ff.): Art. "Maschine", in: Ritter and Gründer (Eds.) (1971 ff.).

Schnädelbach, H. (1983): Philosophie in Deutschland 1831–1933. Frankfurt am Main.

Schneider, I. (1988): Isaac Newton. München.

Schofield, R.E. (1970): Mechanism and Materialism: British Natural Philosophy in an Age of Reason. Princeton (NJ).

Schramm, M. (1985): Natur ohne Sinn? Das Ende des teleologischen Weltbildes. Graz/Wien/Köln.

Schüller, V. (1994): Das Helmholtz-Liesche Raumproblem und seine ersten Lösungen, in: Krüger (Ed.) (1994).

Schulz, R. (1995): Helmholtz und Gadamer. Provokation und Solidarität. Über den Ursprung der philosophischen Hermeneutik im Geist der Naturwissenschaft, in: Philosophia naturalis 32, p. 141 ff.

Schurz, G. (Ed.) (1988): Erklären und Verstehen in der Wissenschaft. Wien/München.

Schwertschlager, J. (1883): Kant und Helmholtz erkenntniss-theoretisch verglichen. Freiburg.

Scott, W.L. (1970): The Conflict Between Atomism and Conservation Theory. 1644 to 1860. London/New York.

Seeliger, R. (1948): Analogien und Modelle in der Physik, in: Studium Generale 1, p. 125 ff.

Shapin, St. and S. Schaffer (1985): Leviathan and the Air-Pump. Hobbes, Boyle, and the Experimental Life. Princeton (NJ).

Siebeck, H. (1890): Über die Entstehung der Termini natura naturans und natura naturata, in: Archiv für Geschichte der Philosophie 3, p. 370 ff.

Siemens-Helmholtz, E. von (Ed.) (1929): Anna von Helmholtz. Ein Lebensbild in Briefen. 2 Vols. Berlin.

Smith, C. and M.N. Wise (1989): Energy and Empire. A Biographical Study of Lord Kelvin. Cambridge etc.

Snelders, H.A.M. (1980): Christiaan Huygens and the Concept of Matter, in: Bos et al. (Ed.) (1980).

Sommerfeld, A. (1950): Vorlesungen über theoretische Physik I und II. Leipzig/Wiesbaden.

Sorell, T. (1991): Scientism. Philosophy and the Infatuation with Science. London/New York.

Spaemann, R. (1986): Ende der Modernität?, in: P. Koslowski et al. (Eds.), Moderne oder Postmoderne? Weinheim.

Stäckel, P. (1863 ff.): Gauss als Geometer, in: Carl Friedrich Gauss. Werke. Göttingen, Vol. 10.2, Treatise IV.

Stallo, J.B. (1911): Die Begriffe und Theorien der Modernen Physik. Leipzig.

Stegmüller, W. (1972 ff.): Art. "Hypothese", in: J. Speck (Ed.), Handbuch wissenschaftstheoretischer Begriffe. 3 Vols. Göttingen.

Stegmüller, W. (1978 ff.): Hauptströmungen der Gegenwartsphilosophie. 3 Vols. Stuttgart [6]1978 (Vol. I) and [8]1987 (Vols. II and III).

Steinle, F. (1994): Experiment, Speculation and Law. Faraday's Analysis of Arago's Wheel, in: Philosophy of Science Association, Vol. 1, p. 293 ff.

Steinle, F. (1995): Looking for a "Simple Case": Faraday and Electromagnetic Rotation, in: History of Science 33, p. 179 ff.

Steinle, F. (2005): Explorative Experimente. Ampère und die Ursprünge der Elektrodynamik. Stuttgart.

Stichweh, R. (1984): Zur Entstehung des modernen Systems wissenschaftlicher Disziplinen: Physik in Deutschland. 1740–1890. Frankfurt am Main.

Ströker, E. (1967): Denkwege der Chemie. Elemente ihrer Wissenschaftstheorie. Freiburg/München.

Stromberg, W.H. (1989): Helmholtz und Zoellner: 19th-Century Empiricism, Spiritism, and the Theory of Space Perception, in: Journal of the History of the Behavioral Sciences 25, p. 371 ff.

Sudhoff, K. (1922): Rudolf Virchow und die Deutschen Naturforscherversammlungen. Leipzig.

Süßmann, G. (1990): Kennzeichnungen der Räume mit konstanter Krümmung, in: Philosophia Naturalis 27, p. 206 ff.

Szabó, I. (1976): Geschichte der mechanischen Prinzipien. Basel/Stuttgart.

Szabó, I. et al. (1971 ff.): Art. "Hypothese", in: Ritter and Gründer (Ed.) (1971 ff.).

Thackray, A. (1970): Atoms and Power: An Essay on Newtonian Matter-Theory and the Development of Chemistry. Cambridge (MA)/London.

Thomas von Aquin (1256 ff.): De veritate, in: Thomas von Aquin, Opera Omnia, Vol. 14. Paris 1875.

Thomson, W. (1884): Notes of Lectures on Molecular Dynamics, and the Wave-Theory of Light. Baltimore (MD).

Thomson, W. and P.G. Tait (1871 ff.): Handbuch der theoretischen Physik. Trans. by H. Helmholtz and D. Wertheim. Braunschweig 1871 (1. Vol., 1. part) and 1874 (2. part).

Torretti, R. (1979): Philosophy of Geometry from Riemann to Poincaré. Dordrecht etc.

Toulmin, St. and J. Goodfield (1966): The Architecture of Matter. London.

Treue, W. and K. Mauel (Ed.) (1976): Naturwissenschaft, Technik und Wirtschaft im 19. Jahrhundert. Göttingen.

Truesdell, C. (1968): Essays in the History of Mechanics. Berlin etc.

Truesdell, C. (1984): An Idiot's Fugitive Essays on Science, Methods, Training, Criticism, Circumstances. New York etc.

Truesdell, C. (1989): Newtons Einfluss auf die Mechanik des 18. Jahrhunderts, in: K. Hutter (Ed.), Die Anfänge der Mechanik. Berlin etc.
Tuchman, A. (1994): Helmholtz and the German Medical Community, in: Cahan (Ed.) (1994).
Turner, R. St. (1970 ff.): Hermann von Helmholtz, in: C.C. Gillispie (Ed.), Dictionary of Scientific Biography. 16 Vols. New York.
Turner, R. St. (1971): The Growth of Professorial Research in Prussia, 1818 to 1848 – Causes and Context, in: Historical Studies in the Physical Sciences 3, p. 137 ff.
Turner, R. St. (1977): Hermann von Helmholtz and the Empiricist Vision, in: Journal of the History of the Behavioral Sciences 13, p. 48 ff.
Turner, R. St. (1994a): Consensus and Controversy: Helmholtz on the Visual Perception of Space, in: Cahan (Ed.) (1994).
Turner, R. St. (1994b): In the Eye's mind. Vision and the Helmholtz-Hering Controversy. Princeton (NJ).
Tyndall, J. (1870): Faraday und seine Entdeckungen. Eine Gedenkschrift. Ed. and with a foreword by H. Helmholtz. Braunschweig.
Tyndall, J. (1874): Fragmente aus den Naturwissenschaften. With a foreword and additions by H. Helmholtz. Braunschweig.
Uexküll, J. von (1947): Der Sinn des Lebens. Gedanken über die Aufgaben der Biologie, mitgeteilt in einer Interpretation der zu Bonn 1824 gehaltenen Vorlesung des Johannes Müller [...], with a prognosis by Th. von Uexküll. Godesberg.
Vaihinger, H. (1920): Die Philosophie des Als Ob. 4. edition. Leipzig.
Vidoni, F. (1991): Ignorabimus! Emil du Bois-Reymond und die Debatte über die Grenzen wissenschaftlicher Erkenntnis im 19. Jahrhundert. Frankfurt am Main etc.
Virchow, R. (1863): Über den vermeintlichen Materialismus der heutigen Naturwissenschaft, in: Amtlicher Bericht der 38. Versammlung Deutscher Naturforscher und Ärzte. 1864, p. 35 ff.
Vogel, St. (1994): Sensation of Tone, Perception of Sound, and Empirism: Helmholtz's Physiological Acoustics, in: Cahan (Ed.) (1994).
Volkert, K. (1993): On Helmholtz' Paper *Über die thatsächlichen Grundlagen der Geometrie*, in: Historia Mathematica 20, p. 307 ff.
Volkert, K. (1996): Hermann von Helmholtz und die Grundlagen der Geometrie, in: Eckart and Volkert (Eds.) (1996).
Wahsner, R. (1992): Zur Helmholtzschen Kritik des Raum-Apriorismus, in: Wahsner, Prämissen physikalischer Erfahrung. Berlin.
Walch, J.G. (1726): Philosophisches Lexikon. 2. edition. Leipzig.
Warburg, E. et al. (1922): Helmholtz als Physiker, Physiologe und Philosoph. Karlsruhe.
Weizsäcker, C.F.v. (1943): Die Auswirkungen des Satzes von der Erhaltung der Energie in der Physik, in: Weizsäcker, Zum Weltbild der Physik. Stuttgart 1970.
Weizsäcker, C.F.v. (1974): Die Einheit der Natur. München.
Welsch, W. (1988): Unsere postmoderne Moderne. 2. edition. Weinheim.
Werner, F. (1997): Hermann Helmholtz' Heidelberger Jahre (1858–1871). Berlin.
Westfall, R.S. (1971): Force in Newton's Physics. London/New York.
Westfall, R.S. (1977): The Construction of Modern Science. Mechanism and Mechanics. Cambridge etc.
Westman, R.S. (1980): Huygens and the Problem of Cartesianism, in: Bos et al. (Eds.) (1980).
Whitt, L.A. (1990): Atoms or Affinities? The Ambivalent Reception of Daltonian Theory, in: Studies in History and Philosophy of Science 21, p. 57 ff.
Whittaker, E.A. (1958): A History of the Theories of Aether and Elecricity. 2 Vols. London etc.
Wieland, W. (1970): Die aristotelische Physik. Göttingen.
Wien, W. (1919): Hermann von Helmholtz zu seinem 25-jährigen Todestage, in: Die Naturwissenschaften 7, p. 645 ff.
Winckelmann, A.v. (Ed.) (1894 ff.): Handbuch der Physik. Breslau.
Winters, S.M. (1985): Hermann von Helmholtz's Discovery of Force Conservation. Diss. Baltimore (MD).

Wise, M.N. (1981): German Concepts of Force, Energy, and the Electromagnetic Ether: 1845–1880, in: G.N. Cantor and M.S.J. Hodge, Conceptions of Ether. Cambridge.

Wise, M.N. (1983): On the Relation of Physical Science to History in Late Nineteenth-Century Germany, in: L. Graham, W. Lepenies and P. Weingart (Eds.), Functions and Uses of Disciplinary Histories. Vol. VII.

Wittgenstein, L. (1970): Schriften. Frankfurt am Main.

Wolff, M. (1978): Geschichte der Impetustheorie. Untersuchungen zum Ursprung der klassischen Mechanik. Frankfurt am Main.

Wollgast, S. (1973): Naturwissenschaftlicher Materialismus und Pantheismus bei Hermann von Helmholtz, in: Banasiewicz et al. (1973).

Woodruff, A.E. (1968): The Contributions of Hermann von Helmholtz to Electrodynamcis, in: Isis 5, p. 300 ff.

Wright, G.H. von (1974): Erklären und Verstehen. Frankfurt am Main 1991.

Wundt, W. (1866): Die physikalischen Axiome und ihre Beziehung zum Causalprincip. Ein Capitel aus einer Philosophie der Naturwissenschaften. Erlangen.

Zilsel, E. (1976): Die sozialen Ursprünge der neuzeitlichen Wissenschaft. Ed. and trans. by W. Krohn. Frankfurt am Main.

Zöllner, C.F. (1872): Über die Natur der Cometen. Beiträge zur Geschichte und Theorie der Erkenntnis. Leipzig.

Index

References of main importance are printed in italics

Archimedes

NEW STUDIES IN THE HISTORY AND PHILOSOPHY OF SCIENCE AND TECHNOLOGY

1. J.Z. Buchwald (ed.): *Scientific Credibility and Technical Standards in 19th and Early 20th Century Germany and Britain.* 1996 ISBN 0-7923-4241-0
2. K. Gavroglu (ed.): *The Sciences in the European Periphery During the Enlightenment.* 1999 ISBN 0-7923-5548-2; Pb 0-7923-6562-1
3. P. Galison and A. Roland (eds.): *Atmospheric Flight in the Twentieth Century*, 2000 ISBN 0-7923-6037-0; Pb 0-7923-6742-1
4. J.M. Steele: *Observations and Predictions of Eclipse Times by Early Astronomers.* 2000 ISBN 0-7923-6298-5
5. D-W. Kim: *Leadership and Creativity.* A History of the Cavendish Laboratory, 1871–1919. 2002 ISBN 1-4020-0475-3
6. M. Feingold: *The New Science and Jesuit Science: Seventeenth Century Perspective.* 2002 ISBN 1-4020-0848-1
7. F.L. Holmes, J. Renn, H-J. Rheinberger: *Reworking* the Bench. 2003 ISBN 1-4020-1039-7
8. J. Chabás, B.R. Goldstein: *The Alfonsine Tables of Toledo.* 2003 ISBN 1-4020-1572-0
9. F.J. Dijksterhuis: *Lenses and Waves.* Christiaan Huygens and the Mathematical Science of Optics in the Seventeenth Century. 2004 ISBN 1-4020-2697-8
10. L. Corry: *David Hilbert and the Axiomatization of Physics (1898–1918).* From Grundlagen der Geometrie to Grundlagen der Physik. 2004 ISBN 1-4020-2777-X
11. J.Z. Buchwald and A. Franklin (eds.): *Wrong for the Right Reasons.* 2005 ISBN 1-4020-3047-9
12. M. Feingold and V. Navarro-Brotons (eds.): *Universities and Science in the Early Modern Period.* 2006 ISBN 1-4020-3974-3
13. R.R. Hamerla: *An American Scientist on the Research Frontier.* Edward Morley, Community, and Radical Ideas in Nineteenth-Century Science. 2006 ISBN 1-4020-4088-1
14. J. Schickore and F. Steinle (eds.): *Revisiting Discovery and Justification.* Historical and Philosophical Perspectives on the context distinction. 2006 ISBN 1-4020-4250-7
15. K. Nickelsen: *Draughtsmen, Botanists and Nature.* The construction of Eighteenth-Century Botanical Illustrations. 2006 ISBN 1-4020-4819-X
16. R. MacLeod and J.A. Johnson (eds.): *Frontline and Factory.* Comparative Perspectives on the Chemical Industry at War, 1914–1924. 2006 ISBN 1-4020-5489-0
17. G. Schiemann: *Hermann von Helmholtz's Mechanism: The Loss of Certainty.* A Study on the Transition from Classical to Modern Philosophy of Nature. 2008 ISBN 978-1-4020-5629-1

MIX
Papier aus verantwortungsvollen Quellen
Paper from responsible sources
FSC® C105338

FSC
www.fsc.org

If you have any concerns about our products,
you can contact us on
ProductSafety@springernature.com

In case Publisher is established outside the EU,
the EU authorized representative is:
Springer Nature Customer Service Center GmbH
Europaplatz 3, 69115 Heidelberg, Germany

Printed by Libri Plureos GmbH
in Hamburg, Germany